ALGAL BLOOMS AND MEMBRANE BASED DESALINATION TECHNOLOGY

LOREEN OPLE VILLACORTE

Cover image:

Aerial photo of a non-toxic "red tide" algal bloom caused by *Noctiluca scintillans* in New Zealand.
© Miriam Godfrey

ALGAL BLOOMS AND MEMBRANE BASED DESALINATION TECHNOLOGY

DISSERTATION

Submitted in fulfillment of the requirements of
the Board for Doctorates of Delft University of Technology
and
of the Academic Board of the UNESCO-IHE
Institute for Water Education
for
the Degree of DOCTOR
to be defended in public on
Wednesday, 8 January 2014, 12:30 pm
in Delft, the Netherlands

by

Loreen Ople VILLACORTE

Master of Science in Water Supply Engineering
UNESCO-IHE Institute for Water Education

born in Cebu, the Philippines

This dissertation has been approved by the supervisor:
Prof. dr. M. D. Kennedy

Composition of Doctoral Committee:

Chairman	Rector Magnificus TU Delft
Vice-Chairman	Rector UNESCO-IHE
Prof. dr. M.D. Kennedy	UNESCO-IHE/Delft University of Technology, supervisor
Prof. dr. ir. J.C. Schippers	UNESCO-IHE/Wageningen University
Prof. dr. ir. J.S. Vrouwenvelder	Delft University of Technology / KAUST
Prof. dr. ir. W.G.J. van der Meer	Delft University of Technology
Prof. dr. ir. E. Roesink	University of Twente
Dr. F. Hammes	EAWAG Switzerland
Prof. dr. ir. D. Brdanovic	Delft University of Technology, reserve member

CRC Press/Balkema is an imprint of the Taylor & Francis Group, an informa business

Published by:
CRC Press/Balkema
PO Box 11320, 2301 EH Leiden, The Netherlands
e-mail: Pub.NL@taylorandfrancis.com
www.crcpress.com – www.taylorandfrancis.com

ISBN 978-1-138-02626-1 (Taylor & Francis Group)

Acknowledgement

This thesis project was carried out at UNESCO-IHE Institute for Water Education in collaboration with the Wetsus Center of Excellence for Sustainable Water Technology. I would like to express my sincere gratitude to my supervisors, Maria Kennedy and Jan Schippers, for their guidance, encouragement, inspiration, patience and trust in shaping and completing this thesis research to its current form. Thanks are due to the Wetsus Center of Excellence for Sustainable Water Technology for their financial support and for the unique opportunity of having productive interactions with both the academia and the water industry. I am also grateful for the valuable support and advice of Gary Amy, Harvey Winters and Hans Vrouwenvelder of the Water Desalination and Reuse Center (WDRC-KAUST) during the course of this project.

I would like to express my appreciation to Rinnert Schurer and Evides Waterbedrijf for their assistance in the sampling, monitoring and analyses in Jacobahaven and de Biesbosch. I am grateful for the support and valuable feedbacks from the R & D group in Pentair X-flow (Bastiaan Blankert, Remon Dekker, Frederik Spenkelink, Ferry Horvath, Stefan van Hof, etc.,). It has been a learning experience interacting with both Evides and Pentair as it has given me a good sense of reality of how scientific research can be applied in practice.

Nine master students and interns contributed to this research, namely: Ramesh Duraisamy, Barun Karna, Muna Gharaibeh, Vaidotas Kisielius, Zanele Nyambi, Dino Berenstein, Katarzyna Micor, Shalane Mari and Heather Ross Oropeza. An acknowledgement is not enough to describe the important contribution of Yuli Ekowati and Helga Calix. Their skills and coolness in performing some of the most challenging experiments in this project are really admirable.

The completion of this thesis may not be possible without the technical/non-technical support of my colleagues/peers (in alphabetical order): Abdulai Salifu, Andrea Radu, Andrew Maeng, Anique Karsten, Arie Zwijnenburg, Assiyeh Tabatabai, Bernadeth Lohmann, Caroline Plugges, Changwon Ha, Chantal Groenendijk, Chol Abel, Claudia Dreszer, Don van Galen, Elodie Loubineaud, Emmanuelle Prest, Erik Roesink, Evan Spruijt, Ferdi Battes, Florian Beyer, Frank Wiegman, Fred Kruis, Giuliana Ferrero, Helga Calix, Herman Smit, Hilde Prummel, Jan Bahlman, Jelmer Dijkstra, Joe Woolley, Jolanda Boots, Juma Hamad, Koos Baas, Laurens Welles, Lute Broens, Lyzette Robbemont, Maikel Maloncy, Mark van Loosdrecht, Mieke

Kliejn, Nirajan Dhakal, Paula van den Brink, Paul Buijs, Peter Heerings, Peter Stroo, Saaed Baghoth, Sergio Salinas-Rodriguez, Sheng Li, Stefan Huber, Tanny van der Klis, Tarek Waly, Thomas Neu and Victor Yangali. I would also like to thank all the participants of the Wetsus research theme "Biofouling" for the fruitful discussions over the last 4 years.

Perhaps more people have directly or indirectly supported me over the last 4 years. I sincerely apologize to those who are not mentioned in this acknowledgement.

Special thanks to my friends and extended family in the Netherlands, who have been very accommodating all through these years, namely: Els and Joost, Trudy and Geert Jan, Lucille and Arjen, Henk and Ruben, the International Student Church of Delft and many more. I am very grateful to my family in the Philippines for their never-ending support even from afar. I am indebted to my parents for pushing all the odds to send me to university, a precious opportunity they did not have themselves. This important milestone is dedicated to them.

And finally, to my loving wife Helga and my sweet little daughter Elsi, this would have been a very difficult journey without your daily inspiration, patience and understanding. Thanks for letting me see the most beautiful side of life. I always feel very blessed to be surrounded by you.

Loreen Ople Villacorte

5 December 2013
Delft, the Netherlands

Summary

Seawater desalination technology has been rapidly growing in terms of installed capacity (~80 million m³/day as of 2013), plant size and global application. However, seawater reverse osmosis (SWRO) systems are prone to operational problems such as membrane fouling. To minimize fouling, pre-treatment of seawater is necessary upstream of the SWRO system. The pretreatment in most SWRO plants, specifically in the Middle East, comprise coagulation followed by granular media filtration (GMF) pre-treatment. In recent years, however, low pressure membranes such as ultrafiltration (UF) are increasingly being used as a preferred alternative to GMF. As more extra large SWRO plants (>500,000 m³/day) are expected to be installed over the coming years, frequent chemical cleaning (>1/year) of the SWRO will be unfeasible and installing a reliable pre-treatment system will be even more important.

An emerging threat to SWRO is the seasonal proliferation of microscopic algae in seawater called algal blooms. These natural phenomena can potentially occur in most coastal areas of the world where SWRO plants are installed. Recent severe algal bloom outbreaks in the Middle East region in 2008 and 2013 have caused clogging of GMF pre-treatment systems which also resulted in inacceptable quality (silt density index, SDI>5) of GMF effluent. The latter eventually led to temporary shutdown of several SWRO plant mainly due to concerns of irreversible fouling of the downstream SWRO membranes. The poor performance of GMF during algal blooms has shifted the focus of the desalination industry to UF as a main pre-treatment for SWRO. Thus, an extensive investigation on the impact of algal blooms on both UF and RO membranes is required.

The goal of this study is to understand the adverse impact of seawater algal bloom on the operation of UF and RO membrane systems. Furthermore, this study aims to develop/improve methods to investigate the characteristics and the membrane fouling potential of algae and algal organic matter (AOM). The ultimate goal is to provide engineers/operators with a better understanding as well as reliable assessment tools to develop robust processes and effective operation strategies to maintain stable operation in membrane-based desalination plants during an algal bloom.

The potential problems which may occur in membrane-based desalination plants (UF pre-treatment followed by SWRO) during treatment of algal bloom impaired waters are: (1) particulate fouling in UF due to accumulation of algal cells and their detritus, (2) organic fouling in UF or RO due to accumulation of AOM, and (3) biological fouling in RO initiated and/or enhanced by AOM. These potential issues were addressed in this study by performing both theoretical and experimental analyses.

The first step to a better understanding of the effects of algal bloom on membrane filtration was to perform characterization studies with batch cultures of common bloom-forming marine (*Chaetoceros affinis, Alexandrium tamarense*) and freshwater (*Microcystis sp.*) algae. The bloom forming algae showed different growth patterns and AOM production at different phases of their life cycles. The AOM which the algae produced wa extracted and characterized using various microscopic, choromatographic and spectrophotometric techniques. The main finding is that AOM comprise mainly biopolymers (e.g., polysaccharides, proteins) while the remaining fractions comprise refractory organic matter (e.g., humic-like substances) and/or low molecular weight biogenic substances. Polysaccharides with associated sulphate and fucose functional groups were ubiquitous among biopolymers produced by the 3 algal species. Furthermore, these biopolymers were capable of adhering to clean UF and RO membranes but the adhesion was much stronger on membranes already fouled by algal biopolymers. Some indications show that this adhesive fraction of AOM is mainly made up of transparent exopolymer particles (TEP).

Further investigations were performed to monitor the presence of TEP in membrane-based desalination systems. For this purpose, the two existing methods to measure TEPs were first modified or further improved in terms of their applicability in studying different pre-treatment processes and membrane fouling. Various improvements were introduced mainly to minimize the interference of dissolved salts in the sample, to decrease the lower limit of detection, to measure both particulate and colloidal TEPs (down to 10 kDa) and to develop a reproducible calibration method with a standard polysaccharide (Xanthan gum). Successful application of the two methods (with modifications and improvements) was demonstrated in monitoring TEP accumulation in algal cultures and TEP removal in the pre-treatment processes of 4 RO plants.

Long term (~4 years) TEP monitoring in the raw water of an SWRO plant in the Netherlands showed spikes in the TEP concentration typically coinciding with the spring algal bloom (March-May), which were mainly dominated by diatoms and/or *Phaeocystis*. Further monitoring of the fate of TEP through the treatment processes of 4 RO plants showed substantial reductions in TEPs and biopolymer concentrations and membrane fouling potential (in terms of the modified fouling index-ultrafiltration, MFI-UF) after UF (with and without coagulation), coagulation–sedimentation-filtration and coagulation-dissolved air flotation-filtration treatments. On the other hand, lab-scale membrane rejection experiments demonstrated that complete removal of algal-derived biopolymers may be possible by using UF membranes with lower molecular weight cut-off ($\leq$10 kDa). Overall, results from both lab-scale and full-scale plant experiments showed significant correlations between TEP concentration and MFI-UF, an indication that TEPs likely have a major role in the fouling of UF pre-treatment systems and possibly in SWRO systems, if not effectively removed by the pre-treatment process.

The role of algae in the fouling of capillary UF membranes was investigated based on theoretical and experimental analyses. Results show that membrane fouling of UF during filtration of algal bloom impacted water is mainly due to accumulation of TEPs on membrane pores and surfaces, rather than the accumulation of algal cells. This was corroborated by the correlation ($R^2>0.8$) between MFI-UF and TEP concentration and not with algal cell concentration ($R^2<0.5$).

Additional investigations were performed to verify if algal cells can cause severe plugging in capillary membranes. Theoretically, algal cell transport through an inside-out driven capillary UF is mainly influenced by inertial lift, which may cause plugging problems as it enhances deposition of algal cells at the dead-end section of the capillary. Hydraulic calculations suggest that severe plugging due to significant accumulation of large algal cells (>25 μm) may cause substantial increase in membrane and cake resistance due to localised increase in flux during constant flux filtration. Such potential problems can be addressed by shortening the period between filtration cycles, lowering the membrane flux and/or introducing a small cross-flow (bleed) at the dead-end of the capillary.

Fouling in capillary UF membranes by biopolymers from bloom-forming marine algae were further investigated in relation to the solution chemistry of the feed water. Lowering the pH of the feed water resulted to an increase in UF membrane fouling potential of biopolymers. Varying the ionic strength in the feed water led to substantial variations in fouling potential. The role of mono- and di- valent cations on the fouling potential of biopolymers was also investigated by varying the cation composition of the feed water matrix. Among the major cations abundant in seawater, Ca^{2+} ions demonstrated a significant influence on the fouling propensity of AOM in terms of MFI-UF and non-backwashable fouling rate. Furthermore, the effect of solution chemistry on the fouling behaviour of algal biopolymers, in terms of hydraulic resistance and backwashability, is substantially different to what was observed in model polysaccharides (e.g., sodium alginate and gum xanthan) traditionally used in UF fouling experiments. This indicates that model polysaccharides such as sodium alginate and gum xanthan are not reliable in simulating the fouling propensity of algal biopolymers in seawater.

The last phase of the study demonstrated the important role of AOM on biofilm development in SWRO system by performing accelerated biological fouling experiments (8-20 days) in cross-flow capillary UF membranes (10 kDa MWCO) and spiral wound RO membranes (using MFS). It was illustrated in these experiments that when biodegradable nutrients (CNP) are readily available in the feed water, the substantial presence of AOM - either in the feedwater or on the membrane/spacers - can accelerate the occurrence of biofouling in SWRO. However, when nutrients are not readily available in the feed water, direct organic fouling by AOM may occur but with a much lower fouling rate than when nutrients (N,C,P) are readily available. These findings were supported by the results of growth potential tests (using flow cytometry) whereby the net growth of bacteria in natural seawater increased with increasing AOM concentration. Consequently, effective removal of AOM from RO feed water is necessary to minimise biofouling in SWRO plants affected by algal blooms.

Overall, this study demonstrates that better analytical parameters and tools are essential in elucidating the adverse impacts of algal bloom in seawater on the operation of membrane-based desalination plants (UF-RO). It also highlighted the importance of developing effective pre-treatment processes to remove AOM from the raw water and reduce the membrane fouling potential of the feed water for downstream SWRO membranes. Since seawater algal blooms generally occur within a short period of time (e.g., from days to few weeks) and are also difficult to predict, it is essential that MFI-UF, TEP and algae concentrations are regularly monitored in the raw and pre-treated water of SWRO plants, so that corrective measures can be implemented in time in the pre-treatment system during the onset of an algal bloom.

Table of Contents

Acknowledgement   3

Summary   5

Chapter 1   General introduction   11

Chapter 2   Marine algal blooms   27

Chapter 3   Seawater reverse osmosis and algal blooms   45

Chapter 4   Characterisation of algal organic matter   71

Chapter 5   Measuring transparent exopolymer particles in   109
                freshwater and seawater

Chapter 6   Fate of transparent exopolymer particles in integrated   147
                membrane systems

Chapter 7   Fouling of ultrafiltration membranes by algal   177
                biopolymers in seawater

Chapter 8   Fouling potential of algae in inside-out capillary UF   213
                membranes

Chapter 9   Biofouling in cross-flow membranes facilitated by algal   247
                organic matter

Chapter 10 General conclusions   279

Samenvatting   287

Abbreviations   291

Publications and awards   293

Curriculum vitae   297

1

General introduction

Contents

1.1	Background	12
1.2	Membrane filtration	14
1.3	Membrane-based desalination	15
1.4	Algal blooms	18
1.5	Membrane fouling by algal organic matter	20
1.6	Transparent exopolymer particles	20
1.7	Aim and scope of the study	21
1.8	Research objectives	22
1.9	Outline of the thesis	22
1.10	Acknowledgements	23
References		24

1.1 Background

The global demand for safe and reliable water supply for agricultural, municipal and industrial use has been steadily growing over the last few decades. The main driving factors are population growth, economic development and depletion of traditional freshwater supplies. Water scarcity is already a long standing issue in several countries in North Africa, the Arabian Peninsula and the Caribbean, where freshwater availability is less than 1000 m³/person/year (Figure 1-1). It is projected that by the year 2030, nearly half of the world population will be living in areas of high water stress (WWAP, 2012). Moreover, future population growth will be mainly in developing countries which are either already experiencing "water stress" or in areas with limited access to safe drinking water (WWAP, 2012). Furthermore, several other countries which overall are still not considered water stressed are currently experiencing localized water scarcity, especially in coastal regions (WWAP, 2012).

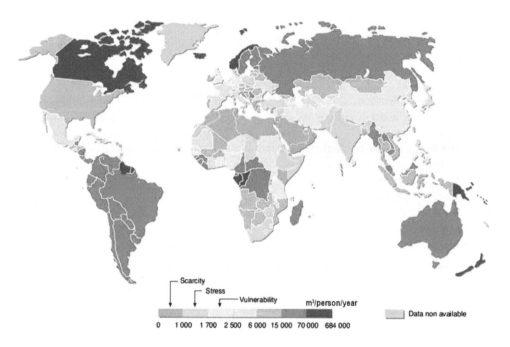

Figure 1-1: Composite map showing global freshwater availability by country in 2007. Source: Philippe Rekacewicz, UNEP/GRID-Arendal via WWAP (2012).

Economic and demographic growths have resulted in over-abstraction of conventional freshwater resources. As of 2012, freshwater abstractions in the Arabian Peninsula, North Africa and South Asia were about 500%, 175% and 45% of their internal renewable water resources, respectively (FAO, 2012). Many countries within these regions have been resorting to seawater desalination to ease the water supply-demand gap. Other measures have also been implemented such as utilising water more efficiently, reducing leakages in public water supply networks and wastewater reuse. These water saving measures are increasingly implemented but their overall contribution to increasing the current water supply are rather limited. Consequently, seawater desalination is often the preferred option to satisfy the demand.

As of 2010, about 44% of the global population and 8 of the 10 largest metropolitan areas in the world are located 150 km from the coastline (UN Atlas of the Oceans, 2010). Therefore, the prospect of widespread application of seawater desalination in the near future is very likely. In fact, it is projected that a cost-effective application of desalination technologies can increase the global clean water supply by about 20% between 2020 and 2030 (WWAP, 2012).

Large scale application of seawater desalination started way back in the 1960's using thermal distillation processes. In the 1970's, reverse osmosis (RO) technology was introduced mainly for brackish water desalination. Subsequent advancement in material science in the 1980's led to the development of more robust composite aromatic polyamide RO membrane which allows extending RO application to seawater desalination (Wilf *et al.*, 2007). Seawater desalination was mainly based on thermal processes (e.g., multi-stage flash and multi-effect distillation) until the last decade when reverse osmosis technology started to dominate the market (Figure 1-2).

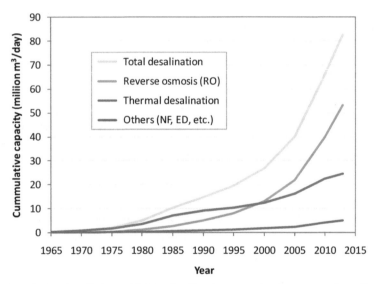

Figure 1-2: Cumulative installed worldwide desalination capacity in terms of applied technology (DesalData, 2013).

The rapid growth of the application of RO desalination technology in recent years is not only driven by the steady increase in water supply demand but also by the declining RO water production cost (Lattemann *et al.*, 2010). It is expected that by 2015, the average global production cost of RO desalinated water will be about 0.5 USD/m^3, which means that large scale application of seawater RO (SWRO) desalination will become more economically attractive and competitive with conventional water treatment processes (GWI, 2007).

1.2   Membrane filtration

In addition to RO, other types of pressure-driven membranes have been applied for drinking and industrial water production. Separation processes using polymeric membranes can cover a wide size spectrum of contaminants, from suspended solids to dissolve organic compounds to inorganic ions (Figure 1-3). The physical and operational characteristics of different categories of membrane filtration processes are presented in Table 1-1.

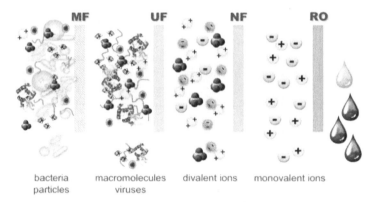

Figure 1-3: Contaminants which can be removed by membrane filtration processes. Adopted from Cath (2010) via The National Academies Press.

Table 1-1: Characteristics of different membrane filtration processes.

Process and abbreviations	Pore size (nm)	MWCO* (kDa)	Pressure (bar)	Materials typically retained
Microfiltration, MF	50-5000	> 500	0.1-2	particles + large colloids + large bacteria
Ultrafiltration, UF	5-50	2 - 500	1-5	as above + small colloids + small bacteria + viruses + organic macromolecules
Nanofiltration, NF	< 1	0.2 - 2	5-20	as above + multi-valent ions
Reverse osmosis, RO	<< 1	< 0.2	10-100	as above + mono-valent ions

* Molecular weight cut-off = molecular weight of solutes with similar weight of which 90% were rejected by the membrane.

Low pressure-driven membranes such as microfiltration (MF) and ultrafiltration (UF) membranes are usually used in treating waters with high turbidity or high particulate matter loading. MF/UF can remove particulate (e.g., bacteria, algae) and some colloidal materials (e.g., biopolymers). Physical sieving is the main rejection mechanism, where water permeates through the membrane due to the applied driving pressure. Over time, the deposits on the membrane can serve as a self-rejecting layer and retain even smaller particles than the pore size of the membrane. UF have finer pores than MF, hence, it is expected to have higher contaminant rejection efficiency as well as higher operating pressure than MF. UF has been shown to be more effective than MF in removing bacteria, viruses and other pathogenic microorganisms in the water (Jacangelo *et al.*, 1995). Both MF and UF are widely applied in surface and wastewater treatment and increasingly as pre-treatment for NF/RO systems (Busch *et al.*, 2009).

Nanofiltration (NF) membrane is mainly applied for softening processes, colour and disinfection-by-product removals. NF membranes have a nominal pore size of approximately <0.001 microns and an operational pressure between 5 and 20 bars. The separation process in NF membranes not only involves physical sieving but also solution diffusion, the Gibbs-Donnan effect, dielectric exclusion and electromigration (Hussain *et al.*, 2008). The membrane removes mainly multi-valent ions and natural organic matter, while allowing some of the mono-valent ions to pass through. The total salt rejection of NF is in the range of 50-70% while organic carbon removal is between 70 and 95% (Schäfer, 2001).

Reverse osmosis (RO) membrane filtration is used primarily in brackish and seawater desalination and for waters contaminated with micro-pollutants. An RO membrane can remove both particulate and dissolved constituents from the water including natural organic matter and both mono- and multi-valent ions. With more stringent regulations regarding water quality, RO is now also being considered for surface water and secondary wastewater effluent treatment, particularly for trace contaminants removal (e.g., pesticides) and disinfection (Hofman *et al.*, 1997; Gomez *et al.*, 2012). The rejection mechanism is mainly solution diffusion through a dense separation layer in the polymer matrix, where water is allowed to pass while most solutes (~99.8% of salt ions) remain and concentrate on the feed side of the membrane. This process requires high pressure to be exerted on the feed side of the membrane due to build-up of osmotic pressure. In seawater, which has around 24 bar of natural osmotic pressure, a feed pressure of 40–70 bars is usually used.

1.3   Membrane-based desalination

Various membrane-based technologies have been developed over the years for water desalination processes, namely: RO, NF, electrodialysis (ED), thermo-osmosis, pervaporation, and membrane distillation. Among these technologies, only RO, NF and ED have been used in large scale applications while the rest are still in the research and developmental stages (Gullinkala *et al.*, 2010). ED desalination process involves moving salt ions selectively through a membrane driven by an electrical potential. RO is currently the dominant desalination technology and is widely applied for both drinking and industrial water production while ED is mainly used for industrial purposes (Figure 1-4).

It is projected that the global desalination capacity will reach 98 million m³/day by 2015, a large majority of which will be based on RO desalination technology (Latteman *et al.*, 2010). Currently, RO desalination has a global online capacity of 39.4 million m³/day, which is about twice the installed capacity of thermal desalination (Figure 1-4). Thermal desalination plants (i.e., MED, MSF, VC) are mainly installed in oil-rich countries of the Middle East while RO desalination is almost exclusively used in the rest of the world. Almost half (46%) of the RO desalinated waters were from seawater and the rest were mainly from brackish, freshwater and treated wastewater (Figure 1-4). This is a testament to the growing importance of RO desalination in coastal areas in the world where freshwater is a limited commodity or too polluted to be treated by conventional water treatment processes.

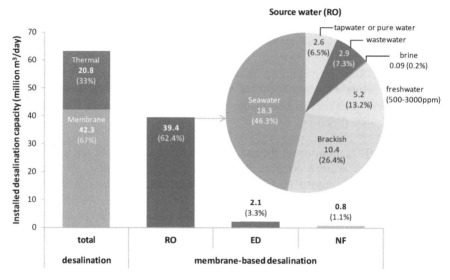

Figure 1-4: The global online desalination capacity (in million m³/day) as of June 2013 with regards to desalination technology and RO source water (inset chart). Primary data from Desaldata (2013).

Figure 1-5 illustrates that high concentration of large RO desalination plants have been installed in the Middle East region, the Mediterranean area (e.g., Spain and Algeria), northern and central Europe (e.g., Germany, Netherlands and UK), the Caribbean, Japan, Korea and the US (e.g., Florida and California).

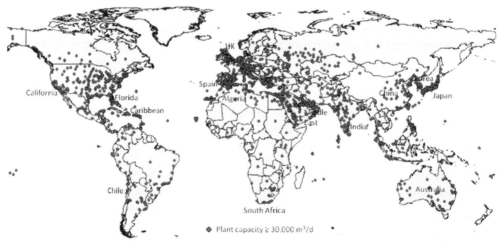

Figure 1-5: Global distribution of large RO plants with installed capacities ≥ 30,000 m³/day. Map processed by N. Dhakal with primary data from DesalData (2012).

Seawater desalination by reverse osmosis is considered to be more energy efficient, more compact and more flexible (modular) compared to other desalination processes. The current water production cost of RO desalination is generally cheaper than thermal desalination processes (GWI, 2007). Such cost is expected to decrease further as more efficient and/or extra large RO systems will be installed in the near future.

Currently, the main "Achilles heel" for the cost-effective application of RO is membrane fouling (Flemming *et al.*, 1997). The accumulation of particulate and organic materials from seawater and biological growth in membrane modules frequently cause operational problems in SWRO. These may result in one or a combination of the following:

1) higher energy cost due to higher operating pressure;
2) higher chemical consumption/cost due to additional chemical pre-treatment (e.g., coagulation) and frequent chemical cleaning of the membranes;
3) higher material cost due to frequent replacement of damaged or irreversibly fouled membranes;
4) lower rate of water production due to longer system downtime during chemical cleaning and membrane replacement; and
5) declining product water quality due to increased salt passage through the membranes.

The above-mentioned problems have increased the necessity of pre-treating the RO feed water with conventional treatment processes, such as granular media filtration or coagulation and sedimentation followed by media filtration, to maintain more stable and more reliable operation. This necessity also paved the way for the development of integrated membrane systems (IMS), in which RO systems are preceded with different pre-treatment processes to remove potential foulants from the RO feedwater (Schippers *et al.*, 2004). Among these pre-treatment processes, low pressure membranes (MF and UF) have been progressively applied in recent years to further reduce membrane fouling in RO/NF systems.

As shown in Figure 1-6, the application of UF pre-treatment for SWRO has been rapidly increasing since 2006. As of 2013, SWRO plants with UF pre-treatment accounts for about 30% of total SWRO capacity. However, this percentage is expected to increase further in the future as UF is currently preferred for its better treatment reliability (in terms of maintaining low SDI/MFI in the RO feed water) and lower chemical consumption than conventional pre-treatment systems (Busch *et al.*, 2009).

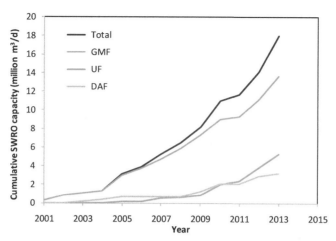

Figure 1-6: Comparison of production capacity of 49 largest SWRO plants installed between 2001 and 2013 in terms of pre-treatment technologies: granular media filtration (GMF), ultrafiltration (UF) and dissolved air flotation (DAF). DAF pre-treatment system is always installed in combination with GMF and/or UF pre-treatments. SWRO capacity based from DesalData (2013).

1.4   Algal blooms

There is growing evidence that algae are a major cause of operational problems in SWRO plants. Many SWRO plants abstract raw water in coastal sources where algal blooms frequently occur (Caron *et al.*, 2010; Richlen *et al.*, 2010; Petry *et al.*, 2007).

An algal bloom is a "population explosion" of naturally occurring microscopic algae, triggered mainly by seasonal changes in temperature, abundance of sunlight and/or high nutrient concentration in the water. Some algal blooms are considered harmful because the causative algal species produce toxic organic compounds which can cause illness/mortalities to humans and/or aquatic organisms. However, some harmful algal blooms do not produce toxic compounds but the algal biomass and algal organic matter (AOM) they produce can accumulate in dense concentrations near the water surface. Bacterial degradation of this organic material can lead to a sudden drop in dissolved oxygen concentration in the water and eventually cause mortalities of aquatic flora and fauna. During the last decades, the number of harmful algal blooms, the type of resources affected and economic losses reported have all increased dramatically (Anderson *et al.*, 2010). Economic losses mainly affect the fishing and aquaculture industry but recently the desalination industry has been affected as well.

The adverse effect of algal blooms on SWRO desalination systems started to gain more attention during the severe "red tide" bloom in the Gulf of Oman between 2008 and 2009 (Figure 1-7). That algal bloom forced several SWRO plants in the region to reduce or shutdown operations due to clogging of pre-treatment systems (i.e., granular media filters) and/or due to unacceptable RO feed water quality (i.e., silt density index, SDI>5) which triggers concerns of irreversible fouling problems in RO membranes (Berktay, 2011; Richlen *et al.*, 2010; Nazzal, 2009; Pankratz, 2008).

Generally, RO suppliers can only guarantee smooth operation with their RO membranes if the feed water has an SDI<5. This incident highlighted a major problem that algal blooms may cause in countries relying largely on SWRO plants for their water supply. Several arid coastal regions in the world (e.g., Chile, California) which are increasingly using SWRO technology for water supply are also vulnerable to this problem (Petry *et al.*, 2007; Caron *et al.*, 2010).

In SWRO plants, granular media filters (GMF) are usually installed to pre-treat seawater before being fed to the RO system. During algal bloom, the GMF can minimise breakthrough of algal cells but a substantial fraction of the AOM can still pass through the pre-treatment systems which can then potentially cause fouling in the downstream RO system. To solve the problem of poor quality of the pre-treated water (GMF effluent), a couple of options have been proposed such as incorporating and/or increasing the dose of coagulant in front of the GMF to improve the effluent water quality. However, an increase in coagulant dosage may further increase the rate of clogging in GMF. Installing a dissolved air flotation (DAF) system in front of the GMF will enable increase in coagulant dosage and improve the effluent quality while reducing clogging problems in GMF. Another option is to install an ultrafiltration (UF) membrane system to replace GMF. UF pre-treatment can guarantee an RO feed water with low SDI even during severe algal bloom. However, some concerns have been expressed regarding the rate of fouling in UF membrane systems (e.g., backwashable and non-backwashable fouling) during algal bloom period (Schurer *et al.*, 2013). To overcome this concern, incorporating in-line coagulation or a DAF system preceding a UF system has been recommended (Anderson and McCarthy, 2012).

Figure 1-7: The massive red tide bloom in the Gulf of Oman as shown in this satellite image acquired by Envisat's MERIS instrument on November 22, 2008 (Credits: C-wams project, Planetek Hellas/ESA). Yellow points indicate location of large SWRO plants in the area. Inset screenshots of online news regarding SWRO plant shutdown due to red-tide in the gulf in 2008 and 2013 (www.arabianbusiness.com).

1.5   Membrane fouling by algal organic matter

Algal blooms can cause fouling problems in both MF/UF and NF/RO systems. During MF/UF treatment of algal bloom-impacted water, particulate and organic materials comprising algae cells and AOM can accumulate to form a cake layer on the surface of the membranes. This cake can cause a substantial increase in the required driving pressure to maintain the permeate flux in the system. NF/RO systems are primarily designed to remove dissolved constituents in the water but they are most vulnerable to spacer clogging problems by particulate material from the feedwater. For this reason, NF/RO systems are generally preceded by a pre-treatment process to minimise particulate and organic fouling potential of the feed water. When MF/UF is applied as pre-treatment for RO, particulate and organic fouling problems during algal blooms is expected to mainly occur in the MF/UF pre-treatment system itself.

So far, a limited number of studies have investigated the effect of algal blooms on the operational performance of MF/UF membrane systems (e.g., Kim and Yoon, 2005; Ladner et al., 2010; Qu et al., 2012; Schurer et al., 2013). Most of these studies have suggested that the accumulation of AOM is the main cause of membrane fouling rather than the algae themselves. However, a synergistic effect between algal cells and AOM may intensify the rate of fouling in UF membranes. More studies are needed to illustrate and to better explain the mechanisms involved in fouling of MF/UF membranes due to accumulation of algae and AOM.

In 2005, Berman and Holenberg reported for the first time that some types of AOM, particularly transparent exopolymer particles (TEPs), can potentially initiate and enhance biofouling in RO systems (Berman and Holenberg, 2005). TEPs are a major component of AOM and are mainly compose of acidic polysaccharides and glycoproteins. They are characteristically sticky, so they can adhere and accumulate on the surface of the membranes and spacers. The accumulated TEPs can serve as a "conditioning layer" – a good platform for effective attachment and initial colonization by bacteria which may then accelerate biofilm formation in RO membranes (Bar-Zeev et al., 2012). Furthermore, TEPs might be partially degradable and may later serve as a substrate for bacteria (Passow and Alldredge, 1994; Alldredge et al., 1993).

The potential problems of TEP accumulation in RO can be more serious than in UF because RO systems are not backwashable and chemical cleaning might not be effective in removing these materials. Nevertheless, the current notion of the role of TEP on biofouling still needs to be verified and their effect on the operation of RO membranes still needs to be demonstrated.

1.6   Transparent exopolymer particles

One of the main bottlenecks of studying TEPs is the lack of reliable methods to measure and monitor these substances in seawater. A number of TEP quantification methods have been introduced in the last two decades, but the application of these methods has been limited due to interference of dissolved salts in seawater (e.g.,

Arruda-Fatibello *et al.*, 2004; Thornton *et al.*, 2007). A critical analysis of the currently available methods identified several questions regarding their reproducibility, calibration, limit of detection and range of TEP sizes taken into account. As TEP is gaining more attention in the SWRO desalination industry, there is an increasing need for a reliable method to measure these substances in marine waters.

Currently, the most widely used method to measure TEP only accounts for TEPs retained on 0.4 μm polycarbonate filters (Passow and Alldredge, 1995). However, TEPs are agglomeration of particulate and colloidal hydrogels which can vary in size from few nanometres to hundreds of micrometres (Passow, 2000; Verdugo *et al.*, 2004). In SWRO plants with MF/UF pre-treatment, only colloidal TEPs are expected to reach the SWRO system. Hence, it is important to measure colloidal TEPs (<0.4μm) as well since these particles cannot be ignored in studying the potential role of TEP in membrane fouling. Furthermore, the fate of colloidal TEPs through different pre-treatment processes is crucial towards developing an effective pre-treatment strategy for mitigating biological fouling in SWRO systems during algal blooms.

1.7   Aim and scope of the study

RO is currently the state-of-the-art seawater desalination technology capable of providing safe and reliable water supply in freshwater scarce coastal areas of the world. A major obstacle for the successful application of this technology is membrane fouling in the RO membrane itself and/or the UF pre-treatment system during algal blooms.

A key to understanding why algal blooms affect the operation of membrane systems is to study their occurrence and growth dynamics of bloom-forming algae as well as the chemical composition, size and membrane fouling potential of AOM, including TEPs.

During algal blooms, both particulate and organic fouling can occur in UF membranes while biofouling are more likely to occur in the RO system. To develop strategies to control operational problems caused by these fouling phenomena, a better understanding of the processes involved is crucial.

This study focuses on elucidating the adverse impacts of seawater algal blooms on the operation of UF and RO membrane systems. Furthermore, various characterisation and monitoring methods are applied while improving the reliability of current methods to measure TEPs in seawater is explored. The ultimate goal is to provide engineers/operators with better monitoring tools and sufficient knowledge to help them develop robust process design and/or effective operation strategy for maintaining stable operation in SWRO plants during algal blooms.

1.8 Research objectives

The specific objectives of this project are the following:

1. Investigate the release, characteristics, membrane rejection and fouling potential of AOM from three common species of bloom-forming algae in marine and freshwater sources.

2. Improve the reliability of current methods to measure TEP in fresh and saline waters, extend the method to cover smaller colloidal TEPs (<0.4 μm) and verify the application of the modified methods in seawater and freshwater.

3. Monitor the presence of TEP in the source water and their fate over the treatment processes of various RO plants by applying the newly developed methods for measuring particulate and colloidal TEPs.

4. Measure the UF membrane fouling propensity of AOM produced by marine bloom forming algae at different solution pH, ionic strength and cation composition.

5. Elucidate the role of algae (with and without AOM) in the fouling of inside-out capillary MF/UF membranes during severe algal bloom situations.

6. Demonstrate the possible role of AOM on biofouling in RO membranes treating algal bloom impaired seawater.

1.9 Outline of the thesis

This thesis is made up of 10 chapters; 2 review articles and six are presenting the results and findings of the different segments of the research. Figure 1-8 illustrate the different topics covered by this study.

- After this introductory chapter is **Chapter 2**, a review of the background knowledge on the characteristics and occurrence of algal bloom and the organic materials they produce.

- **Chapter 3** is a review of the state-of-the-art knowledge on the potential impact of algal blooms on the operation of SWRO plants and the current pre-treatment strategies to mitigate membrane fouling.

- In **Chapter 4**, investigations on the characteristics and membrane retention of AOM released by 3 common species of bloom-forming algae using various characterisation techniques is presented.

- In **Chapter 5**, the development and application of the newly improved methods to measure particulate and colloidal TEPs in seawater and freshwater is described.

- In **Chapter 6**, field studies on the presence of TEP and other biopolymers in the source water and their fate over the different pre-treatment processes of various RO plants is discussed.

- In **Chapter 7**, investigations on the UF membrane fouling potential of AOM from marine algae at different solution pH, ionic strength and cation composition are explained and compared with model polysaccharide foulants.

- In **Chapter 8**, theoretical and experimental studies regarding the membrane fouling and plugging potential of algae in inside-out capillary UF are presented.

- In **Chapter 9**, experimental studies to demonstrate the possible role of AOM on biofouling in seawater RO membranes are discussed.

- And finally, **Chapter 10** provides the summary of conclusions, outlook and recommendations for further research.

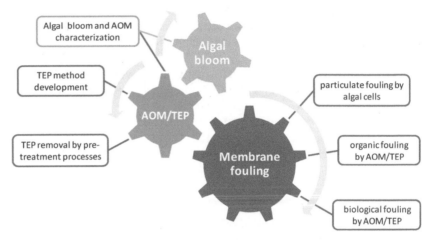

Figure 1-8: A schematic representation of the scope and different segments of the thesis.

1.10 Acknowledgements

Various parts of this project were implemented in collaboration with a number of researchers. The works of the following researchers are reflected in some parts of this thesis (in chronological order):

- Mr. Ramesh Duraisamy from India,
- Mr. Dino Berenstein from Suriname,
- Ms. Helga Calix Ponce from Honduras,
- Mr. Barun Karna from Nepal,
- Ms. Katarzyna Micor from Poland,
- Ms. Shalane Mari from Curacao,
- Ms. Muna Gharaibeh from Jordan,
- Ms. Heather Ross Oropeza from Bolivia,
- Ms. Yuli Ekowati from Indonesia,
- Mr. Vaidotas Kisielius from Lithuania, and
- Ms. Zanele Nyambi from South Africa,

Mr. Duraisamy, Mr. Karna, Ms. Gharaibeh and Ms. Nyambi are alumni of UNESCO-IHE who completed their MSc thesis within the framework of this project as partial fulfilment of their studies in the MSc programme Municipal Water and Infrastructure.

The following are highly acknowledged for their assistance: Rinnert Schurer (Evides) for the pilot plant monitoring studies, Gary Amy and Harvey Winters (KAUST) for the AOM characterisation studies, Thomas Neu (Helmholtz Centre for Environmental Research – UFZ) for the lectin-CLSM analyses, Emmanuelle Prest and Hans Vrouwenvelder (TUDelft) for the flow cytometry studies, Mieke Kleijn and Evan Spruijt (WUR) for the AFM analyses and Arie Zwijnenburg, Paula van den Brink (Wetsus) and all the participants of the Wetsus theme "Biofouling" for the fruitful discussions. The valuable supports of Evides Waterbedrijf in the Jacobahaven UF-RO demonstration plant and Pentair X-flow (Bastiaan Blankert, Ferry Horvath and Frederik Spenkelink and Erik Roesink) in the UF membrane fouling studies are also highly acknowledged.

This work was performed at UNESCO-IHE Institute for Water Education in Delft, the Netherlands (www.unesco-ihe.org) with the financial assistance from Wetsus Center of Excellence for Sustainable Water Technology (www.wetsus.nl). Wetsus is co-funded by the Dutch Ministry of Economic Affairs and Ministry of Infrastructure and Environment, the European Union Regional Development Fund, the Province of Fryslân and the Northern Netherlands Provinces.

References

Alldredge, A. L., Passow, U., & Logan, B. E. (1993). The abundance and significance of a class of large, transparent organic particles in the ocean. Deep-Sea Research I, 40, 1131–1140.

Anderson D.M., and McCarthy S. (2012) Red tides and harmful algal blooms: Impacts on desalination operations. Middle East Desalination Research Center, Muscat, Oman. Online: www.medrc.org/download/habs_and_desaliantion_workshop_report_final.pdf

Anderson, D. M., Reguera, B., Pitcher, G. C. and Enevoldsen, H. O. (2010), "The IOC International Harmful Algal Bloom Program: History and Science Impacts", Oceanography, 23 (3), 72-85.

Bar-Zeev E, Berman-Frank I, Girshevitz O, Berman T. (2012) Revised paradigm of aquatic biofilm formation facilitated by microgel transparent exopolymer particles. PNAS 109(23), 9119-24.

Berktay, A. (2011). Environmental approach and influence of red tide to desalination process in the middle-east region. International Journal of Chemical and Environmental Engineering 2 (3), 183-188.

Berman, T., and Holenberg, M. (2005) Don't fall foul of biofilm through high TEP levels. Filtration & Separation, 42(4), 30-32.

Busch M., Chu R., Rosenberg S. (2009) Novel Trends in Dual Membrane Systems for Seawater Desalination: Minimum Primary Pretreatment and Low Environmental Impact Treatment Schemes In: Proceedings of International Desalination Association World Congress, Dubai, UAE.

Caron , D.A., Garneau, M.E., Seubert, E., Howard M.D.A.,, Darjany L., Schnetzer A., Cetinic, I., Filteau, G., Lauri, P., Jones, B. and Trussell, S. (2010), "Harmful Algae and Their Potential Impacts on Desalination Operations of Southern California", Water Research, 44, pp. 385–416.

Cath, T. (2010) Advanced Water Treatment Engineering and Reuse. Graduate Course Script. Colorado School of Mines, Golden, CO.

DesalData (2013) Worldwide desalination inventory (MS Excel Format), downloaded from DesalData.com (GWI/IDA) on June 6, 2013.

FAO (2012) AQUASTAT database. http://www.fao.org/nr/aquastat.

Flemming, H. C., Schaule, G., Griebe, T., Schmitt, J., and Tamachkiarowa, A. (1997) Biofouling--the Achilles heel of membrane processes. *Desalination*, 113(2), 215-225.

Gomez V., Majamaa K., Pocurull E. and Borrull F. (2012) Determination and occurrence of organic micropollutants in reverse osmosis treatment for advanced water reuse. Water Science & Technology 66 (1), 61–71.

Gullinkala, T., Digman B., Gorey C., Hausman R., Escobar I. C. (2010) Desalination: Reverse osmosis and membrane distillation. In: I.C. Escobar and A.I. Schäfer (eds.) Sustainable Water for the Future: Water Recycling versus Desalination, Sustainability Science and Engineering vol. 2, pp 65– 93.

GWI (2007) Desalination Markets 2007, A Global Industry Forecast (CD ROM), Global Water Intelligence, Media Analytics Ltd., Oxford, UK, 2007, www.globalwaterintel.com.

Hofman, J. A. M. H., Beerendonk, E. F., Folmer, H. C., & Kruithof, J. C. (1997). Removal of pesticides and other micropollutants with cellulose-acetate, polyamide and ultra-low pressure reverse osmosis membranes. Desalination,113(2-3), 209-214.

Hussain, A. A., Nataraj, S. K., Abashar, M. E. E., Al-Mutaz, I. S., & Aminabhavi, T. M. (2008). Prediction of physical properties of nanofiltration membranes using experiment and theoretical models. Journal of Membrane Science,310(1-2), 321-336.

Jacangelo, J. G., Adham, S. S., and Laine, J. M. (1995) Mechanism of Cryptosporidium, Giardia, and MS2 Virus Removal by MF and UF. Journal American Water Works Association, 87(9), 107-121.

Kim, S.H. and Yoon, J.H., (2005), "Optimization of Microfiltration for Seawater Suffering from Red Tide Contamination", Desalination, 182, pp. 315-321.

Ladner, D. A., Vardon, D. R. and Clark M.M. (2010), "Effects of Shear on Microfiltration and Ultrafiltration Fouling by Marine Bloom-forming Algae", Journal of Membrane Science, 356, pp. 33–43.

Lattemann S., Kennedy M.D., Schippers J.C., Amy G. (2010) Global desalination situation. In: I.C. Escobar and A.I. Schäfer (eds.) Sustainable Water for the Future: Water Recycling versus Desalination, Sustainability Science and Engineering vol. 2, pp 7-39.

Nazzal N. (2009) 'Red tide' shuts desalination plant. Gulf News, Dubai, UAE. Available from http://gulfnews.com/news/gulf/uae/environment/red-tide-shuts-desalination-plant-1.59095.

Pankratz, T. (2008), "Red Tides Close Desal Plants", Water Desalination Report, 44, (1).

Passow, U., & Alldredge, A. L. (1994). Distribution, size, and bacterial colonization of transparent exopolymer particles (TEP) in the ocean. Marine Ecology Progress Series, 113, 185–198.

Passow, U., & Alldredge, A. L. (1995). A dye-binding assay for the spectrophotometric measurement of transparent exopolymer particles (TEP). Limnology and Oceanography, 40, 1326–1335.

Petry, M., Sanz M. A., Langlais C., Bonnelye V., Durand J.P., Guevara D., Nardes W. M., Saemi, C. H. (2007), "The El Coloso (Chile) Reverse Osmosis Plant", Desalination , 203, pp. 141–152.

Qu, F., Liang, H., Tian, J., Yu, H., Chen, Z., & Li, G. (2012). Ultrafiltration (UF) membrane fouling caused by cyanobateria: Fouling effects of cells and extracellular organics matter (EOM). Desalination, 293, 30-37.

Richlen M. L., Morton S. L., Jamali E. A., Rajan A., Anderson D. M. (2010), "The Catastrophic 2008–2009 Red Tide in the Arabian Gulf Region, with Observations on the Identification and Phylogeny of the Fish-killing Dinoflagellate Cochlodinium Polykrikoides", Harmful Algae, 9(2), pp. 163-172.

Schäfer, A. (2001). Natural organics removal using membranes: principles, performance, and cost. CRC Press, 406 pp.

Schippers, J. C., Kruithof, J. C., Nederlof, M. M., Hofman, J. A. M. H., and Taylor, J. S. (2004) Integrated Membrane Systems, AWWA Research Foundation and American Water Works Association.

UN Atlas of the Oceans (2010) Uses: Human Settlements on the Coast, Available at http://www.oceansatlas.org, Retrieved 21 October 2013.

Schurer R., Tabatabai A., Villacorte L., Schippers J.C., Kennedy M.D. (2013) Three years operational experience with ultrafiltration as SWRO pre-treatment during algal bloom. Desalination & Water Treatmen 51 (4-6), 1034-1042.

Wilf M., Awerbuch L., Bartels C., Mickley M., Pearce G., Voutchkov N. (2007) The Guidebook to Membrane Desalination Technology, Balaban Desalination Publications, L'Aquila, Italy (2007) 524 pp.

WWAP - World Water Assessment Programme (2012) The United Nations World Water Development Report 4: Managing Water under Uncertainty and Risk. Paris, UNESCO. http://unesdoc.unesco.org/images/0021/002156/215644e.pdf

2

Marine algal blooms

Contents

2.1 Introduction...24
2.2 Causative factors 24
    2.2.1 Natural processes 25
    2.2.2 Anthropogenic loadings..25
2.3 Impact on humans and the environment 26
2.4 Bloom-forming species 27
    2.4.1 Diatoms..29
    2.4.2 Dinoflagellates 30
    2.4.3 Raphidophytes...31
    2.4.4 Cyanobacteria (blue-green algae) 31
    2.4.5 Chlorophytes (green algae) 31
    2.4.6 Haptophytes 31
2.5 Harmful algal blooms..32
2.6 Algal organic matter (AOM)..34
    2.6.1 Transparent exopolymer particles (TEP)...35
    2.6.2 Marine mucilage 37
2.7 Summary...38
References 39

2.1   Introduction

Microscopic algae are one of the most ubiquitous forms of life on earth. These organisms thrive near the well-lit upper layer of the water column (pelagic zone) and obtain energy (for growth and reproduction) from sunlight through the process of photosynthesis. It is estimated that micro-algae are responsible for about half the production of organic matter in the earth's oceans (Field *et al.*, 1998). As a major producer of organic compounds, they essentially sustain the food web in the ocean and therefore have a crucial role in the biogeochemical cycles and formation of marine ecosystems (Falkowski *et al.*, 1998; Thomas *et al.*, 2012).

Micro-algae are very diverse in terms of size, shape and habitat. Their cells cover a wide range of shapes (rods to spheres) and sizes (0.2 to >2000 µm). They can also congregate to form long chains and large floating colonies. In the ocean, various taxa of algae exist. Occasionally, favourable conditions in the water can trigger rapid reproduction of some these algae, eventually resulting in dense concentration of algal biomass in the water. This sudden spike of algae concentration is called an algal bloom.

The reported frequency and severity of algal blooms have been increasing over the years. Some of these blooms showed damaging effects to the fishing and aquaculture industry and recently, they are also considered a major problem to the desalination industry (Caron *et al.*, 2010; Anderson and McCarthy, 2012). Reported operational problems in seawater desalination systems due to algal blooms have been increasing, especially in the Middle East where desalination is extensively applied and is an essential component of the water supply infrastructure (Berktay, 2011; Richlen *et al.*, 2010; Pankratz, 2008). Other areas in the world which rely on seawater desalination for their water supply are also vulnerable to this problem.

The current knowledge about algal blooms in the perspective of marine science, public health and aquaculture/fishing industry has been presented in various literatures (Sellner *et al.*, 2003; Anderson *et al.*, 2012). However, algal blooms are still poorly explained in the perspective of seawater desalination. This chapter presents basic background information about marine algal blooms relevant to membrane-based desalination.

2.2   Causative factors

Algae are opportunistic organisms. They are present in surface waters, usually in small concentrations or in resting stage (cyst or spores), waiting for the right conditions (e.g., sufficient sunlight, temperature and nutrients) to bloom. In due course, the bloom will terminate due to nutrient depletion, growth inhibition by pathogens and parasites and/or grazing by higher organisms. There are two main factors that can trigger an algal bloom: (1) natural processes and (2) anthropogenic loadings.

2.2.1 Natural processes

In addition to solar radiation, the distribution and concentration of algae in a body of water can be greatly influenced by natural physico-chemical variations (e.g., temperature, current, salinity, nutrients load, etc.) in the system coupled with the unique life cycles and behaviours of algal species present (Sellner et al., 2003). Natural phenomena such as storm events can cause increase of river discharges of nutrients to the sea while strong winds can induce mixing and transport of nutrient deposits from the lower water column to the surface where they can be utilised by algae (Smith et al., 1990; Trainer et al., 1998). Coastal upwelling, which is driven by the combination of wind, the Coriolis effects, and the Ekman drift, is a major factor for the transport of nutrients from the bottom of the sea to the surface (Mote and Mantua, 2002; Bakun, 1990). It was also reported that wind-driven dust events carrying iron-rich aerosols from the Sahara Desert may influence the frequency and severity of algal blooms in the Florida coasts, which is on the other side of the Atlantic (Walsh and Steidinger, 2001). A similar scenario may have occurred after dust events around the Yellow Sea (Shi et al., 2012), South China Sea (Wang et al., 2012) and the Persian Gulf region (Hamza et al., 2011; Nezlin et al., 2010).

2.2.2 Anthropogenic loadings

Human activities can trigger algal blooms by increasing nutrient loadings in coastal seawater through river discharge of untreated wastewater and run-off of untreated livestock wastes and residual fertilisers from agricultural areas. Increased incidence of severe algal blooms in populated areas has been shown to have substantial correlation with human population, fertilizer use and livestock production (Anderson et al., 2002; Sellner et al., 2003). Many regions in the world which implemented stricter environmental regulations to limit anthropogenic nutrient discharges to rivers have observed localised reduction in algal blooms, as in the case of the Seto Inland Sea in Japan (Okaichi, 1989).

Over the years, the reported frequencies and severity of algal blooms has been increasing and even spotted in locations where such phenomenon was never reported before. However, such increase may be attributed to the increased scientific attention to the problem and advancement in monitoring techniques (Anderson et al., 2010a). The common public perception about the trigger of algal blooms is environmental pollution caused by human activities, which is not true in most cases (Sellner et al., 2003). The typical global distribution of algae based on chlorophyll-a concentration is shown in Figure 2-1. The high concentrations of algae are located in areas near the mouth of large rivers, lakes, bays and various coastal regions, but do not necessarily occur in areas with high human activity and rather coincide with natural processes such as ocean currents along the peripheries of the ocean gyres and near the Polar Regions.

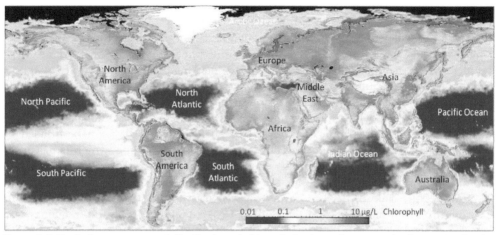

Figure 2-1: Typical average annual distribution of chlorophyll *a* in surface water bodies on Earth. The figure was modified from the composite map generated by Gledhill and Buck (2012) based on the 2009 Aqua MODIS chlorophyll composite (http://oceancolor.gsfc.nasa.gov/cgi/l3).

2.3   Impact on humans and the environment

The impact of algal blooms to the environment and humans ranges from being advantageous to harmful. Most algal blooms are advantageous to the aquatic environment where they are the primary producers of the food chain. Since fish and other sea creatures are directly or indirectly dependent on algae for nourishment, blooms can benefit the local fishing, aquaculture and tourism industry. However, some algae can cause problems when they reach sufficient numbers, due to either their production of toxins or their high biomass concentration (Glibert *et al.*, 2005).

The most obvious consequence of algal bloom is water discoloration. Dense concentration of algal cells can cause reddish, brownish or greenish discoloration of the water. "Red tide" blooms is a familiar type of algal bloom caused by micro-algae with reddish cell wall pigments. Algal biomass and their exudates can aggregate and float on the water surface as scum (mucilage formation) or as foam accumulating along the beaches. Dense biomass and exudates concentration can cause light and oxygen deprivation to other aquatic organisms living below the water surface, resulting in mortalities of aquatic organisms and/or damage of ecosystems (e.g., coral reef). Anoxic conditions may occur due to the high oxygen demand during bacterial decomposition of algal biomass. Some species of algae are known to produce toxic compounds that can alter cellular processes not only in aquatic organisms but also birds, humans and other mammals.

In general, severe forms of algal bloom can have negative consequences to human health, aquatic environment and/or local economies (Figure 2-2). A recent example is the severe red tide bloom in the Middle East Gulf region in 2008-2009 where it caused not only mortalities and damages to the marine ecosystem but also forced various desalination plants (the main source of water supply) in the area to stop or reduce operations due to fouling and odour issues (Pankratz, 2008; Richlen *et al.*, 2010).

Figure 2-2: Clockwise from top left: A "red tide" bloom by *Noctiluca scintillans* in New Zealand (M. Godfrey); green scum formed by toxic *Microcystis* bloom (W. Carmichael); massive cyanobacterial bloom around Gotland Island, Sweden (USGS/NASA); beach foam during *Phaeocystis* bloom in the North Sea (M. Veldhuis); massive fish mortalities in a lagoon in Rio de Janeiro due to suspected oxygen depletion by decomposing algae (C. Simon/AFP/Getty Images); beach closed due to red tide bloom near Sydney, Australia (S. Cocksedge).

2.4   Bloom-forming species

Guiry (2012) estimated that up to ~1 million species of algae exist on Earth, of which only about 44,000 species have so far been properly classified. The Intergovernmental Oceanographic Commission of UNESCO identified about 300 species of micro-algae that were reported to cause blooms in aquatic environments (IOC-UNESCO, 2013). An algal bloom event is often dominated by a group or a species of algae. The duration of an algal bloom event can be for a period of few days to several weeks, depending on the life cycles of causative species. The major groups of algae which are often reported to cause severe blooms are diatoms, dinoflagellates and cyanobacteria. However, some species of haptophytes, raphidophytes and chlorophytes were also often reported in many occasions. Some examples of the common species of bloom-forming algae are described in Table 2-1 and illustrated in Figure 2-3.

Table 2-1: Characteristics of common bloom-forming species of microscopic algae.

Bloom-forming algae	Cell shape $(\mu m)^{(+)}$	Cell size (μm)	Severe bloom $(cells/ml)^{(\#)}$	Potential adverse effect/consequences
Dinoflagellates				
Alexandrium tamarense	RE	25-32	10,000 [a]	toxic bloom, red tide, O_2 depletion
Cochlodinium polykrikoides	RE	20–40	48,000 [b]	toxic bloom, red tide, O_2 depletion
Karenia brevis	RE	20-40	37,000 [c]	toxic bloom, red tide, O_2 depletion
Noctiluca scintillans	Sp	200-2000	1,900 [d]	red/pink/green tide, O_2 depletion
Prorocentrum micans	FE	30-60	50,000 [e]	red/brown tide, O_2 depletion
Diatoms (golden brown)				
Chaetoceros affinis	OC	8-25	900,000 [f]	O_2 depletion, fish gill irritation
Pseudo-nitzschia spp.	0.8*PP	3-100	19,000 [g]	toxic bloom, O_2 depletion
Skeletonema costatum	Cy	2-25	88,000 [h]	O_2 depletion
Thalassiosira spp.	Cy	10-50	100,000 [e]	O_2 depletion
Cyanobacteria (blue-green)				
Anabaena spp. [fw]	Sp	3-12	10,000,000 [i]	toxic bloom, O_2 depletion
Microcystis spp. [fw]	Sp	2-7	14,800,000 [j]	toxic bloom, O_2 depletion
Nodularia spp.	Cy	6-100	605,200 [k]	toxic bloom, O_2 depletion
Haptophytes				
Emiliania huxleyi	Sp	2-6	115,000 [l]	O_2 depletion
Phaeocystis spp.	0.9*Sp	4-9	52,000 [m]	beach foam, O_2 depletion
Raphidophytes				
Chattonella spp.	Co+0.5*Sp	10-40	10,000 [n]	toxic bloom, red tide, O_2 depletion
Heterosigma akashiwo	Sp	15-25	32,000 [h]	toxic bloom, red tide, O_2 depletion
Chlorophytes (green)				
Chlorella vulgaris [fw]	Sp	2-10	145,000 [o]	green tide, O_2 depletion
Scenedesmus spp. [fw]	RE	2-25	820,000 [p]	green tide, O_2 depletion

RE=rotational ellipsoid; S=sphere; FE=flattened ellipsoid; OC=oval cylinder; PP=parallelepiped; Cy=cylinder; Co=cone; O_2=dissolved oxygen; (fw) freshwater algae; (+) Equivalent geometric dimensions of algal cells based on Olenina *et al.* (2006); (#) Maximum recorded concentrations reported in literature.

References: (a) Anderson & Keafer (1985); (b) Kim (2010); (c) Tester *et al.* (2004); (d) Chang (2000) as cited by Fonda-Umani *et al.* (2004); (e) Kim and Yoon (2005); (f) Villacorte *et al.* (2013); (g) Anderson *et al.* (2010b); (h) Shikata *et al.* (2008); (i) Hori *et al.* (2002); (j) Dixon *et al.* (2011a); (k) McGregor *et al.* (2012);(l) Berge (1962); (m) Janse *et al.* (1996); (n) Orlova *et al.* (2002); (o) Plantier *et al.* (2012); (p) Lürling and van Donk (1997).

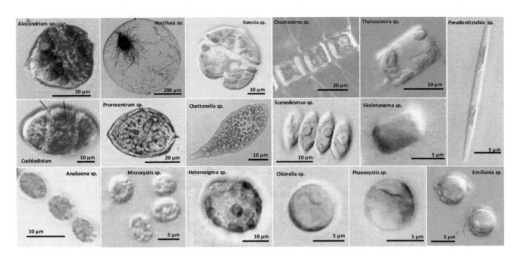

Figure 2-3: Light microscope photographs of common species of bloom forming algae. Photo sources: CCAP, Richlen *et al.* (2010), CCMP, www.algaebase.org, D. Vaulot/www.planktonnet.awi.de, www.shigen.nig.ac.jp.

2.4.1 Diatoms

Diatoms are yellowish brown algae known for their characteristic "glassbox" shape. Each cell is encased inside two overlapping siliceous frustules or exoskeleton. Their dense frustules are rather heavy, but diatom cells can remain buoyant by pumping light-weight ions into their cell vacuole. Since diatoms rely heavily on photosynthesis for their energy, they thrive near the surface of the water. However when nutrients are depleted, they tend to sink to the bottom of the water body where they lay dormant until they migrate back to the surface when sufficient nutrients and light are again available. Diatoms reproduce asexually by binary fission.

In temperate regions, diatom blooms normally occur at the start of spring season when nutrients, light and temperature conditions are just sufficient, but it is generally difficult to predict in other regions. Commonly reported bloom-forming species of marine diatoms belong to the genus *Chaetoceros*, *Skeletenoma* and *Thalassiosira*. Since diatoms require silicates for their exoskeleton formation, the termination of a bloom event is often defined by the depletion of silicates (Martin-Jézéquel *et al.*, 2000). In the Oosterschelde bay (SW Netherlands), the first bloom event in spring is usually dominated by diatoms; hence, silicates are the first essential nutrients to be depleted (Figure 2-4). Aside from nutrient depletion, a diatom bloom can also be terminated by predation by other aquatic organisms including dinoflagellates.

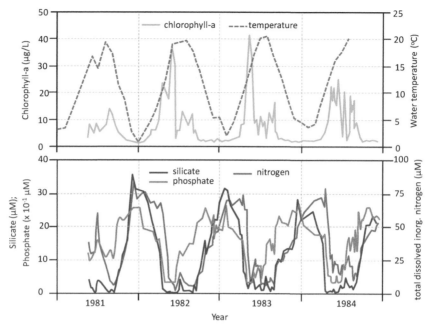

Figure 2-4: Typical drop of essential nutrient concentrations after the spring algal bloom events in the Oosterschelde bay (SW Netherlands). Chlorophyll-a, silicate, phosphate and nitrogen graphs redrawn from Wetsteyn *et al.* (1990) and temperature data adopted from van der Hoeven (1984).

2.4.2 Dinoflagellates

Dinoflagellates are mostly unicellular, rarely colonial, bi-flagellated algae with pigmentation generally reflecting a red color. Dinoflagellate cells range in size from 5 to 2000 μm. Majority of dinoflagellates are photosynthetic while some are heterotrophic. However, some of the photosynthesizing species can show capacities to ingest bacteria and other algae (including other dinoflagellates) as a facultative or a typical feature of their lifestyles (Reynolds, 2006). Among bloom-forming species of algae, dinoflagellates are better adapted to taking advantage of available nutrients. Using their flagellates for mobility, they can swim up and down the water column to photosynthesise during the day and take advantage of higher nutrient levels at deeper portions during the night. Cell reproduction is asexual through binary fission. When conditions are no longer favourable for growth, they form heavy resting cysts which sediment to the seafloor where they lay dormant until the cysts transform to motile cells again the next time growth conditions are favourable (Anderson *et al.*, 1985). Since dinoflagellates have a slower reproduction time than diatoms, they tend to follow diatom blooms and often blooms can stay for a longer period (Smayda and Reynolds, 2003; Smayda and Trainer, 2010). The known bloom-forming species of dinoflagellates include *Noctiluca scintillans*, *Cochlodinium polykrikoides* and various species under the genus *Alexandrium*, *Karenia* and *Prorocentrum*.

2.4.3 Raphidophytes

Raphidophytes is a smaller group of unicellular, bi-flagellated algae with no rigid cell walls. They are photosynthetic autotrophs comprising both freshwater and marine species. Various marine raphidophytes can cause toxic "red tide" blooms. For instance, some species of *Chatonella* has been reported to cause blooms sporadically during summer months in the Seto Inland Sea and the Sea of Japan (Nakamura *et al.* 1989; Orlova *et al.*, 2002). *Heterosigma akashiwo*, which is named after the Japanese word for red tide, is a common bloom-forming raphidophyte in Japanese waters as well as in Brazil, New Zealand and various coastal areas of North America (Shikata *et al.*, 2008; Guiry, 2013).

2.4.4 Cyanobacteria (blue-green algae)

Also known as blue-green algae, cynobacteria are actually a group of bacteria which obtain their energy through photosynthesis. Most cyanobacteria are obligate phototrophs which means they prefer to grow in well-lit portions of the water column. Their cells range in size from 0.5 to 60 μm but they can form colonies of up to few centimetres in diameter. Just like diatoms, cyanobacteria can control their buoyancy, allowing them to migrate vertically through the water column. They are considered as the most successful group of microorganisms on earth and are ubiquitous in both freshwater and marine ecosystems where they can seasonally form blooms. Cyanobacterial blooms can be very intense and can cover a wide surface area (Figure 2-2). Bloom concentration of several million cells/ml is not uncommon, often resulting in green oil paint-like scum formations on the water surface which can cause asphyxia on various aquatic organisms. Moreover, some cyanobacteria under the genus *Microcystis* and *Anabaena* in freshwater and *Nodularia* and *Lyngbya* in brackish/saline waters are not only capable of accumulating in dense concentration but also of synthesising toxic compounds which can affect both aquatic organisms and humans.

2.4.5 Chlorophytes (green algae)

Chlorophytes, or more commonly known as green algae, is a big group of phototrophic organisms ranging from unicellular algae to macroscopic seaweeds. Green algal blooms caused by multiple species of microscopic chlorophytes are a common occurrence in stagnant bodies of freshwater (e.g., lakes, ponds). However in seawater, green algal blooms are mainly dominated by macroscopic chlorophytes (Nelson *et al.*, 2008; Leliaert *et al.*, 2009). The largest recently recorded "green tide" phenomenon occurred in the Yellow Sea (near Qingdao, China) was caused by the seaweed *Ulva prolifera* (Leliaert *et al.*, 2009). Such occurrence has been recurring in the area ever since, possibly due to increased human activities (Liu *et al.*, 2013).

2.4.6 Haptophytes

Some algae under the phylum haptophytes can form massive blooms. A common example is the algae *Emiliania huxleyi*, reported to form widespread blooms (up to 250,000 km² of surface area) in the North Atlantic Ocean, North Sea, Black Sea and Norwegian fiords (Holligan *et al.*, 1993; Berg, 1962; Tyrrell and Merico, 2004). Intense

E. huxleyi blooms can turn a vast area of the ocean to milky colour due to light reflectance of the calcite disks covering the algal cells. *Phaeocystis* is another notable haptophyte which can frequently cause blooms in coastal seawater. For instance, in the coastal areas of the North Sea, foam accumulation resulting from *Phaeocystis* bloom is a regular occurrence during the spring period and in sometimes they caused mortalities in local aquatic fauna (Peperzak and Poelman, 2008).

2.5 Harmful algal blooms

Red tides are often perceived as synonymous to harmful algal blooms (HABs). However, not all red tide blooms are harmful and not all HAB species cause red tides (Anderson, 1994). Red tides are increasingly associated with HABs because various red tide forming dinoflagellates and raphidophytes release toxic compounds that can directly and/or indirectly affect other aquatic organisms and mammals, including humans and other land animals that live around or use the affected body of water as a food source. Other blooms such as those caused by few species of cyanobacteria and diatoms (e.g., *Pseudo-nitzschia*) were also reported to produce toxic compounds. The common marine HAB species and the toxins they produced are presented in Table 2-2.

Table 2-2: Common illness reported due to toxins released by marine HAB species and affected areas.

Syndrome	Toxins	Causative algae	Commonly affected areas	Ref.
Paralytic shellfish poisoning (PSP)	Saxitoxins, Gonyautoxins	*Alexandrium spp. Gymnodinium spp. Pyrodinium spp.*	U.S. west coast, Alaska, New England, Canada, Chile, Europe, South Africa, Asia, Australia, New Zealand	[1],[2]
Neurotoxic shellfish poisoning (NSP)	Brevetoxins	*Kerenia brevis, Karenia brevisulcatum Chatonella spp. Fibrocapsa japonica Heterosigma akashiwo*	U.S. Gulf coast, New Zealand, Japan, Australia	[1],[2],[3]
Diarrhetic shellfish poisoning (DSP)	Okadaic acid	*Dinophysis spp. Prorocentrum lima*	Europe, Japan, Canada (Atlantic coast), South Africa, Chile, Thailand, New Zealand, Australia	[1],[4]
Amnesic shellfish poisoning (ASP)	Domoic acid	*Pseudo-nitzchia spp.*	U.S. west coast, Alaska, Canada (Atlantic coast), Chile, Australia, New Zealand, United Kingdom	[1],[2],[5]
Azaspiracid shellfish poisoning (AZP)	Azaspiracid	*Protoperidinium crassipes*	England, Scotland, Ireland, France, Spain, Morocco, Norway	[1],[2],[6]
Ciguatera fish poisoning (CFP)	Ciguatoxins, Maitotoxins	*Gambierdiscus toxicus*	Hawaii, Gulf of Mexico, Puerto Rico, the Caribbean, Australia, many Pacific islands	[1],[2],[7]

References: [1] Anderson (2012); [2] Wang (2008); [3] Watkins *et al.* (2008); [4] Subba Rao *et al.* (1993); [5] Jeffery *et al.* (2004); [6] Twiner *et al.* (2008); [7] Friedman *et al.* (2008).

The severe consequences of toxic HABs include mortalities of fish, birds, and mammals (including human), respiratory or digestive tract problems, memory loss, seizures, lesions and skin irritation, and damage of coastal resources including submerged aquatic vegetation and benthic fauna (Sellner *et al.*, 2003). Even when concentration of toxin-producing algae in the water is rather low, it may still cause health problems to humans who consumed bivalve molluscs (e.g., mussels, clams, oysters) which have accumulated toxins over time by ingestion of algal cells. Saxitoxins are one of the commonly occurring HAB toxins which can cause paralytic shellfish poisoning (PSP) to mollusc-feeding mammals. They are produced by various species of dinoflagellates under the genus *Alexandrium*, *Gymnodinium* and *Pyrodinium*. The global distribution of HAB events involving these species of algae is presented in Figure 2-5.

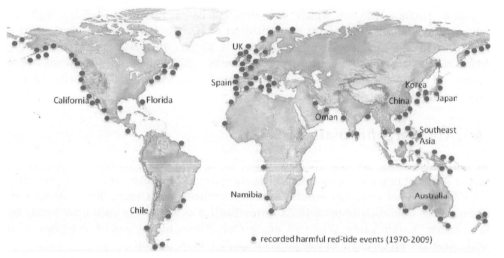

Figure 2-5: Global distribution of "red tide" HAB events based on PSP toxins detected in shellfish or fish. Map modified from Anderson *et al.*, 2012 and the US National Office for Harmful Algal Blooms.

Cochlodinium polykrikoides is another commonly occurring HAB species not listed in Table 2-2. So far, there is no clear consensus among marine scientists regarding the associated toxic mechanism or the chemical nature of toxins produced by the algae. Various studies categorized *Cochlodinium* species as a taxa with multiple toxins which may include neurotoxic, hemolytic, hemagglutinative and zinc-bound PSP toxins (Tang and Gobler, 2009). Harmful blooms of *C. polykrikoides* have been recorded in various parts of the world including East and Southeast Asia, the Middle East and the United States.

Some species of marine and brackish water cyanobacteria under the genus *Nodularia* also produce a toxin known as nodularin (Laamanen *et al.*, 2001). This toxin is a potent hepatotoxin and can cause damage to the liver of mammals who ingested it. Massive *Nodularia* blooms have occurred frequently in the Baltic Sea where it was reported to cover up to an area of more than 60,000 km^2 (Kahru *et al.*, 1994).

Some HABs are not caused by toxin producing species but species which tend to accumulate in dense concentrations on the surface of the water. This can be harmful to aquatic organisms because it can cause light deprivation as well as sudden drop of dissolved oxygen concentration (hypoxia) in the lower water column resulting from excessive cellular respiration and bacterial degradation of dead algal material. Hypoxic conditions induced by cyanobacterial blooms (e.g., *Microcystis*) are often reported in large freshwater lakes (e.g., Lake Taihu, China; Li *et al.*, 2012a). In seawater, anoxic conditions during the sedimentation phase of *Phaeocystis* blooms were also reported to cause mortalities of wild and/or cultured marine fauna in Ireland (Rogers and Lockwood, 1990), the Netherlands (Peperzak and Poelman, 2008), China (Lu and Huang, 1999), and Vietnam (Tang *et al.*, 2004).

About 60-80 species out of the 300 species of bloom-forming algae have been reported to cause HABs, 75% of which are dinoflagellates (Smayda, 1997). High cell concentration is not necessarily an indication of HAB but it is rather dependent on the causative species. For example in South Korea, a HAB alert is raised when *Cochlodinium polykrikoides* concentration exceeds 1,000 cells/mL while during diatom blooms, an alert will only be issued when concentration exceeds 50,000 cells/ml (Kim, 2010; Kim and Yoon, 2005).

2.6 Algal organic matter (AOM)

The natural organic matter (NOM) present in aquatic environment is a mixture of diverse forms of organic compounds originating from both autochthonous (local input) and allochthonous (external input) sources (Leenheer and Croué, 2003). Algae are a major source of autochthonous NOM in the Earth's oceans, accounting for about half the organic matter input (Field *et al.*, 1998). These algae-derived substances are collectively known as algal (or algogenic) organic matter (AOM).

Algal blooms, especially diatom blooms, often represent the highest annual pulses of AOM production in the global ocean (Burrell, 1988). Algal blooms (harmful or non-harmful) produce various forms and differing concentrations of AOM comprising polysaccharides, proteins, lipids, nucleic acids and other dissolved organic substances (Fogg, 1983; Bhaskar and Bhosle, 2005; Decho, 1990; Myklestad, 1995). Among the major components of AOM, Myklestad (1995) highlighted the significance of extracellular polysaccharides as they may comprise > 80% of AOM production. A significant fraction of these exopolysaccharides are highly surface-active and sticky (e.g., transparent exopolymer particles, TEP), which has been suspected to play a major role in the aggregation dynamics of algae during bloom events (Myklestad, 1995; Mopper *et al.*, 1995).

There are two types of AOM, namely: (1) organic substances released during the metabolic activity of algae known as extracellular organic matter (EOM) and (2) substances released through autolysis and/or during the process of cell decay called intracellular organic matter (IOM). Algal cells excrete EOM mostly in response to low nutrient stress and other unfavourable conditions (e.g. light, pH and temperature) or invasion by bacteria or viruses (Fogg, 1983; Leppard, 1993; Myklestad, 1999). EOM substances can be either discrete or remained attached (bound) to the algal cell as coatings. Discrete EOMs often contain mainly polysaccharides and tend to be more hydrophilic while bound EOM contain more proteins and tend to be more hydrophobic (Qu et al., 2012). On the other hand, IOMs comprise mainly low molecular weight polymers released from the interior of compromised, dying or deteriorating cells, which sometimes carry toxins, and taste and odour compounds (Dixon et al., 2010; Li et al., 2012b). Considering the conditions of how they are released, the contribution of IOM to the total AOM production is expected to increase during the stationary-death phase of an algal bloom.

2.6.1 Transparent exopolymer particles (TEP)

A substantial fraction of AOM can be made up of high molecular weight, hydrophilic, anionic muco-polysaccharides and glycoproteins. In the field of oceanography and limnology, such substances are collectively known as transparent exopolymer particles (TEP). The term "TEP" was first introduced by Alldredge et al. (1993) to describe gel-like transparent substances in seawater which can be collected by microfiltration (i.e., 0.4µm polycarbonate filter) and visualised and counted by staining with Alcian blue dye. The dye is a strongly cationic compound which has specific affinity to form bonds with anionic sugar moieties (e.g., carboxyl, sulphate and phosphate groups), yielding an insoluble non-ionic blue precipitate (Ramus, 1977; Scott and Dorling, 1965). More methods were further developed based on this technique to routinely investigate TEPs in saline and freshwater environments (e.g., Passow and Alldredge, 1995; Arruda-Fatibello et al., 2004; Thornton et al., 2007). Ever since the first method was introduced, various studies have shown that TEPs may have a crucial role in abiotic particle formation, aggregation, sedimentation, food web structure and carbon cycling in the ocean and some inland bodies of water (see review by Passow, 2002a).

In natural waters, TEPs mainly originate from exudates of phyto- and bacterio-planktons (Passow, 2002a,b), but may also originate from larger aquatic organisms such as macroalgae, oysters, mussels, scallops and sea snails (McKee et al. 2005; Heinonen et al. 2007; Thornton, 2004). During algal bloom, TEPs are mainly from algal-derived IOM and/or EOM (Figure 2-6). Generally, TEPs are highly heterogeneous, both physically and chemically. Their volume and stability largely depends on environmental conditions while their chemical composition is known to be highly variable depending on the species releasing them and the prevailing growth conditions (Passow, 2002a).

TEPs are found in most marine and freshwater environments, with reported concentrations up to ~900,000 TEP/ml (Passow, 2002a). The observed size range of TEPs is typically 5 to 200 µm and they oftentimes formed an integral part of marine snow-sized aggregates >500 µm (Alldredge et al., 1993; Passow and Alldredge, 1994). Although TEPs were operationally defined as Alcian blue stainable substances retained on 0.40 µm filters, they can form from much smaller colloidal polymers (TEP-precursors), perhaps fibrils as small as 1–3 nm in diameter and up to 100s of nm in length (Leppard *et al.*, 1977; Passow, 2000).

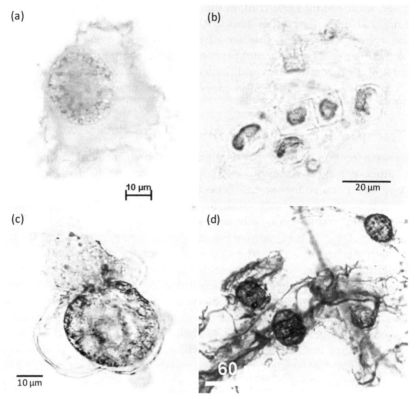

Figure 2-6: Alcian blue stained transparent exopolymers particles (TEP) released by bloom-forming algae: (a) excretions by Lepidodinium chlorophorum (Claquin *et al.*, 2008); (b) excretions by *Chaetoceros affinis* (Villacorte *et al.*, 2013) (c) release from a broken *Gonyaulax fragilis* cell (Pompei *et al*, 2003); (d) excretions by *Gonyaulax hyalina* (MacKenzie *et al.*, 2002).

TEPs are important in many aspects of particle dynamics in aquatic systems and have long been associated with the coagulation and sedimentation of particles in surface waters (Passow, 2002a; Passow *et al.*, 1994). They are known to be highly flexible and sticky, which might explain their tendency to aggregate into large flocs and to adhere to other materials (Mopper *et al.*, 1995). Their relative stickiness is reported to be 2–4 orders of magnitude higher than most suspended particles in natural waters (Passow, 2002a). TEPs can aggregate with and facilitate sedimentation of solid materials in natural waters such as carbonates, clays, algae, bacteria, and detrital matter (Alldredge *et al.*, 1993; Kiørboe and Hansen, 1993; Thornton, 2002). However, when the TEP to solid particle ratio in the aggregate is high, the sinking of the aggregate can be retarded due to the low specific gravity (~0.86) of TEPs (Azetsu-Scott and Passow, 2004). Such characteristic may also provide a vehicle for the upward flux of biological and chemical components in the marine environment, including bacteria, phytoplankton, organic carbon, and reactive trace elements.

Because of their adhesive characteristic, TEPs can accumulate on solid-liquid interfaces and facilitate adsorption of suspended particles, including bacteria. The adsorbed and suspended TEPs can be colonized, degraded and may later serve as a substrate for bacteria (Passow and Alldredge, 1994; Alldredge *et al.*, 1993). Recently, Berman and co-workers proposed a "revised paradigm of aquatic biofilm formation facilitated by TEPs" emphasising the important role of TEPs in the conditioning and bacterial colonisation of surfaces exposed to seawater (Berman and Holenberg, 2005; Bar-Zeev *et al.*, 2012a).

2.6.2 Marine mucilage

Marine mucilage is a phenomenon characterized by the appearance of a sporadic but massive accumulation of gelatinous material at and below the water surface (Leppard, 1995). It is generally a result of excessive production of EOM during an algal bloom in response to low nutrient (P, N, Si) stress and/or invasion by pathogens (Mingazzini and Thake, 1995). Severe mucilage events occasionally occur in the North Sea, Adriatic Sea and other parts of the Mediterranean region but proliferation of smaller mucilage aggregates such as "marine snow" has been commonly reported in oceanic and marine systems (Mingazzini and Thake, 1995; Lancelot, 1995; Rinaldi *et al.*, 1995; Gotsis-Skretas, 1995). In the North Sea, the colony-forming *Phaeocystis* is identified as the main culprit of mucilaginous phenomena (Lancelot, 1995). In the Adriatic Sea, it is mainly attributed to EOM production by diatoms (e.g., *Nitzschia closterium, Chaetoceros affinis* and *Skeletonema costatum*) but some cyanobacteria and benthic macroalgae (e.g., *Tribonema marinum* and *Acinetospora crinite*) may have been likely involved as well (Innamorati, 1995; Mingazzini and Thake, 1995).

Marine mucilages have been reported in different forms: marine snow (>0.5 mm diameter), strings (2-15 cm long), tapes and clouds of up to several kilometres long (Figure 2-7). The mucilaginous aggregates are a heterogeneous consolidation of inorganic particles, biogenic debris/exudates (including TEPs) and dead and living organisms, including healthy eukaryotes and prokaryotes (Alldredge and Silver, 1988; Leppard, 1995). The aggregates are generally unstable and tend to change in size, shape and colour (becoming progressively darker with age) over time. Weather and wave conditions in the sea can dictate the formation and termination of the phenomenon, as a storm event can disperse the mucilage aggregates over a short period of time (Mingazzini and Thake, 1995).

Figure 2-7: Clockwise from top left: a trail of surface mucilage mats in the northern Adriatic Sea (S. Fonda Umani); heavy mucilage blobs accumulating near the seafloor (M. Cornello/NG); a marine snow aggregate containing organic materials and brown coloured phytoplankton cells (Lampitt, 2001); a marine biologist investigating a mucilage blob (*N. Caressa /NG)*; oceanographers using nets to collect samples from falling marine snow (http://www.waterencyclopedia.com/Mi-Oc/Ocean-Biogeochemistry.html).

2.7 Summary

Seasonal proliferation of marine microscopic algae or algal blooms can potentially occur in many coastal areas of the world. This phenomenon is mainly triggered by natural processes but human activities have been shown to increase their reported frequency and severity. The recorded algal cell size, algal population and potential consequences during algal blooms can vary substantially, owing mainly to the high diversity of species which caused them. Some types of blooms are considered harmful because of their capability to produce toxins and/or their high biomass concentration.

Harmful and non-harmful blooms can produce varying concentrations and different forms of algogenic organic matter (AOM) which are either actively exuded by living algal cells and/or released through lyses of compromised or deteriorating cells. Polysaccharide substances generally comprise a large majority of AOM production during algal bloom. A substantial fraction of these substances are acidic, highly sticky and surface active, commonly known as transparent exopolymer particles (TEP). These materials are essential components of mucilaginous aggregates (e.g., marine snow, sea foam) which can form in both the surface and lower part of the water column during the senescent stage of the bloom. TEPs have been identified as a potential initiator/promoter of biofilm in aquatic systems and membrane filtration systems. Hence, TEPs may be the key to understanding the adverse impact of algal blooms in membrane-based desalination systems.

References

Alldredge, A. L., & Silver, M. W. (1988). Characteristics, dynamics and significance of marine snow. Progress in Oceanography, 20(1), 41-82.

Alldredge, A. L., Passow, U., & Logan, B. E. (1993). The abundance and significance of a class of large, transparent organic particles in the ocean. Deep-Sea Research I, 40, 1131–1140.

Anderson C.R., Sapiano M.R.P., Prasad M.B.K., Long W., Tango P.J., Brown C.W. and Murtugudde R. (2010b) Predicting potentially toxigenic Pseudo-nitzschia blooms in the Chesapeake Bay. Journal of Marine Systems, 83: 127 – 140.

Anderson D. (1994) Red tides. Scientific American 271, 52-58.

Anderson D. (2012) Harmful algae, Online access http://www.whoi.edu/redtide/species/by-syndrome

Anderson D.M., and McCarthy S. (2012) Red tides and harmful algal blooms: Impacts on desalination operations. Middle East Desalination Research Center, Muscat, Oman. Online: www.medrc.org/download/habs_and_desaliantion_workshop_report_final.pdf

Anderson DM, Glibert PM, Burkholder JM (2002) Harmful algal blooms and eutrophication: nutrient sources, composition, and consequences. Estuaries 25: 704–726.

Anderson, D. M., Coats, D. W, Tyler, M. A. (1985). Encystment of dinoflagellate Gyrodinium uncatenum: temperature and nutrient effects. J. Phycol. 21:200-206

Anderson, D.M. and Keafer, B.A. (1985) Dinoflagellate cyst dynamics in coastal and estuarine waters. pp. 219-224, In: D.M. Anderson, White, A.W., Baden, D.G. (eds) Toxic dinoflagellates. Proc. 3rd Int'l. Conf., Elsevier, New York.

Anderson, D.M., B. Reguera, G.C. Pitcher, and H.O. Enevoldsen (2010a) The IOC International Harmful Algal Bloom Program: History and science impacts. Oceanography 23(3), 72–85.

Anderson, DM and Cembella, AD and Hallegraeff, GM (2012) Progress in understanding harmful algal blooms: paradigm shifts and new technologies for research, monitoring and management, Annual Review Marine Science 4, 143-176.

Arruda-Fatibello S, Vieira AAH, Fatibello Filho O (2004) A rapid spectrophotometric method for the determination of transparent exopolymer particles (TEP) in freshwater. Talanta, 62(1):81-85.

Azetsu-Scott, K., and Passow, U. (2004) Ascending marine particles: Significance of transparent exopolymer particles (TEP) in the upper ocean. Limnol. Oceanogr, 49(3), 741-748.

Bakun A. (1990) Global Climate Change and Intensification of Coastal Ocean Upwelling. Science 247, 198-201.

Bar-Zeev E, Berman-Frank I, Girshevitz O, and Berman T (2012) Revised paradigm of aquatic biofilm formation facilitated by microgel transparent exopolymer particles. PNAS 109(23):9119–9124.

Berge G (1962) Discoloration of the sea due to Coccolithus huxleyi bloom. Sarsia 6, 27–40.

Berktay, A. (2011). Environmental approach and influence of red tide to desalination process in the middle-east region. International Journal of Chemical and Environmental Engineering 2 (3), 183-188.

Berman T, Holenberg M (2005) Don't fall foul of biofilm through high TEP levels. Filtration & Separation 42(4):30-32.

Bhaskar, P. V., & Bhosle, N. B. (2005). Microbial extracellular polymeric substances in marine biogeochemical processes. Current Science, 88(1), 45-53.

Burrell, D.C. (1988). Carbon flow in fjords. Oceanogr. Mar. Biol. Ann. Rev. 26: 143-226.

Caron , D.A., Garneau, M.E., Seubert, E., Howard M.D.A.,, Darjany L., Schnetzer A., Cetinic, I., Filteau, G., Lauri, P., Jones, B. and Trussell, S. (2010), "Harmful Algae and Their Potential Impacts on Desalination Operations of Southern California", Water Research, 44, pp. 385–416.

Chang, F. H. (2000) Pink blooms in the springs in Wellington Harbour. Acquacult. Update, 24, 10–12.

Claquin P., Probert I., Lefebvre S., Veron B. (2008) Effects of temperature on photosynthetic parameters and TEP production in eight species of marine microalgae. Aquat Microb Ecol 51, 1-11.

Decho, A.W. (1990) Microbial exopolymer secretions in ocean environments: their role(s) in food webs and marine processes. Oceanography and Marine Biology: an Annual Review 28, 73-153.

Dixon M.B., Richard Y., Ho L., Chow C.W.K., O'Neill B.K. and Newcombe G. (2011a). A coagulation-powdered activated carbon-ultrafiltration - multiple barrier approach for removing toxins from two australian cyanobacterial blooms. Journal of Hazardous Materials, 186(2-3), 1553-1559.

Dixon, M.B., Falconet, C., Ho, L., Chow, C.W.K., O'Neill, B.K., Newcombe, G. (2010). Nanofiltration for the removal of algal metabolites and the effects of fouling. Water Science & Technology - WST 61 (5), 1189-1199.

Falkowski P.G., Barber R.T., and Smetacek V. (1998). Biogeochemical Controls and Feedbacks on Ocean Primary Production. Science 281 (5374), 200-206.

Field C.B., Behrenfeld M.J., Randerson J.T., and Falkowski P. (1998) Primary Production of the Biosphere: Integrating Terrestrial and Oceanic Components. Science 281, 237-240.

Fogg, G. E. (1983) The ecological significance of extracellular products of phytoplankton photosynthesis. Bot. Mar., 26, 3–14.

Fonda-Umani S., Beran A., Parlato S., Virgilio D., Zollet T., De Olazabal A., Lazzarini B., and Cabrini M. (2004) Noctiluca scintillans Macartney in the Northern Adriatic Sea: long-term dynamics, relationships with temperature and eutrophication, and role in the food web, J. Plankton Res. 26(5), 545-561.

Friedman MA, Fleming LE, Fernandez M, Bienfang P, Schrank K, Dickey R, Bottein M.Y., Backer L., Ayyar R., Weisman R., Watkins S., Granade R., and Reich A. (2008) Ciguatera fish poisoning: treatment, prevention and management. Mar Drugs 6, 456–479.

Gledhill M and Buck KN (2012) The organic complexation of iron in the marine environment: a review. Front. Microbio. 3:69.

Glibert, P.M., Anderson D.M., Gentien P., Granéli E., and Sellner K.G. (2005) The global, complex phenomena of harmful algal blooms. Oceanography 18(2):136–147.

Gotsis-Skretas, O. (1995). Mucilage appearances in greek waters during 1982-1994. Science of the Total Environment, 165, 229-230.

Guiry, M. D. (2012) How many species of algae are there? Journal of Phycology, 48: 1057–1063.

Guiry, M. "Algaebase". Retrieved 27 October 2013; Online access www.algaebase.org.

Hamza W., Enan M.R., Al-Hassini H., Stuut J.-B., de-Beer D. (2011) Dust storms over the Arabian Gulf: a possible indicator of climate changes consequences. Aquatic Ecosystem Health & Management 14 (3), 260-268.

Holligan P.M. Fernandez E., Aiken J., Balch W.M., Boyd P., Finch M., Groom S.B., Malin G., Muller K., Purdie D.A., Robinson C., Trees C.C., Turner S.M., van der Wal P. (1993) A biogeochemical study of the coccolithophore Emiliania huxleyi in the north Atlantic. Global Biogeochem Cycles 7(4): 879-900.

Hori, K., Ishii, S., Ikeda, G., Okamoto, J., Tanji, Y., Weeraphasphong, C., & Unno, H. (2002). Behavior of filamentous cyanobacterium anabaena spp. in water column and its cellular characteristics. Biochemical Engineering Journal,10(3), 217-225.

Innamorati, M. (1995). Hyperproduction of mucilages by micro and macro algae in the tyrrhenian sea. Science of the Total Environment, 165, 65-81.

IOC-UNESCO (2013) What are harmful algae? Retrieved 30 September 2013, Online access http://hab.ioc-unesco.org/

Janse, I., Van Rijssel, M., Gottschal, J.C., Lancelot, C., Gieskes, W.W.C. (1996) Carbohydrates in the North Sea during spring blooms of Phaeocystis: A specific fingerprint. Aquat. Microb. Ecol. 10, 97–103.

Jeffery B., Barlow T., Moizer K., Paul S., Boyle C. (2004) Amnesic shellfish poison, Food and Chemical Toxicology 42(4), 545–557.

Kahru, M., U. Horstmann, and O. Rud (1994) Satellite detection of increased cyanobacterial blooms in the Baltic Sea: natural fluctuation or ecosystem change? Ambio 23:469–472.

Kim H.-G. (2010) An Overview on the Occurrences of Harmful Algal Blooms (HABs) and Mitigation Strategies in Korean Coastal Waters In: A. Ishimatsu and H.-J. Lie (eds.) Coastal Environmental and Ecosystem Issues of the East China Sea, pp. 121–131.

Kim S.H. and Yoon J.S. (2005) Optimization of microfiltration for seawater suffering from red-tide contamination. Desalination 182(1–3), 315–321.

Kiørboe T. and Hansen J.L.S. (1993) Phytoplankton aggregate formation: observations of patterns and mechanisms of cell sticking and the significance of exopolymeric material. J. Plankton Res. 15(9): 993-1018.

Laamanen, M. J., Gugger, M. F., Lehtimäki, J. M., Haukka, K., & Sivonen, K. (2001). Diversity of toxic and nontoxic nodularia isolates (cyanobacteria) and filaments from the baltic sea. Applied and Environmental Microbiology, 67(10), 4638-4647.

Lampitt R.S. (2001) Marine Snow. pp. 1667–1675, In: J.H. Steele (ed) Encyclopedia of Ocean Sciences, Academic Press, Oxford.

Lancelot, C. (1995). The mucilage phenomenon in the continental coastal waters of the north sea. Science of the Total Environment, 165, 83-102.

Leenheer J.A. and Croué J.P. (2003) Characterizing Aquatic Dissolved Organic Matter: Understanding the unknown structures is key to better treatment of drinking water. Environmental Science & Technology 37 (1), 18A-26A.

Leliaert, F., Zhang, X., Ye, N., Malta, E.-j., Engelen, A. H., Mineur, F., Verbruggen, H. and De Clerck, O. (2009), Research note: Identity of the Qingdao algal bloom. Phycological Research, 57: 147–151.

Leppard, G. G. (1995). The characterization of algal and microbial mucilages and their aggregates in aquatic ecosystems. Science of the Total Environment, 165, 103-131.

Leppard, G. G., Massalski, A., & Lean, D. R. S. (1977). Electron-opaque microscopic fibrils in lakes: Their demonstration, their biological derivation, and their potential significance in the redistribution of cations. Protoplasma, 92, 289–309.

Leppard, G.G., 1993. In: Rao, S.S. (Ed.), Particulate Matter and Aquatic Contaminants. Lewis, Chelsea, MI, pp. 169– 195.

Li, H., Xing, P. and Wu, Q. L. (2012a), Characterization of the bacterial community composition in a hypoxic zone induced by Microcystis blooms in Lake Taihu, China. FEMS Microbiology Ecology, 79: 773–784.

Li, L., Gao, N., Deng, Y., Yao, J., & Zhang, K. (2012b). Characterization of intracellular & extracellular algae organic matters (AOM) of microcystic aeruginosa and formation of AOM-associated disinfection byproducts and odor & taste compounds. Water Research, 46(4), 1233-1240.

Liu, D., Keesing, J. K., He, P., Wang, Z., Shi, Y., & Wang, Y. (2013). The world's largest macroalgal bloom in the yellow sea, china: Formation and implications. Estuarine, Coastal and Shelf Science, 129, 2-10.

Lu D., Huang W. (1999) Phaeocystis bloom in southeast China coastal water 1997. Harmful Algae News 19, 9.

Lürling, M. and Van Donk, E. (1997) Morphological changes in Scenedesmus induced by infochemicals released in situ from zooplankton grazers. Limnology and Oceanography, 42, 783-788.

MacKenzie, L., Sims, I., Beuzenberg, V. and Gillespie, P. (2002) Mass accumulation of mucilage caused by dinoflagellate polysaccharide exudates in Tasman Bay, New Zealand, Harmful Algae 1 (2002) 69–83.

Martin-Jézéquel V, Hildebrand M, Brzezinski M (2000) Silicon metabolism in diatoms: implications for growth. J Phycol 36:791-720.

McGregor, G.B.; Stewart, I.; Sendall, B.C.; Sadler, R.; Reardon, K.; Carter, S.; Wruck, D.; Wickramasinghe, W. First Report of a Toxic *Nodularia spumigena*(Nostocales/ Cyanobacteria) Bloom in Sub-Tropical Australia. I. Phycological and Public Health Investigations. *Int. J. Environ. Res. Public Health* 2012, *9*, 2396-2411.

McKee MP, Ward JE, MacDonald BA, Holohan BA (2005) Production of transparent exopolymer particles by the eastern oyster Crassostrea virginica. Marine Ecology Progress Series 183:59-71.

Mingazzini, M., and Thake, B. (1995). Summary and conclusions of the workshop on marine mucilages in the adriatic sea and elsewhere. Science of the Total Environment, 165, 9-14.

Mopper, K., Zhou, J., Sri Ramana, K., Passow, U., Dam, H. G., & Drapeau, D. T. (1995). The role of surface-active carbohydrates in the flocculation of a diatom bloom in a mesocosm. Deep-Sea Research Part II, 42(1), 47-73.

Mote, P.W., and Mantua N.J. (2002) Coastal upwelling in a warmer future. Geophys. Res. Lett., 29 (23), 2138.

Myklestad, S. M. (1995). Release of extracellular products by phytoplankton with special emphasis on polysaccharides. Science of the Total Environment, 165, 155-164.

Myklestad, S. M. (1999). Phytoplankton extracellular production and leakage with considerations on the polysaccharide accumulation. Annali Dell'Istituto Superiore Di Sanita, 35(3), 401-404.

Naidu, G., Jeong, S., Vigneswaran, S., & Rice, S. A. (2013). Microbial activity in biofilter used as a pretreatment for seawater desalination. Desalination, 309, 254-260.

Nakamura, Y., Umemori, T., Watanabe, M. (1989). Chemical environment for red tides due to Chattonella antiqua Part 2. Daily monitoring of the marine environment throughout the outbreak period. J. oceanogr. Soc. Japan 45 116-128

Nelson, T.A., Haberlin, K., Nelson, A.V., Ribarich, H., Hotchkiss, R., Van Alstyne, K.L., Buckingham, L., Simunds, D.J., Fredrickson, K., 2008. Ecological and physiological controls of species composition in green macroalgal blooms. Ecology 89, 1287-1298.

Nezlin, N. P., Polikarpov, I. G., Al-Yamani, F. Y., Subba Rao, D. V., & Ignatov, A. M. (2010). Satellite monitoring of climatic factors regulating phytoplankton variability in the arabian (persian) gulf. Journal of Marine Systems, 82(1-2), 47-60.

Okaichi T (1989) Red tide problems in the Seto Inland Sea, Japan. In: Okaichi T, Anderson DM, Nemoto T (eds) Red tides. Biology, environmental science and toxicology. Elsevier, New York, pp 137–144.

Olenina, I., Hajdu, S., Edler, L., Andersson, A., Wasmund, N., Busch, S., Göbel, J., Gromisz, S., Huseby, S., Huttunen, M., Jaanus, A., Kokkonen, P., Ledaine, I. and Niemkiewicz, E. (2006) Biovolumes and size-classes of phytoplankton in the Baltic Sea, HELCOM Balt. Sea Environ. Proc. 106, ISSN 0357-2994.

Orlova T.Y., Konovalova G.V., Stonik I.V., Selina M.S., Morozova T.V. and Shevchenko O.G. (2002) Harmful algal blooms on the eastern coast of Russia, Taylor, F.J.R. and Trainer, V.L. (Eds.). Harmful Algal Blooms in the PICES Region of the North Pacific. PICES Sci. Rep. No. 23, 152 pp.

Pankratz, T. (2008), "Red Tides Close Desal Plants", Water Desalination Report, 44 (44).

Passow, U. (2000) Formation of transparent exopolymer particles (TEP) from dissolved precursor material. Marine Ecology Progress Series 192, 1-11.

Passow, U. (2002a) Transparent exopolymer particles (TEP) in aquatic environments, Progress In Oceanography 55(3-4), 287-333.

Passow, U. (2002b) Production of transparent exopolymer particles (TEP) by phyto-and bacterioplankton. Marine Ecology Progress Series 236, 1-12.

Passow, U., & Alldredge, A. L. (1994). Distribution, size, and bacterial colonization of transparent exopolymer particles (TEP) in the ocean. Marine Ecology Progress Series, 113, 185–198.

Passow, U., & Alldredge, A. L. (1995). A dye-binding assay for the spectrophotometric measurement of transparent exopolymer particles (TEP). Limnology and Oceanography, 40, 1326–1335.

Passow, U., Alldredge, A. L., & Logan, B. E. (1994). The role of particulate carbohydrate exudates in the flocculation of diatom blooms. Deep-Sea Research I, 41, 335–357.

Peperzak, L., & Poelman, M. (2008). Mass mussel mortality in the netherlands after a bloom of phaeocystis globosa (prymnesiophyceae). Journal of Sea Research, 60(3), 220-222.

Plantier S., Castaing J.-B., Sabiri N.-E., Massé A., Jaouen P. & Pontié M. (2012): Performance of a sand filter in removal of algal bloom for SWRO pre-treatment, Desalination and Water Treatment 51 (7-9), 1838-1846.

Pompei, M., Mazziotti, C., Guerrini, F., Cangini, M., Pigozzi, S., Benzi, M., Palamidesi, S., Boni, L., Pistocchi, R. (2003) Correlation between the presence of Gonyaulax fragilis (Dinophyceae) and the mucilage phenomena of the Emilia-Romagna coast (northern Adriatic Sea), Harmful Algae 2 (2003) 301–316

Qu, F., Liang, H., He, J., Ma, J., Wang, Z., Yu, H., & Li, G. (2012). Characterization of dissolved extracellular organic matter (dEOM) and bound extracellular organic matter (bEOM) of microcystis aeruginosa and their impacts on UF membrane fouling. Water Research, 46(9), 2881-2890.

Ramus, J. (1977). Alcian Blue: A quantitative aqueous assay for algal acid and sulfated polysaccharides. Journal of Phycology, 13, 348–445.

Reynolds, C. S. (2006) The ecology of phytoplankton. Cambridge University Press.

Richlen M. L., Morton S. L., Jamali E. A., Rajan A., Anderson D. M. (2010), "The Catastrophic 2008–2009 Red Tide in the Arabian Gulf Region, with Observations on the Identification and Phylogeny of the Fish-killing Dinoflagellate Cochlodinium Polykrikoides", Harmful Algae, 9(2), pp. 163-172.

Rinaldi, A., Vollenweider, R. A., Montanari, G., Ferrari, C. R., & Ghetti, A. (1995). Mucilages in italian seas: The adriatic and tyrrhenian seas, 1988-1991. Science of the Total Environment, 165, 165-183.

Rogers S.I. and Lockwood S.J. (1990) Observations on coastal fish fauna during a spring bloom of Phaeocystis pouchetii in the eastern Irish Sea. J. Mar. Biol. Assoc. U.K., 70, 249–253.

Scott J.E. and Dorling J. (1965). Differential staining of acid glycolsaminoglycans (mucopolysaccharides) by Alcian blue in salt solutions. Histochemie 5 (3), 221-233.

Sellner, K.G.,Doucette, G.J., Kirkpatrick, G.J., (2003). Harmful algal blooms: causes, impacts and detection. Journal of Ind. Microbiol. Biotechnol. 30, 383-406.

Shi, J.-H., Gao H.-W., Zhang J., Tan S.-C., Ren J.-L., Liu C.-G., Liu Y., and Yao X. (2012), Examination of causative link between a spring bloom and dry/wet deposition of Asian dust in the Yellow Sea, China, J. Geophys. Res., 117, D17304.

Shikata T., Nagasoe S., Matsubara T., Yoshikawa S., Yamasaki Y., Shimasaki Y., Oshima Y., Jenkinson I.R., Honjo T. (2008). Factors influencing the initiation of blooms of the raphidophyte Heterosigma akashiwo and the diatom Skeletoma costatum in a port in Japan. Limnology & Oceanography 53, 2503-2518.

Smayda TJ, Trainer VL (2010) Dinoflagellate blooms in upwelling systems: Seeding, variability, and contrasts with diatom bloom behaviour. Progress in Oceanography 85: 92–107.

Smayda, T. J. (1997) Harmful algal blooms: their ecophysiology and general relevance to phytoplankton blooms in the sea. Limnol. Oceanogr., 42, 1137–1153.

Smayda, T. J., & Reynolds, C. S. (2003). Strategies of marine dinoflagellate survival and some rules of assembly. Journal of Sea Research, 49(2), 95-106.

Smith J.C., Cormier R., Worms J., Bird C.J., Quilliam M.A., Pocklington R., Angus R., Hanic L. (1990) Toxic blooms of the domoic acid containing diatom Nitzschia pungens in the Cardigan River, Prince Edward Island. In: Graneli E, Sundström B., Edler L., Anderson D.M. (eds) Toxic marine phytoplankton. Elsevier, New York, pp 227–232.

Subba Rao DV, Y Pan, V Zitko, G Bugden and K Mackeigan (1993) Diarrhetic shellfish poisoning (DSP) associated with a subsurface bloom of Dinophysis norvegica in the Bedford Basin, eastern Canada. Mar. Ecol. Prog. Ser. 97: 117-126.

Tang D.L., Kawamura H., Doan-Nhu H., Takahashi W. (2004) Remote sensing oceanography of a harmful algal bloom off the coast of southeastern Vietnam J. Geophys. Res., 109, 1–7, C3014.

Tang, Y. Z., & Gobler, C. J. (2009). Characterization of the toxicity of cochlodinium polykrikoides isolates from northeast US estuaries to finfish and shellfish. Harmful Algae, 8(3), 454-462.

Tester P.A., Wiles K., Varnam S.M., Velez Ortega G., Dubois A.M., and Arenas Fuentes V. (2004) Harmful Algal Blooms in the Western Gulf of Mexico: Karenia brevis Is Messin' with Texas and Mexico! pp. 41-43. Steidinger, K. A., J. H. Landsberg, C. R. Tomas, and G. A. Vargo (Eds.). Harmful Algae 2002. Florida Fish and Wildlife Conservation Commission, Florida Institute of Oceanography, and Intergovernmental Oceanographic Commission of UNESCO.

Thomas M.K., Kremer C.T., Klausmeier C.A., and Litchman E. (2012). A Global Pattern of Thermal Adaptation in Marine Phytoplankton. Science 338 (6110), 1085-1088.

Thornton D.O. (2002). Diatom aggregation in the sea: mechanisms and ecological implications. European Journal of Phycology 37, 149-161.

Thornton DCO, Fejes EM, DiMarco SF, Clancy KM (2007) Measurement of acid polysaccharides in marine and freshwater samples using alcian blue. Limnology and Oceanography: Methods, 5(2007):73–87.

Thornton DCO (2004) Formation of transparent exopolymeric particles (TEP) from macroalgal detritus. Marine Ecology Progress Series 282:1-12.

Trainer, V. L., Adams, N. G., Bill, B. D., Anulacion, B. F. and Wekell, J. C. (1998), Concentration and dispersal of a Pseudo-nitzschia bloom in Penn Cove, Washington, USA. Nat. Toxins 6, 113–125.

Twiner M.J., Rehmann N., Hess P. and Doucette G.J. (2008) Azaspiracid Shellfish Poisoning: A Review on the Chemistry, Ecology, and Toxicology with an Emphasis on Human Health Impacts, Mar Drugs 6(2), 39–72.

Tyrrell T, Merico A (2004) Emiliania huxleyi: Bloom observations and the conditions that induce them. Coccolithophores: From the Molecular Processes to Global Impact, eds Thierstein HR, Young JR (Springer, Heidelberg, Germany), pp 75–97.

van der Hoeven (1984) Observations of surface water temperature in the Netherlands: series from KNMI-RWS (in Dutch). Scientific report W.R. 84-5, ISSN 0169-1651.

Villacorte, L.O., Ekowati, Y., Winters, H., Amy, G.L., Schippers, J.C. and Kennedy, M.D. (2013) Characterisation of transparent exopolymer particles (TEP) produced during algal bloom: a membrane treatment perspective. Desalination & Water Treatment 51 (4-6), 1021-1033.

Walsh, J. J., and Steidinger K. A. (2001), Saharan dust and Florida red tides: The cyanophyte connection, J. Geophys. Res., 106(C6), 11597–11612.

Wang D.-Z. (2008) Neurotoxins from Marine Dinoflagellates: A Brief Review, Mar Drugs 6(2), 349–371.

Wang, S.-H., Hsu N. C., Tsay S.-C., Lin N.-H., Sayer A. M., Huang S.-J., and Lau W. K. M. (2012), Can Asian dust trigger phytoplankton blooms in the oligotrophic northern South China Sea? Geophys. Res. Lett., 39, L05811.

Watkins S.M., Reich A., Fleming L.E. and Hammond R. (2008) Neurotoxic Shellfish Poisoning, Mar Drugs 6(3), 431–455.

Wetsteyn, L. P. M. J., Peeters, J. C. H., Duin, R. N. M., Vegter, F., & de Visscher, P. R. M. (1990). Phytoplankton primary production and nutrients in the oosterschelde (the netherlands) during the pre-barrier period 1980-1984.Hydrobiologia, 195(1), 163-177.

Seawater reverse osmosis and algal blooms

Contents

3.1   Introduction 46
3.2   SWRO pre-treatment options..47
      3.2.1   Granular media filtration (GMF) 48
      3.2.2   Dissolved Air Flotation (DAF)..49
      3.2.3   Ultrafiltration (UF)..51
3.3   Impact of algal blooms on UF and SWRO..53
      3.3.1   Particulate fouling...53
      3.3.2   Organic fouling 54
      3.3.3   Biological fouling...55
3.4   Fouling potential indicators..56
      3.4.1   Algae concentration...57
      3.4.2   Biopolymer concentration..58
      3.4.3   Transparent Exopolymer Particles (TEP)...59
      3.4.4   Modified Fouling Index (MFI) 60
      3.4.5   Biological fouling potential 61
3.5   Proposed strategies to control algae and AOM fouling..61
3.6   Future pre-treatment challenges 63
3.7   Summary and outlook...64
References 65

This chapter is an extended version of:

Villacorte L.O., Tabatabai S.A.A., Amy G., Schippers J.C. and Kennedy M.D., (2014) Impact of algal blooms on seawater reverse osmosis: monitoring, membrane fouling and pre-treatment options. *In preparation for Desalination.*

3.1   Introduction

Reverse osmosis (RO) is currently the leading and preferred seawater desalination technology (DesalData, 2013). Currently, the main drawback for the cost-effective application of RO is membrane fouling (Flemming *et al.*, 1997; Baker and Dudley, 1998). The accumulation of particulate and organic material from seawater and biological growth in membrane modules frequently cause operational problems in seawater reverse osmosis (SWRO) plants. To reduce the (in)organic load of colloidal and particulate matter reaching RO membranes and to minimize or delay associated operational problems, pretreatment systems are generally installed upstream of the RO membranes. Most SWRO plants, especially in the Middle East, install coagulation followed by granular media filtration (GMF) to pre-treat seawater. However, in recent years, low pressure membrane filtration is increasingly being used as SWRO pre-treatment.

Over the years, it is becoming more evident that microscopic algae are a major cause of operational problems in SWRO plants (Caron *et al.*, 2010). The adverse effect of algae on SWRO started to gain more attention during a severe algal bloom incident in the Gulf of Oman in 2008-2009 (Figure 3-1). This bloom forced several SWRO plants in the region to reduce or shutdown operation due to clogging of pre-treatment systems (mostly GMF) and/or due to low RO feed water quality (i.e., high silt density index, SDI>5). The latter triggers concerns of irreversible fouling problems in RO membranes (Berktay, 2011; Richlen *et al.*, 2010; Nazzal, 2009; Pankratz, 2008). This incident highlighted a major problem that algal blooms may cause in countries relying largely on SWRO plants for their water supply, and underlines the significance of adequate pretreatment in such systems.

Figure 3-1: A massive red tide bloom in the Gulf of Oman spreading to the Persian Gulf shown in this satellite image acquired by Envisat's MERIS instrument on November 22, 2008 (Credits: C-wams project, Planetek Hellas/ESA). Yellow points indicate locations of major SWRO plants in the area. Inset screenshots of online news regarding SWRO plant shutdown due to red-tide in the gulf area in 2008 and 2013 (www.arabianbusiness.com).

The high particle loading in seawater during an algal bloom in combination with the high filtration rate (5-10 m/h) in the granular filters can cause rapid and irreversible clogging of the GMF. Furthermore, a substantial fraction of algal-derived organic matter (AOM) can pass through GMF, which can potentially cause fouling in downstream RO membranes (Guastalli et al., 2013).

In 2005, Berman and Holenberg reported that some AOM, particularly transparent exopolymer particles (TEPs), can potentially initiate and enhance biofouling in RO systems (Berman and Holenberg, 2005). TEPs are characteristically sticky, so they can adhere and accumulate on the surface of RO membranes and spacers. The accumulated TEPs may serve as a "conditioning layer" – a platform for effective attachment and initial colonization by bacteria - which may then accelerate biofilm formation in RO membranes (Bar-Zeev et al., 2012). Furthermore, TEPs may be partially degradable and may later serve as a substrate for bacterial growth (Passow and Alldredge, 1994; Alldredge et al., 1993).

To solve the problem of high AOM concentration in the GMF filtrate, a few options have been proposed such as incorporating and/or increasing coagulant dosage prior to GMF to improve effluent water quality. However, an increase in coagulant dosage may further increase the rate of clogging in the filter. The addition of coagulant at high dosage results in the creation of large flocs that are captured on the surface of the filters rather than being filtered through the media. This shifts the filtration mechanism from depth filtration to surface blocking, which at filtration rates of typically 5 - 10 m/h gives rise to significant head loss in these systems. Installing a floc removal step such as sedimentation or dissolved air flotation (DAF) in front of the GMF will enable increase in coagulant dose and improve effluent quality while reducing clogging problems in GMF. Another option is to install ultrafiltration (UF) or microfiltration (MF) membrane systems to replace GMF. UF/MF pre-treatment can guarantee an RO feed water with low SDI even during severe algal bloom. However, concerns have been expressed regarding the rate of fouling in UF/MF membrane systems (e.g., backwashable and non-backwashable fouling) during algal bloom periods (Schurer et al., 2013). To overcome this concern, a DAF system preceding a UF/MF system has been recommended (Anderson and McCarthy, 2012).

This chapter reviews state-of-the-art pre-treatment options for SWRO, the effect of algal blooms on membrane filtration systems (e.g., MF/UF and RO), indicators for quantifying these effects and promising pretreatment options for RO desalination of algal bloom-impacted seawater.

3.2   SWRO pre-treatment options

Currently, most SWRO plants are equipped with one or more pre-treatment systems. These mainly include granular media filtration (GMF), dissolved air flotation (DAF) and/or ultrafiltration (UF).

3.2.1 Granular media filtration (GMF)

Conventional pre-treatment systems for SWRO were developed based on existing technology and most commonly consist of conventional granular media filtration (GMF). Single or dual stage granular media filters consisting of sand and anthracite (garnet is sometimes used) is typically applied in conventional pre-treatment systems, in gravity or pressurized configuration. Sand and anthracite (0.8-1.2mm/2-3mm) filter beds are superior to single media filtration in that they provide higher filtration rates, longer runs and require less backwash water. Anthracite/sand/garnet beds have operated at normal rates of approximately 12 m/h and peak rates as high as 20 m/h without loss of effluent quality. In SWRO pre-treatment, the primary function of GMF is to reduce high loads of particulate and colloidal matter (i.e., turbidity).

GMF relies on depth filtration to enhance RO feed water quality. However, when high concentrations of organic matter or turbidity loads are encountered, coagulation is required to ensure that RO feed water of acceptable quality is produced (SDI<5). Coagulation is applied either in full scale or inline mode in these systems. The most commonly applied coagulant in SWRO pre-treatment is ferric salts (i.e., ferric chloride or ferric sulphate).

Poor removal of algae can lead to clogging of granular media filters and short filter runs. While diatoms are well-known filter clogging algae, other algae types can clog filters including green algae, flagellates, and cyanobacteria (Edzwald, 2010). During the severe red tide bloom event in the Gulf of Oman and Persian Gulf in 2008-09, conventional pre-treatment systems were not able to maintain production capacity and treated water quality at high algal cell concentrations of approximately 27,000 cells/mL (Richlen *et al*, 2010). Operation of the media filters was characterized by rapid clogging rates and deteriorating quality of pre-treated water. As a consequence frequent backwashing was required resulting in increased downtime of the system such that the required pre-treatment capacity could no longer be maintained. At the Fujairah plant in UAE, filter runs were reduced from 24 to 2 hours. Furthermore, deteriorating quality of the pre-treated water, i.e. SDI > 5, led to the increased dosage of coagulant to enhance treated water quality. Increasing coagulant dose may lead to higher clogging rates of media filters. Coagulation enlarges particulate and colloidal matter in water and can therefore shift filtration mechanism from standard blocking (depth filtration) to surface blocking (cake filtration). As filtration rates are relatively high (5-10 m/h) in media filters, cake filtration can result in exponential head loss in the filters.

Reducing filtration rates of the media filters during such extreme events can enhance operation. Reducing the rate of filtration by 50% will result in much lower clogging rates, e.g. by a factor 2-4 depending on the size and characteristics of the foulants (e.g., algae). However, reducing filtration rates will require increased surface area of the media filters. This implies significant investment costs and larger foot print of the treatment plant. Another way to enhance operation of GMF during such extreme events is to provide a clarification step, e.g. sedimentation or flotation, after coagulation/flocculation to reduce the load of particulate/colloidal matter (including coagulated flocs) on the media filters.

Flotation is more robust than sedimentation as it can handle large concentrations of suspended matter (e.g. algae). Currently, flotation preceding media filtration is proposed as the solution for algal blooms. Flotation is able to reduce the algal concentration to a large extent, protecting media filters from rapid clogging, reduced capacity, and breakthrough. However a coagulant dose of 1-2 mg Fe^{3+}/L or higher is usually required to render the process effective. Furthermore, coagulant might be required upstream of the media filters to ensure an acceptable SDI in the effluent. Installing flotation units in front of media filtration might be cheaper than sedimentation units, as the surface loading rates in high rate DAF systems can reach 30 m/h (Edzwald, 2010). Consequently, flotation may require much lower footprint than sedimentation. However the process scheme will require flocculation basins, air saturation and sludge treatment facilities.

3.2.2 Dissolved Air Flotation (DAF)

DAF is a clarification process that can be used to remove particles prior to conventional media filtration or MF/UF systems. Raw water is dosed with a coagulant, typically at concentrations lower than those applied for sedimentation, followed by two-stage tapered flocculation. Removal is achieved by injecting the feed water stream with water that has been saturated with air under pressure and then releasing the air at atmospheric pressure in a flotation tank. As the pressurized water is released, a large number of micro-bubbles are formed (approximately 30-100 µm) that adhere to coagulated flocs and suspended matter causing them to float to the water surface where they may be removed by either a mechanical scraper or hydraulic means, or a combination thereof. Clarified water (sub-natant) is drawn off the bottom of the tank by a series of lateral draw-off pipes (see Figure 3-2). Conventional DAF systems operate at nominal hydraulic loading rates of 5-15 m/h. More recent DAF units are developed for loadings of 15-30 m/h and greater. As a result, DAF requires a smaller footprint than sedimentation.

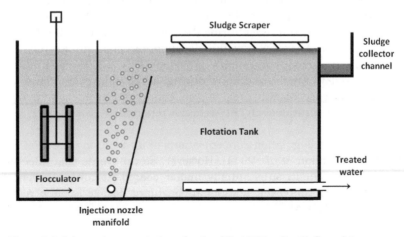

Figure 3-2: Schematic presentation of a simplified DAF unit with flocculator

DAF is more effective than sedimentation in removing low density particles from water and is therefore a suitable treatment process for algal bloom-impacted waters. Gregory and Edzwald (2010) reported 90-99% removal by DAF of algal cells for different algae types compared to 60-90% by sedimentation. A review paper on separation of algae by Henderson *et al.* (2008) reports DAF removals of 96% to about 99.9% when pretreatment and DAF are optimized. Several DAF plants in the Netherlands and Great Britain are primarily used for treatment of algal-laden waters (van Puffelen *et al.*, 1995; Longhurst and Graham, 1987; Gregory, 1997).

In SWRO pre-treatment, DAF prior to dual-stage GMF was tested by Degrémont during early pilot testing for the Taweelah SWRO plant in Abu Dhabi, UAE (Rovel, 2003). DAF was suggested to enhance the robustness of the pre-treatment scheme in case of oil spills or algal bloom events, or in case high coagulant concentrations were required during turbidity spikes. Algal cell concentrations were reportedly below 100 cells/mL during this period, which is far below concentrations observed during severe bloom conditions. Sanz *et al.* (2005) demonstrated the effectiveness of DAF coupled with coagulation prior to dual stage GMF in producing RO feed water with SDI < 4 (typically less than 3) when treating seawater containing various algae, including red tide species. The paper does not specify the initial total cell concentration of the various species. The authors reported more than 99% removal of total algae after DAF and first stage filtration units.

The severe red tide event in 2008-2009 in the Gulf of Oman that led to the shutdown of several desalination plants in the region redirected the attention of the desalination industry to DAF as part of SWRO pre-treatment schemes. DAF is now being regularly used in new SWRO plants in the Persian Gulf upstream of granular media filtration or UF. Although the Fujairah 2 desalination plant was still under construction during that period, Veolia reported that their pilot plant fitted with a DAF unit in the pre-treatment system, continued to operate throughout the red tide bloom (Pankratz, 2008). Expansion of the Fujairah plant, which is under construction, will have DAF as an essential part of the pre-treatment scheme (WaterWorld, 2013). Degrémont reported > 99% removal of algal cells during pilot testing of coagulation/AquaDAF™ prior to GMF in Al-Dur, Bahrain (Le Gallou *et al.*, 2011). However, real bloom conditions were not encountered during the pilot phase with algal cell counts reaching only 200 cells/mL. The Al-Shuwaikh desalination plant in Kuwait equipped with DAF/UF as pre-treatment consistently provided SDI < 2.5 for good quality feed water and <3.5 for deteriorated conditions during a red tide event (Park *et al.*, 2013). However, in this case as well, real bloom conditions as measured by cell counts, chlorophyll-a concentrations, or TEP were not reported.

In most DAF units, coagulation concentrations of up to 20 mg/L as $FeCl_3$ are reported (Rovel, 2003; Le Gallou *et al.*, 2011). However, storm events that affect water quality are reported to result in substantially higher coagulant concentrations for DAF (Le Gallou *et al.*, 2011). Moreover, additional coagulant dosage is often required in GMF units downstream of DAF.

3.2.3 Ultrafiltration (UF)

Over the last decade, the application of UF has been considered as a more reliable alternative to conventional granular media filtration (with and without coagulation) as a pretreatment process for RO systems. UF membranes have been tested and applied at pilot and commercial scale as pre-treatment for SWRO (Wolf et al., 2005; Halpern et al., 2005; Gille and Czolkoss, 2005; Brehant et al., 2002; Glueckstern et al., 2002; Wilf and Schierach, 2001) and offer several advantages over conventional pre-treatment systems; namely, lower footprint, constant high permeate quality (in terms of SDI), higher retention of large molecular weight organics, lower overall chemical consumption, etc. (Wilf and Schierach, 2001; Pearce, 2007). Successful piloting has led to the implementation of UF pre-treatment in several large (>100,000m^3/day) SWRO plants, with a total installed capacity of 3.4 million m^3/day as of 2011. Projections predicted annual installed capacities exceeding 2 million m^3/day in the coming years (Busch et al., 2010).

UF membranes are generally more effective in removing particulate and colloidal matter from seawater than GMF. Hence, they are expected to be more reliable in maintaining an RO feed water with low fouling potential even during an algal bloom period. However, MF/UF membranes were also reported to experience some degree of fouling during algal blooms (Schurer et al., 2012; 2013). So far, a few studies have investigated the effect of algal blooms on the operational performance of UF membranes (Kim and Yoon, 2005; Ladner et al., 2010, Schurer et al., 2012; 2013). These studies agree on the notion that large macromolecules (e.g., polysaccharides and proteins) produced by these algae are the main causes of membrane fouling, and more so than the algal cell themselves. High concentrations of sticky AOM substances (e.g., TEP) present during an algal bloom can impair UF operation (Figure 3-3) by attaching to the membrane surface and pores resulting in permeability decline (CEBs as frequent as once in 6 hours). Under such conditions, operators resorted to coagulation to stabilize operation (Schurer et al, 2012; 2013). With optimized coagulation conditions, operation was stabilized at relatively low doses of ferric (approximately 0.5 mg Fe^{3+}/L) during the bloom period.

Although extensive operational experience on algal blooms of different types and severities is not available, one can extrapolate the results of the existing pilot study and propose that inside-out pressure driven UF membranes are more capable of handling algal bloom events than conventional GMF. This may be attributed to significant differences in hydraulic operational parameters of the two systems. An overview of operational parameters for media filtration and ultrafiltration is presented in Table 3-1. Filtration flux rates in GMF can be up to 100 times higher than flux rates in UF systems. Total filtered volume prior to backwash may be 2000 times higher for GMF compared with UF membranes. Hence, low filtration rates and high backwash frequencies favour overall enhancement of UF systems performance. However, coagulation is required to stabilize hydraulic performance during periods of severe algal bloom.

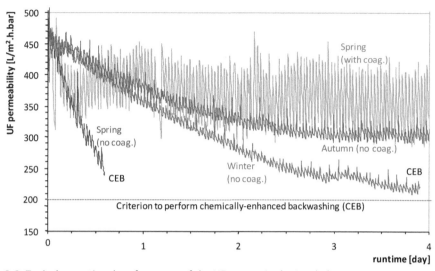

Figure 3-3: Typical operational performance of the UF system in the Jacobahaven seawater UF-RO plant during the bloom (spring) and non-bloom (autumn and winter) seasons. In-line coagulation pre-treatment was implemented during the spring season to stabilise performance of the UF. Graph was redrawn from Schurer *et al.* (2012).

Table 3-1: Operational parameters for ultrafiltration and media filtration typically applied in SWRO pre-treatment (adapted from Schippers, 2012).

	Ultrafiltration	Granular media filtration
Pores [µm]	0.02	150
Filtration rate [L/m²h]	50 – 100	5,000 – 10,000
Run length [hrs]	1	24
Backwash rate : Filtration rate	2.5	2.5 - 5
Backwash time [min]	1	30
Filtered volume/m² per cycle [L]	50 – 100	120,000 – 240,000
Pressure loss [bar]	0.2 – 2	0.2 – 2

Coagulation is commonly applied in inline mode in UF systems for SWRO pre-treatment. Inline coagulation is the application of a coagulant without removal of coagulated flocs through a clarification step. Inline coagulation is also commonly characterized by the absence of a flocculation chamber. Hence, in most applications, inline coagulation is achieved by dosing the coagulant prior to a static mixer or the feed pump of UF membranes to provide adequate mixing. Flocculation is generally not required in UF applications, as enlarging particle size is not an objective and pin-sized flocs are sufficient to enhance UF operation. However, if not optimized, coagulation can deteriorate long-term UF operation. Ferric species (monomers, dimers, trimers, etc.) that are small enough to enter UF pores, can irreversibly foul UF membranes. If low grade coagulants are used, ferrous iron can reach UF membranes and adsorb on the membrane surface or within the pores. Fouling by manganese may also occur for coagulants of low grade. This may result in slow irreversible fouling of UF membranes that can only be removed with specialized cleaning solutions with proprietary recipes. Some of the chemicals that are known to reduce irreversible fouling of UF membranes by iron and manganese are solutions based on ascorbic and oxalic acids of a certain ratio.

3.3   Impact of algal blooms on UF and SWRO

Caron *et al.* (2010) pointed out two potential impacts of algal blooms in membrane-based seawater desalination facilities: (1) significant treatment challenge to ensure the desalination systems are effectively removing algal toxins from seawater and (2) operational difficulties due to increased total suspended solids and organic content resulting from algal biomass in the raw water. The latter is expected to be a major challenge in membrane-based desalination plants (UF pre-treatment preceding the SWRO) considering that majority of algal blooms does not produce toxic compounds. Furthermore, it has been shown that common HAB toxins can be effectively removed by NF (>90%) or RO (>99%) membranes (Laycock *et al.*, 2012; Dixon *et al.*, 2011).

3.3.1 Particulate fouling

High algae biomass in raw water can cause operational problems in membrane systems. During filtration of algal bloom-impacted waters, particulate materials comprising algal cells, their detritus and AOM can accumulate to form a heterogeneous and compressible cake layer on the surface of the membranes which may eventually cause a substantial decrease in overall membrane permeability. RO/NF systems are primarily designed to remove dissolved constituents in the water but they are most vulnerable to spacer clogging problems by particulate matter. For this reason, the majority of RO systems are preceded by a pretreatment process to minimise particulate fouling potential of the feed water. Nevertheless, common pretreatment processes such as granular (dual) media filters may not be reliable to prevent particulate fouling during algal bloom (Berktay, 2011; Nazzal, 2009; Anderson and McCarthy, 2012). The product water of granular media filters can be highly variable over time, with reported algae and biopolymer (algal-released organic macromolecules) removal efficiencies in the range of 48-90% and 17-47%, respectively (Plantier *et al.*, 2012; Salinas-Rodriguez *et al.*, 2009).

Capillary MF/UF membranes may also suffer fibre plugging problems when high concentrations of algae cells are transported to and deposited at the dead-end side of the capillary (Figure 3-4), eventually blocking a section of the feed channel and resulting in a loss of effective filtration area (Heijman *et al.*, 2005; 2007; Lerch *et al.*, 2007; Panglisch, 2003). To maintain constant water production, the flux on the remaining active membrane area will increase, resulting in higher trans-membrane pressure (TMP). Furthermore, conventional hydraulic cleaning (backwashing) may no longer be effective in removing accumulated material in the plugged portions of the capillary.

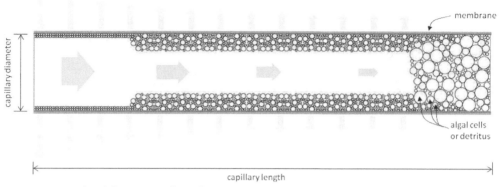

Figure 3-4: Graphical illustration of capillary UF membrane plugging (axial fouling) due to substantial accumulation of algal cells and their detritus from the feedwater. Figure modified from Panglisch (2003).

3.3.2 Organic fouling

During algal blooms, organic fouling in UF membranes often occurs when AOM are abundant in the feed water. AOM produced by common bloom-forming species of algae largely comprise of high molecular weight biopolymers (polysaccharides and proteins) which often includes the sticky TEPs (Myklestad, 1995; Villacorte et al., 2013). TEPs are hydrophilic materials which can absorb/retain water up to ~99% of their dry weight while allowing some water to pass through (Verdugo et al., 2004; Azetsu-Scott and Passow, 2004). This means that they can bulk-up to more than 100 times their solid volume and can easily squeeze through and fill-up the interstitial voids between the accumulated solid particles (e.g., algal cells) on the surface of the membrane (Figure 3-5). It is therefore expected that accumulation of these materials can provide substantial resistance to permeate flow during membrane filtration. As TEP can be very sticky, it may strongly adhere to the surface and pores of UF membranes. Consequently, hydraulic cleaning (backwashing) may no longer be effective in adequately restoring initial membrane permeability (Figure 3-5). This scenario has been reported in recent studies (e.g., Villacorte et al., 2010a,b; Schurer et al., 2012; 2013; Qu et al., 2012a,b), signifying that AOMs could not only cause decrease in hydraulic performance but also non-backwashable or physically irreversible fouling in dead-end UF systems.

Accumulation of AOM materials on RO membranes may result in decrease of normalized flux and feed channel pressure drop. Considering the high operating pressure used in RO, the direct impact of TEP accumulation on operational performance is expected to be much less remarkable than in UF systems. However, the accumulated sticky substances may initiate or promote particulate and biological fouling by enhancing the deposition of bacteria and other particles from the feed water to the membrane and spacers (Berman and Holenberg, 2005; Winters and Isquith, 1979).

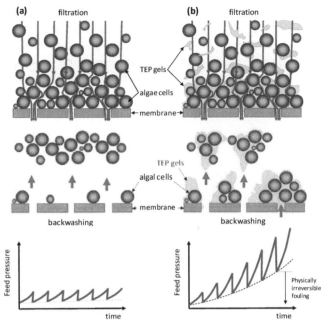

Figure 3-5: Graphical representation of the potential role of TEP in the fouling UF membranes during severe algal bloom: filtration during algal bloom with (a) low and (b) high concentrations of TEP.

3.3.3 Biological fouling

Bacteria has been shown to adhere, accumulate and multiply in RO systems which eventually resulted in the formation of a slimy layer of dense concentrations of bacteria and their extracellular polymeric substances known as biofilm. When the accumulation of biofilm reaches a certain threshold that operational problems are encountered in the membrane system, it is considered as biological fouling or biofouling (Flemming, 2002). An operational problem threshold can be a remarkable (e.g., >15%) decrease of normalised membrane flux, increase in net driving pressure and/or increase in feed channel pressure drop. In the Middle East, about 70% of the seawater RO installations were reported to be suffering from biofouling problems (Gamal Khedr, 2000). Generally, biofouling is only a major problem in NF/RO systems because periodic backwashing and chemical cleaning in dead-end UF systems allows regular dispersion or removal of most of the accumulated bacteria from the membrane; thus, inhibiting the formation of a biofilm.

TEPs produced during algal blooms can initiate and enhance biofouling in RO systems. Because TEPs are characteristically sticky, they can adhere and accumulate on the surface of the membranes and spacers. The accumulated TEPs can serve as a "conditioning layer" – a good platform for effective attachment and initial colonization of bacteria - where bacteria can utilize effectively biodegradable nutrients from the feed water (Berman and Holenberg, 2005; Winters and Isquith, 1979). Furthermore, TEPs can be partially degraded and may later serve as a substrate for bacteria growth (Passow and Alldredge, 1994; Alldredge et al., 1993). Recently, Berman and co-workers proposed a "revised paradigm" of aquatic biofilm formation facilitated by TEPs (Berman and Holenberg, 2005; Berman, 2010; Berman *et al.*, 2011; Bar-Zeev *et al.*, 2012a). As illustrated in Figure 3-6, colloidal and particulate TEPs and protobiofilms (suspended TEPs with extensive microbial outgrowth and colonization) in surface water can initiate, enhance and possibly accelerate biofilm accumulation in RO membranes.

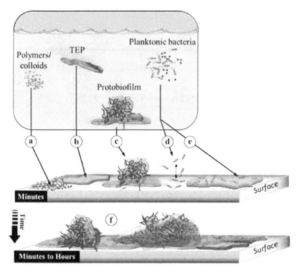

Figure 3-6: Schematic illustration of the possible involvement of (a) colloidal biopolymers, (b) TEPs, and (c) protobiofilm in the initiation of aquatic biofilms. A number of planktonic bacteria (first colonizers) can attach (d) reversibly on clean surfaces or (e) irreversibly on TEP-conditioned surfaces. When nutrients are not limited in the water, (f) a contiguous coverage of mature biofilm can develop within a short period of time (minutes to hours). Figure and description adopted from Bar-Zeev *at al.* (2012a).

Since bacteria requires nutrients for energy generation and cellular biosynthesis, essential nutrients such as biodegradable organic carbon, phosphates and nitrates can be the main factors dictating the formation and growth of biofilm. During the peak of an algal bloom, some of these essential nutrients can be limited (e.g., phosphate) due to algal uptake (Figure 3-6). However when the bloom reaches the death phase, algal cells starts to disintegrate and release some of these nutrients. Hence, biofouling initiated/enhanced by TEPs and AOM may occur within a period of time after the termination of an algal bloom.

3.4   Fouling potential indicators

Monitoring the membrane fouling potential of raw and pre-treated water is important in SWRO plants, especially during algal bloom periods, in order to develop preventive/corrective measures for minimizing the potential adverse impacts to membrane filtration. Various indicators have been proposed to assess the magnitude of the bloom and effectiveness of the pre-treatment systems. The most relevant indicators/parameters are discussed in the following sections.

3.4.1 Algae concentration

The magnitude of algal blooms is mainly measured either in terms of cell count or chlorophyll-a concentration. Bloom-forming algae of different species can vary substantially in terms of cell size and chlorophyll-a content. Hence, relationship between these two parameters also varies. As presented in Chapter 2 (Table 2-1), typical bloom concentrations are higher for smaller algae than larger algae. To compensate for the size differences, cell concentration can be expressed in terms of volume fraction (total cell volume per volume of water sample) instead of cell number per volume of water.

Ideally, the pre-treatment systems of an SWRO plant should effectively remove algal cells to prevent clogging in RO channels. Table 3-2 shows removal efficiencies of selected pre-treatment processes reported in literature.

Table 3-2: Reported removal efficiencies of algae for various treatment processes.

Treatment	Source water	Remarks	Removal (%)	Ref.
Granular media filtration	E. Mediterranean Sea	rapid sand filter (no coag.)	76 ± 13	[1]
	E. Mediterranean Sea	coag. + mixed bed filter	90 ± 8	[2]
	E. Mediterranean Sea	coag. [1 mg $Fe_2(SO_4)_3$] + RSF	79 ± 8	[3]
	W. Mediterranean Sea	press. GMF (anthracite-sand)	74	[4]
	Algae-spiked seawater	dual media filter (no coag.)	48 - 90	[5]
Sedimentation	Lake water	coag. = 20-24 mg Fe^{3+}/L	96	[6]
	algae-spiked freshwater	coag. = 12 mg Al_2O_3/L	90	[7]
Dissolved air flotation (DAF)	W. Mediterranean Sea	coag. (0-6 mg $FeCl_3$/L) + DAF	75	[4]
	lake water	coag. = 7-12 mg Fe^{3+}/L	96	[6]
	algae-spiked freshwater	coag. = 12 mg Al_2O_3/L	96	[7]
	algae-spiked freshwater	coag. = 0.5-4 mg Al_2O_3/L	90 - 100	[8]
	algal culture	coag. = 0.7-3 mg Al^{3+}/L	98	[9]
Microfiltration	Algal cultures	no coagulation	>99	[10]
Ultrafiltration	W. Mediterranean Sea	PVDF; pore size = 0.02μm	99	[4]
	algae-spiked freshwater	PVC; nom. pore size = 0.01μm	100	[11]
Cartridge filters	E. Mediterranean Sea	Disruptor® media	60	[12]
	Lake Kinneret	Amiad™ AMF; 2-20μm	90 ± 6	[13]
	Lake Kinneret	Disruptor® media	65	[12]
	River Jordan	Disruptor® media	85	[12]
	Treated wastewater	Disruptor® media	70	[12]

Note: Removal efficiency calculated based on chlorophyll-a concentration or cell count.
References: [1] Bar-Zeev et al. (2012b); [2] Bar-Zeev et al. (2009); [3] Bar-Zeev et al. (2013); [4] Guastalli et al. (2013) [5] Plantier et al (2012); [6] Vlaski (1997) ; [7] Teixeira & Rosa (2007) ; [8] Teixeira et al. (2010); [9] Henderson et al. (2009) ; [10] Castaing et al. (2011) ; [11] Zhang et al. (2011) ; [12] Komlenic et al. (2013) ; [13] Eschel et al. (2013).

Algae removal in granular media filters (GMF) is highly variable (48-90%) as compared to more stable and much higher removal efficiencies by MF/UF membranes (>99%). High algal removals (>75%) were also reported for sedimentation and DAF treatments. Cartridge filters, which are typically installed after the pre-treatment processes and before the SWRO system, have comparable removal with GMF.

Although algal cell and chl-a concentrations are the main indicators of an algal bloom, these parameters are often not sufficient indicators of the fouling potential of the water. Different bloom-forming species of algae can behave differently in terms of AOM production and at which stage of their life cycle AOM materials are released. More advanced parameters that better indicate the concentration of AOM in feed water and the fouling potential attributed to the presence of AOM are discussed in the following sections.

3.4.2 Biopolymer concentration

Liquid chromatography-organic carbon detection (LC-OCD) is a semi-quantitative method for the measurement and fractionation of organic carbon. This method can be used to quantify the presence of AOM in algal bloom impacted waters. Using this technique, AOM can be fractionated based on molecular size. The high molecular weight fraction are classified as biopolymers while the low molecular weight fractions (<1 kDa) are further sub-classified into humic-like substances, building blocks, acids and neutrals (Huber *et al.*, 2011). Considering that the high molecular weight AOM are likely to deposit/accumulate in the RO system, measuring the biopolymer fraction of organic matter in the water is a promising indicator of organic and biological fouling potential of algal bloom impacted waters. The reported biopolymer removal efficiencies of different treatment processes are presented in Table 3-3.

Table 3-3: Reported removal efficiencies of various treatment processes based on biopolymer concentrations measured using LC-OCD.

Treatment	Source water	Remarks	Removal (%)	Ref.
Granular media filtration	W. Mediterranean Sea	coag. + dual media filter	47	[1]
	W. Mediterranean Sea	press. GMF (anthracite-sand)	18	[2]
	E. Mediterranean Sea	coag. + single media filter	32	[1]
	Estuarine (brackish)	coag. + continuous sand filter	17	[1,3]
Microfiltration	W. Mediterranean Sea	PVDF; nom. pore size= 0.1μm	36	[1]
	W. Mediterranean Sea	PVDF	14	[4]
	Red Sea	ceramic; pore size = 0.08μm	20	[5]
	Oman Gulf	ceramic; pore size = 0.08μm	30	[5]
	Algal cultures	PC; nom. pore size = 0.1μm	52 - 56	[6]
Ultrafiltration	North Sea	MWCO = 300 kDa	50	[4]
	Red Sea	Ceramic; pore size = 0.03μm	40	[5]
	W. Mediterranean Sea	PVDF; pore size = 0.02μm	41	[2]
	Seawater (Sydney)	MWCO = 17.5 kDa	81.3	[7]
	Estuarine (brackish)	no coagulation	70	[1,3]
	River water	PVDF; pore size = 0.02μm	86	[8]
	River water	PVDF; pore size = 0.01μm	59	[9]
	Algal cultures	PES; MWCO = 100kDa	65 - 83	[6]
	Algal cultures	RC; MWCO = 10kDa	83 - 95	[6]
	Algae-spiked seawater	ceramic; pore size = 0.03μm	60	[5]
Subsurface intake	W. Mediterranean Sea	beachwell	70	[1]
	N. Pacific ocean	infiltration gallery	75	[4]

References: [1] Salinas-Rodriguez *et al.* (2009); [2] Guastalli *et al.* (2013); [3] Villacorte *et al.* (2009a); [4] Salinas-Rodriguez (2011); [5] Dramas & Croué (2013); [6] Villacorte *et al.* (2013); [7] Naidu *et al.* (2013) [8] Hallé *et al.* (2009); [9] Huang *et al.* (2011).

Biopolymers in the water can be substantially reduced (>50%) by UF and sub-surface intake (e.g., beach well) treatments while granular media filtration typically remove less than 50% of biopolymers.

3.4.3 Transparent Exopolymer Particles (TEP)

Transparent exopolymer particles (TEP) are a major component of the high molecular weight fraction (biopolymers) of algal organic matter. As discussed in Section 3.3, these materials can potentially cause severe fouling in UF and RO systems. Over the last two decades, various methods were developed to measure TEP by microscopic enumeration or by spectrophometric measurements (Alldredge *et al.*, 1993; Passow and Alldredge, 1995; Arruda-Fatibello *et al.*, 2004; Thornton *et al.*, 2007). The most widely used and accepted method was introduced in 1995 by Passow and Alldredge. This method is based on retention of TEP on 0.4 µm polycarbonate membrane filters and subsequent staining with Alcian blue dye. The reported $TEP_{0.4\mu m}$ reduction by different pre-treatment processes are summarised in Table 3-4.

Table 3-4: Reported removal efficiencies of $TEP_{0.4\mu m}$ for various treatment processes.

Treatment	Source water	Remarks	Removal (%)	Ref.
Granular media filtration	E. Mediterranean Sea	rapid sand filter (no coag.)	51 ± 27	[1]
	E. Mediterranean Sea	coag. + mixed bed filter	27 ± 19	[2]
	E. Mediterranean Sea	coag. [1mg $Fe_2(SO_4)_3$] + RSF	17 ± 28	[3]
	Estuarine (brackish)	coag. + continuous sand filter	65	[4,5]
	River water	coag. (8 ml/L PACl) + RSF	25	[5]
	River water	rapid sand filter (no coag.)	~100	[6]
	Treated wastewater	coag. (10 mg Al^{3+}/L) + RSF	70	[7]
Sedimentation + media filtration	Lake water	coag. = 15 mg Fe^{3+}/L	70	[5]
Ultrafiltration	Estuarine (brackish)	no coagulation	100	[4,5]
	Lake water	no coagulation	100	[5]
	River water	inline coag. = 3 mg Fe^{3+}/L	100	[5]
	River water	inline coag. = 0.3 mg Fe^{3+}/L	100	[5,8]
	Treated wastewater	no coagulation	100	[7]
	Treated wastewater	no coagulation	>95	[6]
Cartridge filter	E. Mediterranean Sea	Disruptor® media	59	[9]
	Lake Kinneret	Amiad™ AMF; 2-20µm	47 ± 21	[10]
	Lake Kinneret	Disruptor® media	63	[9]
	River Jordan	Disruptor® media	82	[9]
	Treated wastewater	Disruptor® media	74	[9]

Note: $TEP_{0.4\mu m}$ measurement based on Passow and Alldredge (1995).
References: [1] Bar-Zeev *et al.* (2012b); [2] Bar-Zeev *et al.* (2009); [3] Bar-Zeev *et al.* (2013); [4] Salinas-Rodriguez *et al.* (2009); [5] Villacorte *et al.* (2009a); [6] van Nevel *et al.* (2012); [7] Kennedy *et al.* (2009); [8] Villacorte *et al.* (2009b); [9] Komlenic *et al.* (2013) ; [10] Eschel *et al.* (2013).

Although they are operationally defined by marine biologists as particles larger than 0.4 µm, TEPs are not solid particles, but rather agglomeration of particulate and colloidal hydrogels which can vary in size from few nanometres to hundreds of micrometres (Passow, 2000; Verdugo *et al.*, 2004). Hydrogels are highly hydrated and may contain more than 99% of water, which means they can bulk-up to more than

100 times their solid volume (Azetsu-Scott and Passow, 2004; Verdugo *et al.*, 2004). Majority of these materials were formed abiotically through spontaneous assembly of colloidal polymers in water (Chin *et al.*, 1998, Passow, 2000). A substantial fraction of colloidal TEPs (<0.4µm) are not covered by the current established measurement methods (e.g., Passow and Alldredge, 1995). Consequently, new TEP methods were recently developed to cover this previously neglected smaller fraction (see Chapter 5).

3.4.4 Modified Fouling Index (MFI)

There are two established methods to measure the particulate/colloidal fouling potential of water: the silt density index (SDI) and the modified fouling index (MFI). Currently, the silt density index (SDI) is the most widely used method to measure the fouling potential of feed water in SWRO plants. It is based on measurements using membrane filters with 0.45 micrometer pores at a pressure of 210 kPa (30 psi). Although this simple technique is widely used in practice, it has been known for many years that SDI has no reliable correlation with the concentration of particulate/colloidal matter (Alhadidi *et al.*, 2013). Hence, it is often insufficient in predicting the fouling potential of SWRO feed water.

A more reliable approach to measure the membrane fouling potential of RO feed water is the modified fouling index (MFI). Unlike SDI, MFI is based on a known membrane fouling mechanism (i.e., cake filtration). This index was developed by Schippers and Verdouw (1980) whereby they demonstrated the linear correlation between the MFI and colloidal matter concentration in the water. Initially, MFI was measured using membranes with 0.45 or 0.05 µm pore sizes and at constant pressure. However, it was later found that particles smaller than the pore size of these membranes most likely play a dominant role in particulate fouling. In addition, it became clear that the predictive value of MFI measured at constant pressure was limited. For these reasons, MFI test measured at constant flux (with ultra-filtration membranes) was eventually developed over the last decade (Boerlage *et al*, 2004; Salinas-Rodriguez *et al.*, 2012).

A comparison in the reduction of fouling potential as measured by MFI-UF in UF and GMF pre-treatment systems is presented in Table 3-5. In general, UF membranes are superior over GMF in terms of MFI-UF reduction

Table 3-5: Reported reduction in particulate/colloidal fouling potential as measured by MFI-UF for various pre-treatment processes (Source: Salinas-Rodriguez, 2011).

Treatment	Source water	Remarks	Reduction (%)
Granular media filtration	W. Mediterranean Sea	anthracite-sand; 1.5 mgFe^{3+}/L	19
	N. Mediterranean Sea	anthracite-sand; 2 mg Fe^{3+}/L	37
Ultrafiltration	W. Mediterranean Sea	PVDF, 0.03µm	66
	W. Mediterranean Sea	PVDF, 0.02µm	52
	N. Mediterranean Sea	PVDF; 0.01µm	68
	North Sea	PES 300 kDa; 0.5 mg Al^{3+}/L	88

3.4.5 Biological fouling potential

Measuring biological fouling potential of RO feed water is rather complicated. Over the years, multiple parameters have been proposed as indicators of biofouling potential, namely: adenosinetriphosphate (ATP), assimilable organic carbon (AOC) and biodegradable dissolved organic carbon (BDOC) (Vrouwenvelder and van der Kooij, 2001; Amy *et al.*, 2011). So far, these parameters are mainly applied in non-saline waters and still not extensively used in seawater RO plants. Furthermore, inline monitors such as the biofilmmonitor and membrane fouling simulator (MFS) have been introduced to measure biofilm formation rate (Vrouwenvelder and van der Kooij, 2001; Vrouwenvelder *et al.*, 2006). Meanwhile, Liberman and Berman (2006) proposed a set of tests to determine the microbial support capacity (MSC) of water samples, namely chlorophyll-a, TEP, bacterial activity, total bacterial count, inverted microscope observations of sedimented water samples, biological oxygen demand (BOD), total phosphorous and total nitrogen. More investigations are needed to assess the reliability of these parameters/monitors to predict the biofouling potential of algal bloom impaired seawater.

3.5   Proposed strategies to control algae and AOM fouling

Reverse osmosis plants operating with direct/open source intake require extensive pre-treatment of the raw water to maintain or prolong reliable performance and membrane life. To mitigate the adverse effects of algal blooms, a reliable pre-treatment should continuously produce high quality RO feed water while maintaining stable operation. For example in GMF and UF pre-treatment systems, a stable operation is based on the ability of the system to maintain acceptable backwash frequency at minimum chemical and energy requirement.

Following the 2008-2009 red tide outbreak in the Gulf of Oman which led to the shutdown of several desalination plants in the region, an expert workshop was held in Oman in 2012 on the impacts of red tides and HABs on desalination operation. During this workshop, DAF and UF were highly recommended as possible alternatives to GMF for maintaining reliable operation in RO plants during severe algal bloom situations (Anderson and McCarthy, 2012). Installing a sub-surface intake (e.g., beach wells) instead of an open intake has been recently considered as a pre-treatment option for SWRO. Figure 3-7 illustrates the different proposed process schemes based on existing technology as measures for mitigating the adverse effects of algal blooms in SWRO desalination systems.

Raw water abstraction in SWRO plants can be through an open or a sub-surface intake structure. For areas prone to algal blooms, sub-surface intake such as beach wells is preferred as it can serve as a natural slow sand filter which can allow substantial removal of algae and algal organic matter (Missimer *et al.*, 2013). Consequently, less-extensive pre-treatment processes are needed to maintain stable operation in the SWRO plant. However, sub-surface intakes are not applicable in some coastal locations where the geology of the area makes it unfeasible to install such intakes.

Figure 3-7: The different pre-treatment options and process schemes to minimise fouling in SWRO system during algal bloom.

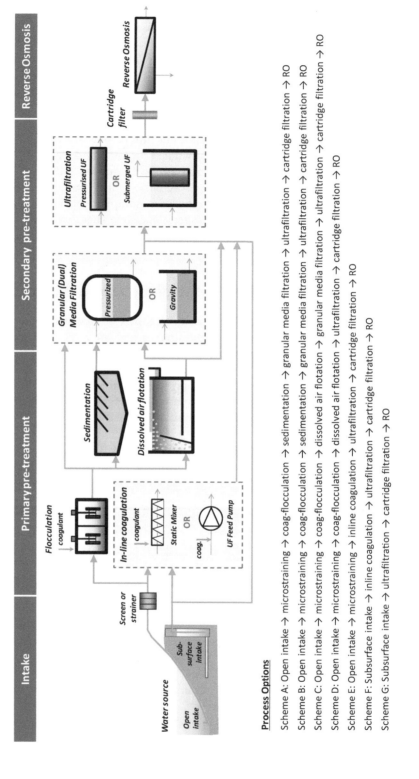

Process Options

Scheme A: Open intake → microstraining → coag-flocculation → sedimentation → granular media filtration → ultrafiltration → cartridge filtration → RO

Scheme B: Open intake → microstraining → coag-flocculation → sedimentation → granular media filtration → ultrafiltration → cartridge filtration → RO

Scheme C: Open intake → microstraining → coag-flocculation → dissolved air flotation → granular media filtration → ultrafiltration → cartridge filtration → RO

Scheme D: Open intake → microstraining → coag-flocculation → dissolved air flotation → ultrafiltration → cartridge filtration → RO

Scheme E: Open intake → microstraining → inline coagulation → ultrafiltration → cartridge filtration → RO

Scheme F: Subsurface intake → inline coagulation → ultrafiltration → cartridge filtration → RO

Scheme G: Subsurface intake → ultrafiltration → cartridge filtration → RO

For SWRO plants operating with an open intake, primary and secondary pre-treatment systems are installed to ensure acceptable RO feed water quality and stable operation during algal blooms. Primary pre-treatment typically includes microstraining/screening to remove large suspended materials (>50 µm), followed by coagulation and sedimentation or DAF. Secondary pre-treatment typically comprises granular (dual) media filtration and/or UF. Granular media filtration requires a full coagulation-flocculation step prior to filtration. On the other hand, UF can operate with or without coagulation and a flocculation or floc removal step is not necessarily required during coagulation. Schurer *et al.*, 2013 demonstrated that UF is capable of maintaining stable operation during algal blooms when preceded with in-line coagulation without additional primary pre-treatment. Other operational measures such as decreasing membrane flux and applying a forward flush cleaning may also improve the performance of UF during severe algal bloom situations.

3.6   Future pre-treatment challenges

Driven by the increasing global demand as well as the economy of scale on the cost of desalinated water, it is projected that more large-scale RO plants (>500,000 m³/day) will be installed in the near future (Figure 3-8; Kurihara & Hanakawa, 2013). If pre-treatment systems in these plants are ineffective during algal bloom, it can likely result in severe organic/biological fouling in SWRO which often requires extensive chemical cleanings (i.e., CIP) to restore membrane permeability. A high frequency (>1 per year) chemical cleaning is a serious problem in large RO plants as it is a laborious and time-consuming process while it also shortens the lifetime of the RO membranes. Usually, part of the RO plant needs to be shut down during the cleaning process, which can result in lower overall plant water production. Since a high cleaning frequency is unfeasible for future large and extra-large SWRO plants (~1,000,000 m³/day), it is essential that future pre-treatment systems is reliable in maintaining an RO feed water with very low fouling potential.

To minimise the cleaning frequency of SWRO plants affected by algal bloom, the development of new generation of pre-treatment technology should focus on complete removal of algae and their AOM as well as limiting the concentration of nutrients in RO feed water. Removal of algae and AOM (e.g., TEPs) in itself can eliminate organic fouling and may substantially delay biological fouling in SWRO as there is no "conditioning layer" to jump-start biofilm development. On the other hand, removal of essential nutrients (e.g., phosphate, AOC) from the RO feed water can further control biological growth in the system. Combining the two treatment strategies can potentially eliminate organic/biological fouling in seawater RO systems, even in algal bloom situations.

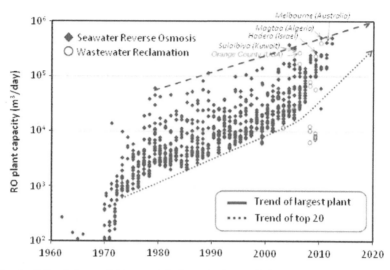

Figure 3-8: Productivity of the top 20 largest RO plants in the world from 1960-2012 and projected plant size for year 2020 (adopted from Kurihara and Hanakawa, 2013).

3.7 Summary and outlook

The recent severe algal bloom outbreaks in the Middle East have resulted in temporary closure of various seawater desalination installations in the region, mainly due to breakdown of pre-treatment systems and/or as a drastic measure to prevent irreversible fouling problems in the downstream RO systems. The major issues which may occur in SWRO plants during an algal bloom are: (1) particulate/organic fouling of pre-treatment systems (e.g., GMF, MF/UF) by algal cells, their detritus and/or AOM, and (2) biological fouling of NF/RO initiated and/or enhanced by AOM. As membrane-based desalination is expected to grow in terms of production capacity and global application, many coastal areas in the world will potentially face a similar scenario in the future if these problems are not addressed accordingly.

To tackle serious operational problems of membrane-based desalination plants, several pre-treatment strategies are currently being proposed. SWRO plants that are fitted with conventional granular media filtration, can gain significant benefits in terms of capacity and product water quality, if preceded by DAF. As a consequence, coagulant consumption will increase as a result of coagulant dosage prior to DAF. Therefore, particularly in countries with stringent legislation, such schemes should be foreseen with coagulant-rich sludge handling and/or treatment facilities. If conditions allow, UF pre-treatment should be incorporated in SWRO plant design as pre-treatment. The advantages of UF over conventional pre-treatment systems are well known to the membrane and desalination societies. Significant additional benefits of UF pre-treatment systems can be gained during periods of severe algal bloom in terms of maintaining pre-treatment capacity and RO feed water quality. Moreover, coagulant consumption is significantly lower in these systems as compared to conventional pre-treatment systems. The future of UF pre-treatment of SWRO lies in the development of tight UF membranes (low MWCO) that can deliver high quality RO feed water (very low SDI or MFI) at minimal coagulant consumption and high output rates.

References

Alhadidi, A., Blankert, B., Kemperman, A. J. B., Schurer, R., Schippers, J. C., Wessling, M., & van der Meer, W. G. J. (2013). Limitations, improvements and alternatives of the silt density index. Desalination and Water Treatment, 51(4-6), 1104-1113.

Alldredge, A. L., Passow, U., & Logan, B. E. (1993). The abundance and significance of a class of large, transparent organic particles in the ocean. Deep-Sea Research I, 40, 1131–1140.

Amy GL, Salinas-Rodriguez SG, Kennedy MD, Schippers JC, Rapenne S, Remize P-J, Barbe C, Manes CLDO, West NJ, Lebaron P, Kooij DVD, Veenendaal H, Schaule G, Petrowski K, Huber S, Sim LN, Ye Y, Chen V and Fane AG (2011). Water quality assessment tools. In: Drioli, E., Criscuoli, A. & Macedonio, F. (eds.) Membrane-Based Desalination - An Integrated Approach (MEDINA). IWA Publishing, New York, pp 3-32.

Anderson D.M., and McCarthy S. (2012) Red tides and harmful algal blooms: Impacts on desalination operations. Middle East Desalination Research Center, Muscat, Oman. Online: www.medrc.org/download/habs_and_desaliantion_workshop_report_final.pdf

Arruda-Fatibello, S. H. S., Henriques-Vieira, A. A. and Fatibello-Filho, O. (2004) A rapid spectrophotometric method for the determination of transparent exopolymer particles (TEP) in freshwater. Talanta 62(1), 81-85.

Azetsu-Scott, K., and Passow, U. (2004) Ascending marine particles: Significance of transparent exopolymer particles (TEP) in the upper ocean. Limnol. Oceanogr, 49(3), 741-748.

Baker, J. S. and Dudley, L. Y. (1998) Biofouling in membrane systems—A review. Desalination 118(1-3), 81-89.

Bar-Zeev E, Berman-Frank I, Girshevitz O, and Berman T (2012a) Revised paradigm of aquatic biofilm formation facilitated by microgel transparent exopolymer particles. PNAS 109(23):9119–9124.

Bar-Zeev, E., Belkin, N., Liberman, B., Berman, T., & Berman-Frank, I. (2012b). Rapid sand filtration pretreatment for SWRO: Microbial maturation dynamics and filtration efficiency of organic matter. Desalination, 286, 120-130.

Bar-Zeev, E., Belkin, N., Liberman, B., Berman-Frank, I., & Berman, T. (2013). Bioflocculation: Chemical free, pre-treatment technology for the desalination industry. Water Research, 47(9), 3093-3102.

Bar-Zeev, E., Berman-Frank, I., Liberman, B., Rahav, E., Passow, U., & Berman, T. (2009). Transparent exopolymer particles: Potential agents for organic fouling and biofilm formation in desalination and water treatment plants. Desalination and Water Treatment, 3(1-3), 136-142.

Berktay, A. (2011). Environmental approach and influence of red tide to desalination process in the middle-east region. International Journal of Chemical and Environmental Engineering 2 (3), 183-188.

Berman T. and Holenberg M. (2005) Don't fall foul of biofilm through high TEP levels. Filtration & Separation 42(4):30-32.

Berman, T. (2010). Biofouling: TEP - a major challenge for water filtration. Filtration and Separation, 47(2), 20-22.

Berman, T., Mizrahi, R., & Dosoretz, C. G. (2011). Transparent exopolymer particles (TEP): A critical factor in aquatic biofilm initiation and fouling on filtration membranes. Desalination, 276(1-3), 184-190.

Boerlage, S. F. E., Kennedy, M. D., Aniye, M. p., Abogrean, E. M., El-Hodali, D. E. Y., Tarawneh, Z. S., and Schippers, J. C. (2000) Modified Fouling Index - ultrafiltration to compare pretreatment processes of reverse osmosis feedwater. Desalination, 131(1-3), 201-214.

Boerlage, S. F. E., Kennedy, M., Tarawneh, Z., De Faber, R., & Schippers, J. C. (2004). Development of the MFI-UF in constant flux filtration. Desalination, 161(2), 103-113.

Brehant, A., Bonnelye, V. & Perez, M., 2002. Comparison of MF/UF pre-treatment with conventional filtration prior to RO membranes for surface seawater desalination. Desalination, 144, pp.353-360.

Caron , D.A., Garneau, M.E., Seubert, E., Howard M.D.A.,, Darjany L., Schnetzer A., Cetinic, I., Filteau, G., Lauri, P., Jones, B. and Trussell, S. (2010), "Harmful Algae and Their Potential Impacts on Desalination Operations of Southern California", Water Research, 44, pp. 385–416.

Castaing J.-B., Masse A., Sechet V., Sabiri N.-E., Pontie M., Haure J., Jaouen P. (2011). Immersed hollow fibres microfiltration (MF) for removing undesirable micro-algae and protecting semi-closed aquaculture basins. Desalination, 276(1-3), 386-396.

Chin W.-C., Orellana M.V., and Verdugo P. (1998) Spontaneous assembly of marine dissolved organic matter into polymer gels. Nature 391, 568–572.

DesalData (2013) Worldwide desalination inventory (MS Excel Format), downloaded from DesalData.com (GWI/IDA) on June 6, 2013.

Dixon, M.B., Falconet, C., Ho, L., Chow, C. W.K., O'Neill, B.K., & Newcombe, G. (2011). Removal of cyanobacterial metabolites by nanofiltration from two treated waters. Journal of Hazardous Materials, 188(1-3), 288-295.

Dramas L. & Croué J.-P. (2013): Ceramic membrane as a pretreatment for reverse osmosis: interaction between marine organic matter and metal oxides, Desalination and Water Treatment, 51 (7-9), 1781-1789.

Edzwald, J.K. (2010) Dissolved air flotation and me. Water Research, 44, 2077-2106.

Eshel G., Elifantz H., Nuriel S., Holenberg M., Berman T. (2013) Microfiber filtration of lake water: impacts on TEP removal and biofouling development. Desalination and Water Treatment 51 (4-6), 1043-1049.

Flemming, H. C. (2002) Biofouling in water systems–cases, causes and countermeasures. Applied Microbiology and Biotechnology 59, 629-640.

Flemming, H. C., Schaule, G., Griebe, T., Schmitt, J., and Tamachkiarowa, A. (1997) Biofouling--the Achilles heel of membrane processes. Desalination, 113(2), 215-225.

Gamal Khedr, M. (2000) Membrane fouling problems in reverse osmosis desalination applications. Desalination and Water Reuse 10, 8-17.

Gille, D. & Czolkoss, W., 2005. Ultrafiltration with multi-bore membranes as seawater pre-treatment. Desalination, 182(1-3), pp.301-307.

Glueckstern, P., Priel, M. & Wilf, M., 2002. Field evaluation of capillary UF technology as a pre-treatment for large seawater RO systems. Desalination, 147, pp.55-62.

Gregory R., Edzwald J.K. (2010) Sedimentation and flotation, J.K. Edzwald (Ed.), Water Quality and Treatment (sixth ed.), McGraw Hill, New York (2010) (Chapter 9).

Gregory, R. (1997) Summary of General Developments in DAF for Water Treatment since 1976. Proceedings Dissolved Air Flotation Conference. The Chartered Institution of Water and Environmental Management, London, 1-8.

Guastalli, A. R., Simon, F. X., Penru, Y., de Kerchove, A., Llorens, J., and Baig, S. (2013). Comparison of DMF and UF pre-treatments for particulate material and dissolved organic matter removal in SWRO desalination. Desalination,322, 144-150.

Hallé, C., Huck, P. M., Peldszus, S., Haberkamp, J., & Jekel, M. (2009). Assessing the performance of biological filtration as pretreatment to low pressure membranes for drinking water. Environmental Science and Technology,43(10), 3878-3884.

Halpern, D.F., McArdle, J. & Antrim, B., 2005. UF pre-treatment for SWRO: pilot studies. Desalination, 182, pp.323-332.

Heijman, S. G. J., Kennedy, M. D., & van Hek, G. J. (2005). Heterogeneous fouling in dead-end ultrafiltration. Desalination, 178(1-3), 295-301.

Heijman, S. G. J., Vantieghem, M., Raktoe, S., Verberk, J. Q. J. C., & van Dijk, J. C. (2007). Blocking of capillaries as fouling mechanism for dead-end ultrafiltration. Journal of Membrane Science, 287(1), 119-125.

Henderson, R. K., Parsons, S. A., and Jefferson, B. (2009). The potential for using bubble modification chemicals in dissolved air flotation for algae removal. *Separation Science and Technology, 44*(9), 1923-1940.

Henderson, R., Parsons, S. A., & Jefferson, B. (2008). The impact of algal properties and pre-oxidation on solid-liquid separation of algae. *Water Research, 42*(8-9), 1827-1845.

Huang, G., Meng, F., Zheng, X., Wang, Y., Wang, Z., Liu, H., & Jekel, M. (2011). Biodegradation behavior of natural organic matter (NOM) in a biological aerated filter (BAF) as a pretreatment for ultrafiltration (UF) of river water.Applied Microbiology and Biotechnology, 90(5), 1795-1803.

Huber S.A., Balz A., Abert M. and Pronk W. (2011) Characterisation of aquatic humic and non-humic matter with size-exclusion chromatography – organic carbon detection – organic nitrogen detection (LC-OCD-OND). Water Research 45:879-885.

Kennedy, M. D., Muñoz Tobar, F. P., Amy, G., & Schippers, J. C. (2009). Transparent exopolymer particle (TEP) fouling of ultrafiltration membrane systems. Desalination and Water Treatment, 6(1-3), 169-176.

Kim S.H. and Yoon J.S. (2005) Optimization of microfiltration for seawater suffering from red-tide contamination. Desalination 182(1–3), 315–321.

Komlenic R., Berman T., Brant J. A., Dorr B., El-Azizi I., Mowers H. (2013) Removal of polysaccharide foulants from reverse osmosis feedwater using electroadsorptive cartridge filters. Desalination and Water Treatment 51 (4-6), 1050-1056.

Kurihara, M., & Hanakawa, M. (2013). Mega-ton Water System: Japanese national research and development project on seawater desalination and wastewater reclamation. Desalination, 308(0), 131-137.

Ladner, D. A., Vardon, D. R. and Clark M.M. (2010), "Effects of Shear on Microfiltration and Ultrafiltration Fouling by Marine Bloom-forming Algae", Journal of Membrane Science, 356, pp. 33–43.

Laycock, M. V., Anderson, D. M., Naar, J., Goodman, A., Easy, D. J., Donovan, M. A., Li, A., Quilliam, M.A., Al Jamali, E., Alshihi, R., Alshihi, R. (2012) Laboratory desalination experiments with some algal toxins. Desalination, 293, 1-6.

Le Gallou, S., Bertrand, S., Madan, K.H. (2011) Full coagulation and dissolved air flotation: a SWRO key pre-treatment step for heavy fouling seawater. In: Proceedings of International Desalination Association World Congress, Perth, Australia.

Lerch, A., Uhl, W., & Gimbel, R. (2007). CFD modelling of floc transport and coating layer build-up in single UF/MF membrane capillaries driven in inside-out mode. Water Science and Technology: Water Supply 7 (4), 37-47.

Liberman, B., and Berman, T. (2006) Analysis and monitoring: MSC - a biologically oriented approach. Filtration & Separation, 43(4), 39-40.

Longhurst, S.J., Graham, N.J.D., (1987) Dissolved air flotation for potable water treatment: a survey of operational units in Great Britain. *The Public Health* Engineer, 14 (6), 71-76.

Missimer T.M., Ghaffour N., Dehwah H.A., Rachman R., Maliva R.G., Amy G. (2013) Subsurface intakes for seawater reverse osmosis facilities: Capacity limitation, water quality improvement, and economics. Desalination 322, 37–51.

Myklestad, S. M. (1995). Release of extracellular products by phytoplankton with special emphasis on polysaccharides. Science of the Total Environment, 165, 155-164.

Naidu, G., Jeong, S., Vigneswaran, S., & Rice, S. A. (2013). Microbial activity in biofilter used as a pretreatment for seawater desalination. *Desalination, 309*, 254-260.

Nazzal N. (2009) 'Red tide' shuts desalination plant. Gulf News, Dubai, UAE. Available from http://gulfnews.com/news/gulf/uae/environment/red-tide-shuts-desalination-plant-1.59095.

Panglisch, S. (2003) Formation and prevention of hardly removable particle layers in inside-out capillary membranes operating in dead-end mode. Water Science and Technology: Water Supply 3, (5-6), 117-124.

Pankratz, T. (2008), "Red Tides Close Desal Plants", Water Desalination Report, 44 (44).

Park, K.S., Mitra, S.S., Yim, W.K., & Lim, S.W. (2013) Algal bloom - critical to designing SWRO pretreatment and pretreatment as built in Shuwaikh, Kuwait SWRO by Doosan. Desalination and Water Treatment. 51, (31-33), 1-12.

Passow and Alldredge, 1995 Passow, U. and Alldredge, A. L. (1995) A Dye-Binding Assay for the Spectrophotometric Measurement of Transparent Exopolymer Particles (TEP). Limnology and Oceanography 40(7), 1326-1335.

Passow, U. (2000) Formation of transparent exopolymer particles (TEP) from dissolved precursor material. Marine Ecology Progress Series 192, 1-11.

Passow, U., & Alldredge, A. L. (1994). Distribution, size, and bacterial colonization of transparent exopolymer particles (TEP) in the ocean. Marine Ecology Progress Series, 113, 185–198.

Pearce, G. (2009) SWRO pre-treatment: treated water quality. Filtration+Separation, 46(6), 30-33.

Pearce, G.K, (2007) The case for UF/MF pre-treatment to RO in seawater applications. Desalination, 203(1-3), 286-295.

Peperzak, L., & Poelman, M. (2008). Mass mussel mortality in the netherlands after a bloom of phaeocystis globosa (prymnesiophyceae). Journal of Sea Research, 60(3), 220-222.

Plantier S., Castaing J.-B., Sabiri N.-E., Massé A., Jaouen P. & Pontié M. (2012): Performance of a sand filter in removal of algal bloom for SWRO pre-treatment, Desalination and Water Treatment 51 (7-9), 1838-1846.

Qu, F., Liang, H., He, J., Ma, J., Wang, Z., Yu, H., & Li, G. (2012a). Characterization of dissolved extracellular organic matter (dEOM) and bound extracellular organic matter (bEOM) of microcystis aeruginosa and their impacts on UF membrane fouling. Water Research, 46(9), 2881-2890.

Qu, F., Liang, H., Tian, J., Yu, H., Chen, Z., & Li, G. (2012b). Ultrafiltration (UF) membrane fouling caused by cyanobateria: Fouling effects of cells and extracellular organics matter (EOM). Desalination, 293, 30-37.

Richlen M. L., Morton S. L., Jamali E. A., Rajan A., Anderson D. M. (2010), "The Catastrophic 2008–2009 Red Tide in the Arabian Gulf Region, with Observations on the Identification and Phylogeny of the Fish-killing Dinoflagellate Cochlodinium Polykrikoides", Harmful Algae, 9(2), pp. 163-172.

Rovel, J.M. (2003) Why a SWRO in Taweelah - pilot plant results demonstrating feasibility and performance of SWRO on Gulf water? In: Proceedings of International Desalination Association World Congress, Nassau, Bahamas.

Salinas Rodríguez S. G., Kennedy M. D., Schippers J. C., Amy G. L. (2009) Organic foulants in estuarine and bay sources for seawater reverse osmosis – Comparing pre-treatment processes with respect to foulant reductions. Desalination and Water Treatment 9 (1-3), 155-164.

Salinas-Rodriguez S.G. (2011) Particulate and organic matter fouling of SWRO systems: Characterization, modelling and applications. Ph.D. thesis, UNESCO-IHE/TUDelft, Delft.

Sanz, M.A., Guevara, D., Beltrán, F., Trauman, E. (2005) 4 Stages pre-treatment reverse osmosis for South-Pacific seawater: El Coloso plant (Chile). In: Proceedings of International Desalination Association World Congress, Singapore.

Schippers J.C. (2012) personal communication.

Schippers, J.C. and Verdouw, J. (1980) The modified fouling index, a method of determining the fouling characteristics of water. Desalination 32, 137-148.

Schurer R., Tabatabai A., Villacorte L., Schippers J.C., Kennedy M.D. (2013) Three years operational experience with ultrafiltration as SWRO pre-treatment during algal bloom. Desalination & Water Treatmen 51 (4-6), 1034-1042.

Schurer, R., Janssen, A., Villacorte, L., Kennedy, M.D. (2012) Performance of ultrafiltration & coagulation in an UF-RO seawater desalination demonstration plant. Desalination & Water Treatment 42(1-3), 57-64.

Teixeira, M. R., & Rosa, M. J. (2007). Comparing dissolved air flotation and conventional sedimentation to remove cyanobacterial cells of microcystis aeruginosa. part II. the effect of water background organics. Separation and Purification Technology, 53(1), 126-134.

Teixeira, M. R., Sousa, V., & Rosa, M. J. (2010). Investigating dissolved air flotation performance with cyanobacterial cells and filaments. Water Research, 44(11), 3337-3344.

Thornton, D. C. O., Fejes, E. M., DiMarco, S. F. and Clancy, K. M. (2007) Measurement of acid polysaccharides (APS) in marine and freshwater samples using alcian blue. Limnology and Oceanography: Methods 5, 73–87.

van Nevel, S., Hennebel, T., De Beuf, K., Du Laing, G., Verstraete, W., & Boon, N. (2012). Transparent exopolymer particle removal in different drinking water production centers. Water Research, 46(11), 3603-3611.

van Puffelen, J., Buijs, P.J., Nuhn, P.N.A.M. and Hijen, W.A.M. (1995) Dissolved air flotation in potable water treatment: the Dutch experience. *Water Science and Technology*, 31 (3-4), 149-157.

Verdugo, P., Alldredge, A. L., Azam, F., Kirchman, D. L., Passow, U. and Santschi, P. H. (2004) The oceanic gel phase: a bridge in the DOM–POM continuum. Marine Chemistry 92, 67-85.

Villacorte, L. O., Kennedy, M. D., Amy, G. L., and Schippers, J. C. (2009a) The fate of Transparent Exopolymer Particles (TEP) in integrated membrane systems: Removal through pretreatment processes and deposition on reverse osmosis membranes. Water Research 43 (20), 5039-5052.

Villacorte, L. O., Kennedy, M. D., Amy, G. L., and Schippers, J. C. (2009b) Measuring Transparent Exopolymer Particles (TEP) as indicator of the (bio)fouling potential of RO feed water. Desalination & Water Treatment 5, 207-212.

Villacorte, L. O., Schurer, R., Kennedy, M., Amy, G., Schippers, J.C. (2010a) The fate of transparent exopolymer particles in integrated membrane systems: a pilot plant study in Zeeland, The Netherlands. Desalination & Water Treatment 13:109-119.

Villacorte, L.O, Schurer, R., Kennedy, M., Amy, G. and Schippers, J.C. (2010b) Removal and deposition of Transparent Exopolymer Particles (TEP) in seawater UF-RO system. IDA Journal 2 (1), 45-55.

Villacorte, L.O., Ekowati, Y., Winters, H., Amy, G.L., Schippers, J.C. and Kennedy, M.D. (2013) Characterisation of transparent exopolymer particles (TEP) produced during algal bloom: a membrane treatment perspective. Desalination & Water Treatment 51 (4-6), 1021-1033.

Vlaski A., (1997). Microcystis aeruginosa Removal by Dissolved Air Flotation (DAF): Options for Enhanced Process Operation and Kinetic Modelling. Doctoral thesis, IHE/TUD, Delft, the Netherlands, 253 pp.

Vrouwenvelder, J. S., and van der Kooij, D. (2001) Diagnosis, prediction and prevention of biofouling of NF and RO membranes. Desalination, 139(1-3), 65-71.

Vrouwenvelder, J. S., van Paassen, J. A. M., Wessels, L. P., van Dam, A. F., & Bakker, S. M. (2006). The membrane fouling simulator: A practical tool for fouling prediction and control. Journal of Membrane Science, 281(1-2), 316-324.

WaterWorld (2013) Fujairah hybrid desalination plant to expand with dissolved air flotation system. Available at: http://www.waterworld.com/articles/2013/01/fujairah-hybrid-desalination-plant-to-expand-with-dissolved-air-floatation-system.html.

WHO (2007) Desalination for Safe Water Supply, Guidance for the Health and Environmental Aspects Applicable to Desalination, World Health Organization (WHO), Geneva.

Wilf, M. and Schierach, M.K. (2001) Improved performance and cost reduction of RO seawater systems using UF pre-treatment. Desalination, 135, 61-68.

Winters, H., and Isquith, I. R. (1979). In-plant microfouling in desalination. Desalination, 30(1), 387-399.

Wolf, P.H., Siverns, S. & Monti, S., 2005. UF membranes for RO desalination pre-treatment. Desalination, 182, pp.293-300.

Zhang, Y., Tian, J., Nan, J., Gao, S., Liang, H., Wang, M., and Li, G. (2011). Effect of PAC addition on immersed ultrafiltration for the treatment of algal-rich water. Journal of Hazardous Materials, 186(2-3), 1415-1424.

Characterisation of algal organic matter

Contents

Abstract 72
4.1 Background 73
    4.1.1   Bloom-forming algae 74
    4.1.2   Characterisation of organic matter in aquatic systems 74
4.2 Materials and methods 75
    4.2.1   Algal culture 75
    4.2.2   AOM extraction and purification 76
    4.2.3   Characterisation techniques 76
    4.2.4   Membrane rejection experiments 81
    4.2.5   Modified fouling index – ultrafiltration (MFI-UF) 81
4.3 Results and discussion 82
    4.3.1   Characteristics of cultured algae 82
    4.3.2   Characterisation of algal organic matter (AOM) 87
    4.3.3   AOM rejection by MF/UF membranes 98
    4.3.4   Specific membrane fouling potential of biopolymers 100
4.4 Summary and conclusions...101
    References 102
    Annex 4-1: Synthetic seawater and culture media..............................105
    Annex 4-2: Molecular structure of selected sugar functional groups....107

This chapter is based on parts of:

Villacorte L.O., Ekowati Y., Neu T., Kleijn M., Winters H., Amy G., Schippers J.C. and Kennedy M.D. (2013) Characterisation of organic matter released by bloom-forming dinoflagellate, diatom and cyanobacteria. *In preparation for Water Research.*

Villacorte L.O., Ekowati Y., Winters H., Amy G., Schippers J.C. and Kennedy M.D. (2013) Characterisation of transparent exopolymer particles (TEP) produced during algal bloom: a membrane treatment perspective. *Desalination and Water Treatment* 51(4-6), 1021-1033.

Abstract

Algae and the organic compounds they release can seriously affect the operation of low pressure (micro- and ultra-filtration) and high pressure (nanofiltration and reverse osmosis) membrane systems. In this study, the nature and behaviour of algal-derived organic materials (AOM) produced by three common species of bloom-forming algae (AT: Alexandrium tamarense, CA: Chaetoceros affinis and MSp: Microcystis sp.) were characterized. Investigations on the physico-chemical properties of AOM were performed employing various characterisation techniques such as liquid chromatography-organic carbon detection, fluorescence spectroscopy, fourier transform infrared spectroscopy, lectin staining coupled with laser scanning microscopy and force measurement using atomic force microscopy. The observed growth behaviour, peak cell concentration and AOM concentration vary substantially between batch cultures of three algal species. CA produced 2-3 times more transparent exopolymer particles (TEP) than AT and 6-9 times more than MSp. The AOM produced by the 3 species are mainly biopolymers (polysaccharides and proteins) while some refractory organic matter (e.g., humic-like substances) and/or low molecular weight biogenic substances were also observed. Polysaccharides with associated fucose and sulphated functional groups were ubiquitous in the biopolymer fraction. The stickiness of AOM in terms of adhesive and cohesive forces varies between source species (CA>AT≥MSp). In general, cohesion forces between AOM materials are generally stronger than the adhesion forces between AOM and clean UF or RO membranes.

The AOM retention by microfiltration (MF) and ultrafiltration (UF) membranes were measured and compared for membranes with different pore sizes. The biopolymer fraction of AOM was substantially removed by MF/UF membranes but not the low molecular weight fractions (e.g., humic-like substances). MF membranes with pore sizes of 0.4 μm and 0.1μm rejected 14-47% and 42-56% of biopolymers, respectively. UF membranes with molecular weight cut-offs of 100kDa and 10kDa rejected 65-83% and 83-97% of biopolymers, respectively.

To illustrate the adverse effect of algal bloom on membrane filtration, the membrane fouling potential based on the modified fouling index (MFI) was measured in batch cultures and extracted AOM. The membrane fouling potential (MFI-UF) of the 3 batch cultures showed clear correlation with TEP concentration (R^2>0.85) but no clear correlation was observed with algal cell concentration. The specific membrane fouling potentials of AOM varied significantly depending on the source algal species (CA>MSp>AT). AOM fractions between 10-100 kDa demonstrated higher specific membrane fouling potential (MFI per mg C/L biopolymers) than the larger fractions (>100kDa). Hence, removal of smaller biopolymers (down to 10 kDa) might be necessary to prevent fouling in RO systems.

4.1 Background

The presence of algae in source water can cause serious challenges in drinking and industrial water production. Microscopic algae in surface waters seasonally proliferate and generate high concentrations of organic substances (Kirchman *et al.*, 1991). This can cause water discoloration, anoxic conditions, odour and toxicity problems and operational problems in water treatment processes (Petry *et al.*, 2007; Caron *et al.*, 2010; Henderson *et al.*, 2008a; Schurer *et al.*, 2012; 2013).

In recent years, algal blooms have been increasingly affecting the operation of membrane–based water treatment plants and it is now considered as a major threat to the desalination industry (Caron *et al*, 2010; Pankratz, 2008). One remarkable example is the 2008–2009 harmful "red tide" event caused by the dinoflagellate *Cochlodinium polykrikoides* in the Middle East Gulf region, when algal concentrations were reported to be between 11,000 and 21,000 cells/ml (Richlen *et al.*, 2010). At least five seawater desalination plants in the region were forced to reduce (or shutdown) operation due to odour issues, clogging of pre-treatment systems (e.g., granular media filters) or due to concerns that the bloom would irreversibly foul reverse osmosis membranes (Berktay, 2011; Pankratz, 2008). This incident highlighted a major problem that algal blooms may pose in countries relying largely on membrane-based desalination plants for their water supply.

Algae can produce various forms and differing concentrations of organic substances such as polysaccharides, proteins, lipids, nucleic acids and other dissolved organic substances (Fogg, 1983; Bhaskar and Bhosle, 2005; Decho, 1990; Myklestad, 1995). Myklestad (1995) highlighted the significance of studying extracellular polysaccharides as they may comprise more than 80% of algal organic matter (AOM). Depending on the species of algae, AOMs are either produced through extracellular release in response to low nutrient stress or other stressful conditions (e.g., unfavourable light, pH or temperature), invasion by bacteria or viruses and/or produced through disruption and decay of algal cells (Fogg, 1983; Leppard, 1993; Myklestad, 1999). However, various species of algae may also release these organic materials under normal conditions (Fogg, 1983). In seawater, a substantial fraction of AOM are highly sticky and are likely involved in the coagulation of colloidal/particulate materials in aquatic environments, which may result in the formation of mucilaginous aggregates such as marine snow and sea foam (Alldredge *et al.*, 1988; Mopper *et al.*, 1995). These sticky materials have been referred to in various studies as transparent exopolymer particles (TEP; see review by Passow, 2002).

AOMs, specifically TEPs, were recently identified as potential causes of biological fouling in reverse osmosis (RO) systems and organic fouling in ultrafiltration (UF) systems (Berman and Holenberg, 2005; Kennedy *et al.*, 2009; Berman *et al.*, 2011). However, their composition and specific membrane fouling potential are still largely unknown. Moreover, the biopolymer components of AOMs have not been sufficiently studied in terms of their removal by and fouling of MF/UF membranes. Since algal blooms are naturally occurring phenomena which are often unpredictable, a fundamental understanding of the physico-chemical characteristics of AOM produced by common species of bloom-forming algae is a significant step towards developing an effective strategy to minimise their adverse effects on membrane-based water treatment systems.

4.1.1 Bloom-forming algae

The Intergovernmental Oceanographic Commission of UNESCO identified about 300 species of microscopic algae that can cause blooms in surface waters (IOC-UNESCO, 2013). The majority of these species belong to the three major groups of micro-algae (i.e., dinoflagellates, diatoms and cyanobacteria). Three commonly-occurring species, one from each group, were investigated in this study, namely: *Alexandrium tamarense* (dinoflagellates), *Chaetoceros affinis* (diatoms) and *Microcystis sp.* (cyanobacteria).

Alexandrium tamarense (AT) is a marine dinoflagellate identified as one of the major causative species of "red tide" (WHOI, 2013). This alga can produce neurotoxins that can cause paralytic shellfish poisoning to humans and other mammals who consumed bivalve molluscs from affected waters (Anderson *et al.*, 2012). AT cells are spherical in shape between 25-45 μm in diameter. The cells are armoured (thecate plate-covered cells) and use two flagella for their mobility.

Chaetoceros affinis is a marine diatom known to release substantial concentrations of extracellular polysaccharides including TEPs (Myklestad *et al.*, 1972). In temperate regions, this species usually thrive during the early spring season. High concentration of this microalga can cause golden brown discoloration of the water. Algal cells resemble an oval cylinder geometric shape ranging in size from 8-30 μm. The algae tend to form chains where cells are connected at the origin of the two setae located on both ends of the cell.

Microcystis sp. is a common freshwater cyanobacteria (blue-green algae) ubiquitous in lakes, rivers and reservoirs during the summer seasons in temperate regions. *Microcystis* are small spherical algae between 3-6 μm in diameter. During a bloom situation, they are known to generate high concentrations of extracellular polymeric substances and have been associated with taste and odour problems in water sources (Qu *et al.*, 2012; Henderson *et al.*, 2008b; Li *et al.*, 2012). Some species such as the *Microcystis aeruginosa* produce a hepatotoxin called microcystin, making this genus one of the main causes of harmful bloom in freshwater environments.

4.1.2 Characterisation of organic matter in aquatic systems

Over the years, various analytical techniques have been introduced to characterise natural organic matter in aquatic systems, some of which have been adopted to investigate the composition of microbial-derived organic matter.

Alcian blue, a dye known to specifically bind with acidic polysaccharides and glycoprotein, has been widely used to visualise and measure TEPs in seawater and lake water (Alldredge *et al.*, 1993; Passow, 2002). Some studies used staining with fluorochrome labelled lectins to visualise and indentify the carbohydrate components of extracellular polymeric substances in microbial biofilms (Neu, 2000; Zippel & Neu,

2011). Lectins are proteins extracted from various organisms which can bind specifically to one or more carbohydrate functional groups. Some functional groups that make up the organic materials can also be identified using fourier transform infrared (FTIR) spectroscopy (Lee *et al.*, 2006).

Qualitative assessment of the presence of protein-like and humic-like materials based on fluorescence excitation-emission matrices (FEEM) has been employed in various applications including characterisation of AOM (Henderson *et al.*, 2008b; Lee *et al.*, 2006; Li *et al*, 2012; Her *et al.*, 2003). Semi-quantitative techniques such as liquid chromatography - organic carbon detection (LC-OCD) can be used to fractionate organic materials based on size and composition. This technique has been extensively applied in characterising natural organic matter in surface water and their fate through the water treatment processes (Huber *et al.*, 2011; Kennedy *et al.*, 2008; Salinas-Rodriguez *et al.*, 2009; Villacorte *et al.*, 2009; Baghoth *et al.* 2011).

The stickiness of high molecular weight organic matter (e.g., biopolymers) can be studied by measuring interaction forces between organic materials and other surfaces using an atomic force microscope. This technique has been applied to measure the adhesive strength of bacterial polysaccharides on polymeric membranes (Yamamura *et al.*, 2008; Frank & Belfort, 2003) and the elastic character of diatom mucilages (Higgins *et al.*, 2002).

4.1.3 Research objective

The objective of this study is to investigate the release, physico-chemical characteristics, membrane retention and fouling potential of AOM from three species of bloom-forming algae in marine and freshwater sources by applying various characterisation techniques.

4.2 Materials and methods

Three selected algal species were grown in batch cultures to represent an algal bloom situation in freshwater and seawater. The algal organic matter (AOM) produced by the three species were extracted and a series of analyses were performed to identify their characteristics, membrane rejection and fouling potential.

4.2.1 Algal culture

Three strains of algae were acquired from the Culture Collection of Algae and Protozoa (Oban, Scotland), namely: *Alexandrium tamarense* (CCAP 1119/32) *Chaetoceros affinis* (CCAP 1010/27) and *Microcystis sp.* (CCAP 1450/13). *Alexandrium tamarense* and *Chaetoceros affinis* were inoculated in sterilised synthetic seawater spiked with nutrients and trace elements based on the L1 and f/2+Si medium, respectively. The artificial seawater (ASW) was prepared to resemble the typical ion composition of the North Sea (TDS 34 g/L, pH 8±0.2). Microcystis Sp. was grown in sterilized BG-11 medium for freshwater algae. The composition of the prepared media and ASW are presented in Annex 4-1. All algal batch cultures were incubated at 20±2°C under an artificial light source (fluorescent lamp) at 12/12h light/dark regime. CA and MSp cultures were continuously mixed in a shaker while AT was mixed manually 1-2 times per day. Light intensities of the fluorescent lamps were adjusted to 40-50 $\mu mol/m^2.s$ for AT and CA cultures and 10-15 $\mu mol/m^2.s$ for MSp.

The average algal cell concentration in batch cultures was monitored by sampling every 2-4 days and counting the cells using Thoma chamber glass slides and a light microscope (Olympus BX51). Additional samples were taken for TEP measurements (Sections 4.2.3.1), lectin staining (Section 4.2.3.2) and MFI-UF measurements (Sections 4.2.5) on selected days during the exponential and stationary-death phases.

4.2.2 AOM extraction and purification

Water samples (0.5 L) were collected during the exponential and stationary-death phases of the three batch cultures of algae. To extract AOM from the culture, the samples were gently mixed and filtered through polycarbonate filters (Nuclepore PC membranes, Whatman) with <0.2 bar of vacuum. Polycarbonate membranes with different pore sizes were used depending on the lower size range of the algal cells: 10 µm for AT, 5 µm for CA and 1 µm for MSp. AOM larger than the filter pore size which may have been retained on the filters and AOM bound to algal cells were not included in the subsequent analysis. The filtered solutions were analysed using LC-OCD (Section 4.2.3.3), FEEM (Section 4.2.3.4) and AFM (Section 4.2.3.6). The same AOM solutions were used in membrane rejection experiments (Section 4.2.4).

To isolate particulate and colloidal materials from the bulk AOM and excess nutrients from the culture medium, dialysis treatment was performed for AOM samples collected during the stationary/death phase. The samples were dialysed using 3.5 kDa RC membrane sacks (Spectra/Por 3, SpectrumLabs) and ultra-pure water (Milli-Q, Millipore) as draw solution. The draw solution was continuously stirred (using a magnetic stirrer) and replenished 1 to 2 times per day. Each dialysis treatment lasted for 4-6 days to remove most of the dissolved salts (based on electrical conductivity measurement) and low molecular weight organics (based on LC-OCD analysis) out of the membrane sacks. After dialysis, the AOM samples inside the membrane sacks were freeze dried and then analysed using FTIR spectroscopy (Section 4.2.3.5).

4.2.3 Characterisation techniques

4.2.3.1 Transparent exopolymer particles (TEP) measurement

TEP concentration was measured using a protocol slightly modified from the method described by Passow and Alldredge (1995). This method measures the relative concentration of TEP retained on PC filters (Whatman Nuclepore) based on the anion density of acidic polysaccharides and glycoproteins. Alcian blue (AB) staining solution was prepared by dissolving 0.025% (m/v) of Alcian Blue 8GX (Standard Fluka, Sigma-Aldrich) in acetic acid buffer solution (pH 2.5). The AB working solution was pre-filtered through 0.05 µm PC filter prior to staining.

Water samples of known volume were filtered through 0.4 µm PC filters by applying 0.2 bar vacuum. Since AB coagulates in contact with saline water, a rinsing procedure was then performed by filtering 1 ml of milli-Q water through the membrane prior to staining in order to replace the adsorbed saline moisture. One ml staining solution was applied over the membrane, allowed to react for 10 seconds and the excess stain was removed by 0.2 bar vacuum. To further remove excess stain, 2 ml of milli-Q water was filtered through the membrane. The stained membrane was then soaked in 6 ml of 60% H_2SO_4 solution and absorbance (at 787 nm) of acid solution was measured after 2 hours. In this study, no calibration with a standard polysaccharide was performed. Hence, TEP concentrations are presented in terms of abs/cm/L instead of µg Xanthan equivalent per liter (µg X_{eq}/L) used in the original method.

Alcian blue staining is routinely performed at pH 2.5 to ensure both sulphated and carboxylated TEPs are fully stained. Lowering the staining pH down to pH 1.0 will only stain sulphated TEPs (Passow and Alldredge, 1995). In this study, this technique was applied to estimate the abundance of sulphated species of TEP. Applying a similar protocol mentioned above, TEPs were measured using 0.1 µm PC membranes with staining solutions prepared at pH 1.0 and pH 2.5 for samples collected during the exponential and stationary/death phase.

4.2.3.2 Lectin staining and confocal laser scanning microscopy (CLSM)

Six different lectins: AAL, UEA-1, Con-A, HMA, PA-I and IAA (Sigma, Vector, EY labs) labelled with either FITC, TRITC or Alexa 488 fluorochromes were prepared for staining polysaccharides produced by different species of algae. The characteristics of the lectins and their known carbohydrate binding specificities are shown in Table 4-1. Raw samples from algal cultures were concentrated on 0.1 µm polycarbonate filters (Whatman Nuclepore) by vacuum filtration (-0.2 bar). Fluorochrome-labelled lectin solutions were then applied over the filter-retained AOM, incubated in the dark for 20 minutes (room temperature) and then excess lectins were removed by vacuum filtration. The stained samples were carefully rinsed by filtering medium solution (specific to the algae species) through the membrane three times to further remove unbound lectins.

Table 4-1: Characteristics of lectins used in this study.

Lectin	Source	Flourochrome	Sugar specificity*
AAL	Aleuria aurantia	Alexa 488	fucose (Fuc)
UEA-1	Ulex europaeus	TRITC	fucose (Fuc)
Con-A	Canavalia ensiformis	FITC	α-D-mannose (α-Man), α-D-glucose (α-Glc)
HMA	Homarus americanus	Alexa 488	Sialic acid
PA-I	Pseudomonas aeruginosa	Alexa 488	galactose (Gal)
IAA	Iberis amara	Alexa 488	N-Acetyl-D-Glucosamine (GlcNAc)

* Based on Neu (2000) and Zippel & Neu (2011)

Imaging of stained AOM samples was performed at the Helmholtz Centre for Environmental Research (Magdeburg, Germany) using a TCS SP5 confocal laser scanning microscope (Leica) equipped with a super continuum light source and an upright microscope. The system was controlled by the software LAS AF version 2.6.1 7314. A range of water-immersible lenses was available to examine the samples. The data presented was recorded using a 63x NA 1.2 wi objective lens. The CLSM signals were recorded in the green channel (excitation = 494 nm; emission = 505-600 nm) for the fluorescently-labelled lectins and in the far-red channel (excitation = 630 nm; emission = 650-750 nm) to detect autofluorescence of chlorophyll-a containing microorganisms (Neu, 2000; Neu *et al.*, 2004).

4.2.3.3 Liquid chromatography - organic carbon detection (LC-OCD)

AOM samples extracted from algal cultures were sent to DOC-Labor (Karlsruhe, Germany) where it was analysed using liquid chromatography - organic carbon detection (LC-OCD). The relative responses of organic carbon, ultraviolet and organic nitrogen at different retention times were measured with an online organic carbon detector (OCD), UV detector (UVD) and organic nitrogen detector (OND). Organic carbon concentrations of biopolymers, humic substances, building blocks, low molecular weight (LMW) acids, and neutrals were determined based on the chromatogram peaks and their retention times (Table 4-2; Huber *et al.*, 2011). The chromatogram results were processed on the basis of area integration using a customised software program CHROMCalc to fractionate the organic carbon concentration in the sample (DOC-Labor, Karlsruhe). Since AOM may comprise large macromolecules (e.g., TEPs), LC-OCD analyses were performed without 0.45 µm inline filtration of samples. The theoretical maximum chromatographable size without sample pre-filtration is 2 µm, which is based on the pore size of the sinter filters of the column used (S. Huber, *per. com.*).

Protein concentration was estimated by assuming that all organic nitrogen detected by the organic nitrogen detector (OND) between 25 and 42 mins retention times were all bound to proteinic compounds. Typical protein compounds contain 14.5-17.5 % nitrogen and 49.7-55.3 % carbon (Rouwenhorst *et al.*, 1991). So, the C:N ratio of proteinic biopolymers can be estimated as 3:1. The estimated AOM protein concentration in mg C/L was calculated based on this ratio. Furthermore, polysaccharide concentrations were computed by subtracting the calculated protein concentration from the organic carbon concentration of biopolymers.

Table 4-2. Descriptions of organic matter fractions measured by LC-OCD (DOC-Labor; Huber *et al.*, 2011)

Organic fraction	Typical size (Da)	Typical composition
Biopolymers	> 20,000	polysaccharides, proteins, amino sugars, polypeptides
Humic subst.	~ 1000	humic and fulvic acids
Building blocks	300 - 500	weathering and oxidation products of humics
LMW Neutrals	< 350	mono-oligosaccharides, alcohols, aldehydes, ketones, amino acids
LMW acids	< 350	all monoprotic organic acids

To estimate the molecular weight distribution of the biopolymer fraction of AOM, selected samples were analysed using high resolution LC-OCD. This modified technique follows a similar principle mentioned above but using 2 series of columns (HW65S and HW50S) instead of one, doubling the mobile phase retention time and the resolution of the chromatogram. Molecular weight fractionations were defined based on calibration with pullulan (PSS, Germany), a polysaccharide with molecular weight between 0.3 and 700 kDa (S. Huber, *per. com.*).

4.2.3.4 Fluorescence excitation-emission matrix (FEEM)

The fluorescence EEM spectra of AOM-containing water samples were measured using a FluoroMax-3 spectrofluorometer (Horiba Jobin Yvon, Inc., USA) with a 150 W ozone-free xenon arc-lamp as a light source for excitation. FEEM measurements were performed at excitation wavelength range from 240 to 450 nm with 10 nm increments and the emission wavelength range from 290 to 500 nm with 2 nm increments. The slit widths were set at 5 nm for excitation and emission. To keep samples free from algal cells, water samples were first filtered through 10µm, 5µm and 1 µm PC membranes by mild vacuum filtration (-0.2 bar) for AT, CA and MSp, respectively. The filter and filter holder were intensively flushed with ultrapure water to remove organic contaminants before sample filtration. Prior to FEEM analyses, the total organic carbon (TOC) concentrations of the samples were measured (Shimadzu TOC-V$_{CPN}$) and then diluted to set the TOC of each sample to ~1 mg C/L. A three dimensional spectrum data series from FEEM were plotted using MatLab R2007b. Background signals were minimised by subtracting the signals of the blank from the sample EEMs. The typical peaks which can be expected within the limits of the covered EEM spectra (Ex. 240-250nm / Em. 290-500 nm) are shown in Table 4-3.

Table 4-3: Typical EEM peak locations of natural organic matter (Coble, 1996; Salinas-Rodriguez *et al.*, 2009; Leenher and Crue, 2003).

Code	Description	Fluorescence range (nm)	
		Excitation	Emission
H1	Humic-like primary peak	330–350	420–480
H2	Humic-like secondary peak	250–260	380–480
Hm	Marine humic-like	300-330	380-420
P1	Protein-like (tyrosine) peak	270–280	300–320
P2	Protein-like (tryptophan) peak, phenol-like	270–280	320–350

For each samples, the fluorescence Index (FI) was calculated based on the ratio of the fluorescence intensity at Em 500 nm/Ex 450 nm and Em 500 nm/Ex 370 nm (McKnight *et al*, 2001). An FI of between 1.7 and 2.0 indicates that the fluorescent organic materials are autochthonous (microbial origin) while an FI between 1.3 and 1.4 indicates they are allochthonous (terrestrial origin).

4.2.3.5 Fourier Transform Infrared (FTIR) Spectroscopy

FTIR spectroscopy was applied to identify the functional groups present on the AOM produced by the three algae species. For this analysis, freeze-dried AOM samples (described in Section 4.2.2) were analysed using a PerkinElmer ATR-FTIR Spectrum 100 instrument at the Aerospace Engineering laboratory of the Technical University of Delft.

4.2.3.6 Force measurement using atomic force microscopy

To assess the stickiness of algal-derived organic matter, AOM to membrane and AOM to AOM interactions were measured using atomic force microscopy (AFM). In this study, the cohesive and adhesive strengths of AOM were investigated by measuring interactions between polystyrene microspheres coated with AOM against clean and

AOM-fouled membranes. The microsphere is attached to the tip of the AFM cantilever. Forces (F) were derived from the cantilever deflection using Hooke's law while the separation distances between the microsphere and the membrane surface are measured from the scanner position and cantilever deflection.

$$F = -k\Delta z$$ Eq. 1

where k is the cantilever spring constant and Δz the deflection of the cantilever.

Each force measurement generates two force-distance curves: the approach force curve and the retract force curve (Figure 4-1). The approach force curve shows the force interactions of the particle probe as it approaches the membrane surface. The retract force features the adhesion force between the particle probe surface and the stationary surface. Adhesion/cohesion forces are the force needed to separate two surfaces from contact and is defined as the equivalent force at the maximum cantilever deflection in the retraction force curve. Furthermore, the total energy needed to completely separate two surfaces was calculated by integrating the measured negative forces (<0 nN) with respect to separation distance in the retract force curve (Figure 4-1).

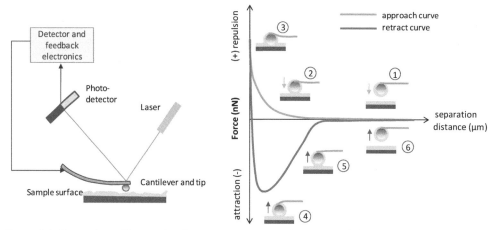

Figure 4-1: Illustration of interaction force measurement using AFM and the generated force-distance curves. The arrows indicate the movement of the particle attached to the cantilever relative to the surface.

Around 10 ml of AOM solution (~5 mgC/L) was poured into a clean plastic petri-dish (50 mm diameter). AFM cantilevers with attached 25 µm polystyrene microsphere tip (Novascan Technologies, Inc.) were submerged in the solution by carefully positioning the cantilever at the bottom of the dish with the microsphere tips facing upwards. The cantilever and tips were submerged for 5 days (4°C) to allow adsorption of AOM. The petri-dish was then allowed to warm up to room temperature (20°C) for >2 hours before AFM force measurements were performed.

Two types of membranes were prepared for this test, namely: polyethersulfone UF (Omega 100 kDa MWCO, Pall Corp.) and polyamide thin-film composite RO membrane (Filmtec BW30, Dow). The UF membranes were first cleaned by soaking them in ultra-pure water for 24 hours and then filtering through >10 ml ultrapure water before the experiment. Small sheets (~1cm x 10cm) of polyamide RO membranes were stored in 1% sodium bisulphite solution for at least 24 hours and then rinsed with ultra-pure water before the experiment. For AOM to AOM interaction measurements, a PES membrane was coated with a thin layer of AOM by filtering 5 ml of AOM solution. Filtration was performed at constant flux (60 L m^{-2}.h^{-1}) using a syringe pump (Harvard pump 33), 30 ml plastic syringe and 25 mm filter holder. The fouled membrane was then placed on the AFM and force measurements with an AOM-coated tip were performed.

All force measurements were performed with ForceRobot 300 (JPK Instruments) at room temperature (~20°C), using a small volume liquid cell, sealed with a rubber ring. In every test, the liquid cell was filled with synthetic seawater solution. The cantilevers were calibrated with respect to their deflection sensitivity from the slope of the constant compliance region in the force curves obtained between the clean, hard surfaces and their spring constant using the thermal noise spectrum of the cantilever deflections (Ralston et al., 2005; Spruijt et al., 2012). The measured cantilever spring constants ranged from 0.18-0.31 N/m although the nominal cantilever spring constant supplied by the manufacturer was higher (0.35 N/m). Force-distance curves were recorded for an approach range of 10 μm, a tip velocity of 16-28 μm/s, 30 seconds surface delay and a sampling rate of 2 kHz. In each cycle of approach and retract, the probe is brought into contact with the surface with an average load force of 5 nN. The force-distance (F-D) curves were recorded in triplicates for each of the six points on the membrane surface within an area of 10 μm x 10 μm. The F-D curves were analyzed using the JPK data processing software to calculate the maximum retract force (adhesion/cohesion) and total energy.

4.2.4 Membrane rejection experiments

MF/UF membranes with different pore sizes were used for AOM rejection experiments, namely: 0.4 and 0.1 μm PC (Whatman) and 100kDa polyethersulfone (PES, Pall) and 10 kDa regenerated cellulose (RC, Millipore). All membranes were soaked for at least 24 hours and/or flushed with milli-Q water to remove possible organic contaminants from the membrane before filtration. Filtration through PC membranes were performed in a vacuum filtration system (-0.2 bar) while filtration through PES and RC membranes were performed at constant flux (60 L/m^2/h) using a syringe/piston pump (Harvard pump 33). Permeate water samples were collected in clean glass tubes for LC-OCD analyses.

4.2.5 Modified fouling index – ultrafiltration (MFI-UF)

The modified fouling index (MFI) was originally developed by Schippers and Verdouw (1980) to measure the membrane fouling potential of the feedwater of reverse osmosis membrane systems based on a cake filtration mechanism at constant pressure. Further improvement of the index, the constant flux MFI-UF (Boerlage et al., 2004;

Salinas-Rodriguez *et al.*, 2012), was applied in this study to measure the membrane fouling potential of algal cultures and AOM.

To better assess the fouling potential of AOM, 10 kDa MWCO UF membranes (Millipore) were used instead of the conventional 100 kDa UF membranes. Filtration was performed at constant flux (60 L m^{-2}.h^{-1}) using a syringe pump (Harvard pump 33), 30 ml plastic syringe and 25 mm filter holder. The increase in trans-membrane pressure (ΔP) was monitored over time for up to 30 mins. The fouling index (I) was then calculated based on the minimum slope of the ΔP versus time plot. Consequently, MFI-UF was calculated by normalising the fouling index to the standard reference conditions set by Schippers and Verdouw (1980) as follows;

$$MFI - UF = \frac{\eta_{20^\circ C} \, I}{2 \, \Delta P_0 A_0^2}$$
Eq. 2

where $\eta_{20^\circ C}$ is the water viscosity at 20°C, ΔP_0 is the standard feed pressure (2 bar) and A$_0$ is standard membrane area (13.8 x10^{-4} m^2).

4.3   Results and discussion

4.3.1 Characteristics of cultured algae

The characteristics of the three algal species - *Alexandrium tamarense* (AT), *Chaetoceros affinis* (CA) and *Microcystis sp.* (MSp) - investigated this study are summarised in Table 4-4.

Table 4-4: Characteristics of bloom-forming algae investigated in this study.

	Alexandrium tamarense	*Chaetoceros affinis*	*Microcystis sp.*
Strain	CCAP 1119/32	CCAP 1010/27	CCAP 1450/13
Type	dinoflagellate	diatom	cyanobacteria
Geometric shape	sphere	oval cylinder	sphere
Typical dimensions	diameter = 25-45 µm	diameter = 8-20 µm	diameter = 4-6 µm
		height = 15-30 µm	
Cell surface area	~3800 µm^2	~1400 µm^2	~50 µm^2
Cell volume	~16500 µm^3	~1920 µm^3	~34 µm^3
Water discoloration	reddish brown	yellowish brown	bluish green
Natural habitat	seawater	seawater	fresh surface waters
Typical bloom period	summer season	spring season	summer season

Batch cultures of the three species were cultivated in the laboratory and monitored for more than 25 days. The changes in cell concentration, TEP concentration and membrane fouling potential (in terms of MFI-UF) in each culture within the monitoring period are presented in Figure 4-2 and discussed in the following sections.

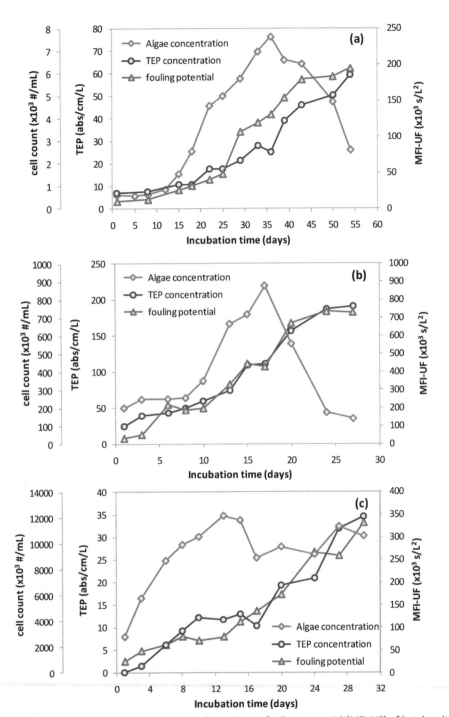

Figure 4-2: Growth curve, TEP concentration and membrane fouling potential (MFI-UF) of batch cultures of (a) AT, (b) CA and (c) MSp.

4.3.1.1 Algal growth

The batch cultures of three algal species showed different growth patterns. *Alexandrium tamarense* (AT) showed a 12-day lag phase, followed by exponential phase and then death phase after day 35. For *Chaetoceros affinis* (CA), an 8-day lag phase was observed, followed by exponential phase and then death phase after day 17. The *Microcystis sp.* culture did not show an apparent lag phase but a growth phase was observed immediately after inoculation until day 12 and then remain in a stationary phase for the rest of the monitoring period (day 30). The maximum cell concentration recorded by the 3 cultures are 7,500 cells/ml, 875,000 cells/ml and 12,000,000 cells/ml for AT, CA and MSp, respectively (Figure 4-2).

The eventual decline of cell concentration in AT (after day 35) and CA (after day 17) cultures may be attributed to depletion of nutrients in the medium (e.g., inorganic carbon, phosphorus, nitrogen, silica and other trace metals). Although *Alexandrium* species can also utilise organic nitrogen and phosphorus from deteriorating cells, they can only take up inorganic carbon (Anderson *et al.*, 2012). It is therefore likely that the death phase of AT was triggered by inorganic carbon depletion in the medium. On the other hand, the rapid decrease of CA after day 17 may be attributed to the depletion of essential nutrients including silicon. The latter is an important limiting nutrient for diatoms like CA as they utilise silicon to synthesize their rigid cell wall (frustules) which are mainly made up of silicates (Martin-Jezequel *et al.*, 2000). Unlike AT and CA, the MSp batch culture did not manifest an apparent death phase within the 30 day monitoring period. This may indicate that MSp were able to utilise nutrients release by deteriorating dead cells.

4.3.1.2 TEP production

All three species of algae produced significant concentrations of TEP (Figures 4-2). CA produced the highest cumulative concentration which was about 3 times that of AT and about 5 times higher than MSp. For AT, TEP release was rather low during the lag phase, followed by rapid increase during the exponential phase which further continued during the stationary-death phase. For CA, TEP production coincided with the increase in cell concentration during the lag and exponential phases and continued to increase more rapidly during the stationary-death phase. Accumulated TEP in MSp culture also increased with cell concentration during the growth and stationary phase followed by a rapid increase in the late stationary phase.

In general, TEPs were produced by both actively growing (exponential phase) and nutrient limited (stationary/death phase) algae. About 60%, 40% and 65% of total accumulated TEP were produced during the stationary/death phase of AT, CA and MSp cultures, respectively. This is an indication that AT and MSp may have released more TEP under nutrient deficient conditions. It is also possible that the majority of these TEPs may have originated from cell wall material and/or intracellular sources as compromised and/or deteriorating cells during the stationary-death phase may have released more TEP-like material. The latter was verified by microscopic examination of Alcian blue stained cells. Figure 4-3a shows a compromised AT cell shedding parts of its cell wall which is stainable with Alcian blue. When the cell has totally deteriorated, what remained were TEP-like material that resembled the form of their thecated cell wall. For MSp, aggregates of TEP-like materials were slightly visible around live cells but these become clearly visible when most of the cells have died and deteriorated (Figure 4-3c). On the contrary, aggregates of TEP were visible around live cells of CA, which may indicate that this species produced most of the TEP extracellularly and not just via release during disruption of algal cells observed in the other two species (Figure 4-3b). This distinct characteristic might be the reason why these species and diatoms in general, are one of the major producers of TEP in the ocean (Passow, 2002).

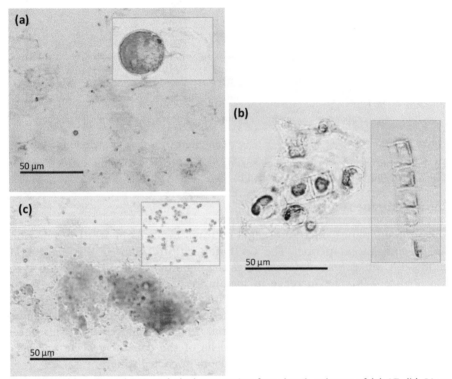

Figure 4-3: Alcian blue stained TEP and algal suspension from batch cultures of (a) AT, (b) CA and (c) MSp. Inset photo of b and c are photos of unstained algal cells.

TEP measurements at different solution pH indicated the abundance of sulphated TEPs in the AOM produced by the 3 algae species. As discussed in Section 4.2.3.1, only sulphated TEPs can be measured if TEP measurement is performed at pH 1.0 while all TEPs (including carboxylated) can be measured at pH 2.5. TEP concentrations in AOM from the three species of algae when measured at pH 2.5 and 1.0 are presented in Figure 4-4.

TEPs produced by AT were mostly sulphated (~93%) during the exponential and death phases while TEPs from CA were at least 99% sulphated in both phases. On the other hand, TEPs from MSp were mostly non-sulphated during the exponential phase and mostly sulphated (~90%) during the stationary phase. Mopper and co-workers reported that sulphated polysaccharides are highly surface-active and are attributed to the high stickiness of TEPs (Mopper *et al.*, 1995; Zhou *et al.*, 1998). Hence, AOM produced by the 3 algal species are likely sticky, especially during the stationary-death phase of the bloom.

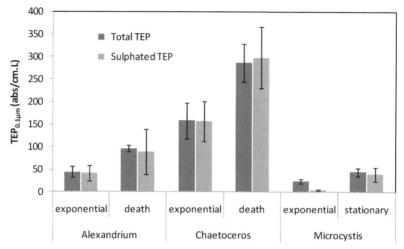

Figure 4-4: TEP concentration in samples originating from different algal cultures measured in terms of total TEP (at pH 2.5) and sulphated TEP (at pH 1.0).

4.3.1.3 Membrane fouling potential

The membrane fouling potential of algal cultures in terms of MFI-UF was monitored during the incubation period. The CA batch culture showed the highest MFI-UF (733,200 s/L^2), which was almost 4 times the maximum MFI-UF of AT and twice that of MSp (Figure 4-2). The increase in membrane fouling potential of the 3 cultures coincided with the increase in TEP concentration. Algal cell concentration did not show a clear relationship with fouling potential (Figure 4-5a) while TEP showed an apparent linear relation with fouling potential (R^2>0.88; Figure 4-5b). This is a strong indication that TEP release and not the cells of the causative algae species has a major role in the fouling potential of algal bloom impacted waters.

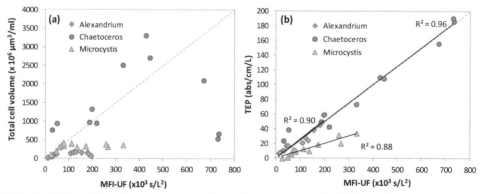

Figure 4-5: Linear relationship between membrane fouling potential (MFI-UF) and (a) total cell volume and (b) TEP concentration.

Due to their particulate sizes (>2μm), algal cells alone are expected to cause lower resistance during dead-end filtration as they form a cake layer that is rather porous compared to TEP gels. Moreover, TEP gels, which are freely suspended or attached to algal cells, are amorphous and are capable of squeezing through the interstitial voids of accumulated algal cells, resulting in the formation of heterogeneous cake layer on the surface of the membrane with a high hydraulic resistance.

4.3.2 Characterisation of algal organic matter (AOM)

The characteristics of AOM extracted from 3 species of algae was investigated using LC-OCD, FEEM, FTIR spectroscopy, lectin staining coupled with CLSM imaging and force measurement with AFM. The results are discussed in the following sections.

4.3.2.1 LC-OCD characterisation

LC-OCD analyses were performed to identify the major components of AOM produced by AT, CA and MSp. The organic carbon detector (OCD) and ultraviolet detector (UVD) chromatograms of AOM samples from 3 algal cultures and their respective growth media (blank) are presented in Figure 4-6.

The AOM from AT comprise mainly biopolymers (e.g., polysaccharides and proteins) and some humic-like substances (Figure 4-6a). The biopolymer fraction showed two distinct chromatogram peaks: 26-31 mins and 31-42 mins. Based on calibration with pullulan, the first peak was estimated to be >1000 kDa and the second peak to be mainly between 100 and 1000 kDa. The biopolymer fraction comprised up to 74% polysaccharides. Aside from biopolymers, building blocks, humic-like substances, LMW acids and LMW neutrals (weakly and uncharged organic compounds) were also identified. Most of these low molecular weight organics may have originated from the culture media (e.g., LMW acids and neutrals). The L1 medium where the AT culture was cultivated contained 3.4 mg/L of EDTA which was used as a chelating agent to minimise precipitation of metals in the medium. LC-OCD analysis of EDTA solution showed a peak appearing at similar elution time as the LMW acid peak (50-57 mins) detected in the AOM samples. During the exponential phase, this acid peak did not show apparent change relative to the blank medium which means no additional acids were produced by algae. However during the death phase, the acid peak shifted to slightly lower retention time and the signal is much higher than what was found in the culture media. The UV detector identified a significant signal at the same retention time which means that the additional OC signal may have originated from aromatic humic-like substances. The AOM concentration increased from 5.3 to 7.7 mg C/L between exponential and stationary phases, which was largely due to the increase of biopolymer concentration (Table 4-5).

For CA, two major peaks were observed (Figure 4-5b), namely: biopolymers (25-40 mins) and low molecular weight (LMW) acids (45-55 mins). Some minor signals were also observed for building blocks and LMW neutrals. The medium on which the alga was inoculated showed a similar acid peak which was relatively stable at different growth phases. This signal also originated from the chelating agent (EDTA) used in preparing the f/2+Si medium. Discounting this signal would indicate that CA produced AOM which were mainly biopolymers with low concentrations of building blocks and LMW organic neutrals. The biopolymer fraction comprised up to about 80% polysaccharides.

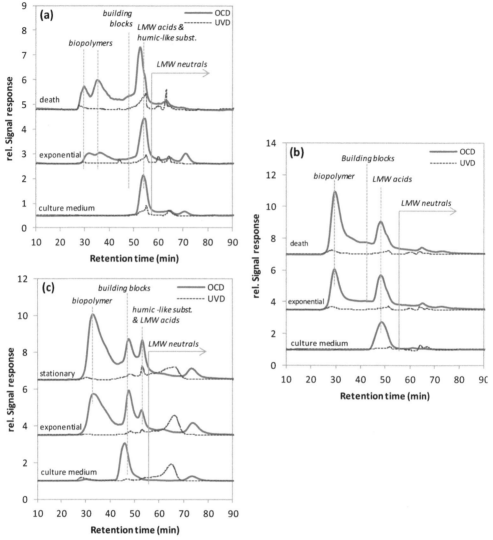

Figure 4-6: LC-OCD chromatograms of blank culture media and AOM from (a) AT, (b) CA and (c) MSp at different growth phases. Offsets in presentation of chromatograms were intentionally made for clarity.

The OCD chromatograms of MSp samples (Figure 4-5c) showed four peak signals: biopolymers (28-42 mins), building blocks (42-51 mins) and LMW acids and humic-like substances (51-55 mins). Contrary to CA, the LMW acid peak of MSp comprised humic-like substances based on the high UV signal peak that was observed at similar retention time as the humic OC peak (Figure 4-5c). The medium OCD chromatogram showed a peak (42-53 mins) close to the retention time of the building block peak of MSp. This peak can be attributed to the presence of citric acid (a metal solubilising agent) in the culture medium. Assuming that the building block peak originated entirely from the medium, MSp produced mainly biopolymers, humic substances, LMW organic acids and neutrals. The concentrations of biopolymers and LMW organics were observed to increase by more than 100% between the exponential and stationary phases. The biopolymer fraction comprised up to about 90% polysaccharides. In general, MSp produced majority of the AOM during the stationary phase. The substantial increase of low molecular weight organics during this phase might be an indication of cell lysis and release of intracellular materials to the AOM pool.

Table 4-5: Composition of AOM extracted from 3 species of algae at different growth phases as measured based on LC-OCD analyses.

Algal species	Alexandrium tamarense		Chaetoceros affinis		Microcystis sp.	
Growth phase	exponential	death	exponential	death	exponential	stationary
COC* (mg C/L)	5.26	7.68	7.08	11.26	1.98	4.43
Biopolymers	1.90	4.10	3.43	6.66	1.01	2.47
Protein	0.51	1.07	0.78	1.30	0.09	0.70
Polysaccharides	1.39	3.03	2.65	5.36	0.92	1.77
Humic-like subst.	0.39	0.51	bdl	bdl	0.24	0.66
Building blocks	0.28	0.37	0.78	1.14	0.28	0.29
LMW neutrals	1.00	0.86	1.50	1.98	0.39	0.80
LMW acids	1.69	1.84	1.37	1.47	0.07	0.21
SUVA (L/mg.m)	1.42	1.14	1.93	1.89	6.42	3.93

* COC = chromatographic organic carbon < 2µm

Using high-resolution LC-OCD analysis and calibration with pullulan, the molecular weight distribution of biopolymers from the 3 algal cultures at the stationary/death phase was estimated based on their elution time (Figure 4-7). The relative weight distribution of biopolymers was found to vary between source species. In general, biopolymers from AT and MSp showed some similarities in terms of molecular weight distribution whereby the dominant fraction is between 100 and 1000 kDa. On the other hand, biopolymers from CA comprise mainly materials larger 1000kDa. The similarities in size distribution between AT and MSp might be attributed to how they release biopolymer materials from their cell walls and intracellular storage materials during deterioration of dying or dead cells. On the other hand, living CA cells actively exuded biopolymers, which are likely stickier and have higher aggregation capability, possibly resulting in accumulation of larger aggregates.

The size distribution of biopolymers might have been influenced by the pre-filtration treatment of the AOM samples whereby AOM samples were extracted using different pore size membranes (see Section 4.2.2). AOM samples from AT, CA and MSp were extracted by retaining algal cells on 10, 5 and 1µm PC filters, respectively. The theoretical maximum size limit covered by the LC-OCD analysis is 2 µm, which means biopolymer samples from AT and CA are comparable but a small fraction of biopolymers (1-2 µm) in the MSp samples were excluded in the analyses. This might be the reason for the lower concentration of large biopolymers (>1000 kDa) in MSp-AOM samples.

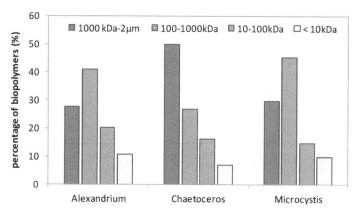

Figure 4-7: LC-OCD molecular weight fractionation of biopolymers produced by 3 species of bloom-forming algae during stationary/death phase.

In summary, LC-OCD analyses illustrated that AOM produced by 3 species of algae are different in terms of composition and molecular weight. AOM from 3 algal species mainly comprise biopolymers while substantial concentrations of LMW organics (e.g., humic-like substances) were also released by 2 algal species (AT and MSp), especially during the stationary-death phase of the bloom.

4.3.2.2 Fluorescence signatures

The EEM spectra of AOM from the three species of algae were analysed to verify the presence of fluorescent organic substances. The analysis was performed after removal of algal cells by filtration (see Section 4.2.3.4), which means the signal of fluorescent organic materials bound to algal cell walls were not measured. Moreover, polysaccharides are non-fluorescent organic substances; hence, the focus of this analysis is to investigate the presence of protein-like and humic/fulvic-like substances in AOM samples.

Both protein-like and humic-like substances were identified in the EEM spectra of AOM from 3 algal species (Figure 4-8). The spectra of AOM from AT identified two peaks which indicates the presence both protein-like and marine humic-like substances. For CA, fluorescence responses mainly originated from protein-like materials while for MSp, the fluorescence responses were from both protein-like and humic/fulvic-like substances. AT and CA mainly produced tryptophan-like proteins (P2) while MSp produced both tyrosine-like (P1) and tryptophan-like proteins. In all identified peaks for each species, a significant increase of fluorescence responses were observed between the exponential and stationary-death phases. The fluorescence indexes (FI) of the AOM samples were between 1.7 and 2.0, confirming their microbial origin.

For AOM from AT and CA, the fluorescence intensity of protein-like peaks was observed to be higher than the humic-like peaks. On the contrary, AOM from MSp showed higher response for the humic-like substances. The EEM spectra of MSp-AOM were similar to what was reported in *hypereutrophic* waters (Lee *et al.*, 2006) and *Microcystis aeruginosa* cultures (Qu *et al.*, 2012; Li *et al.*, 2012). Moreover, previous studies have shown that a significant portion of protein-like materials may be tightly or loosely bound to MSp cells and not as free AOM (Qu *et al.*, 2012), which means the total fluorescence intensity of protein-like materials may have been much higher than what was actually measured. The EEM peak intensities of AOM from 3 species were observed to increase over the incubation period (Figure 4-8).

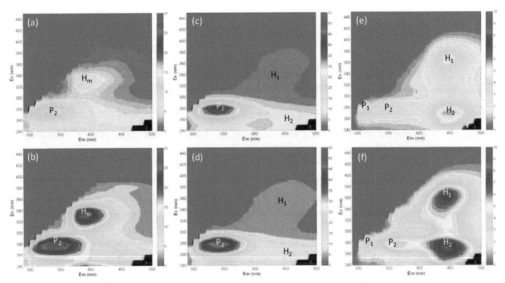

Figure 4-8: Typical EEM spectra of algae cell-free AOM samples from AT (a,b), CA (c,d) and MSp (e,f) collected during the exponential (a,c,e) and stationary growth phases (b,d,f). Legend: H_1= primary humic-like peak; H_2 = secondary humic-like peak; H_m = marine humic-like peak; P_1 = tyrosine-like peak; P_2 = tryptophan-like peak.

In general, the FEEM spectra supported the results of LC-OCD analyses regarding the presence of proteins in AOM released by the three algal species and the presence of humic-like like substances in AOM released by AT and MSp.

4.3.2.3 FTIR spectra

FTIR spectroscopy of purified and freeze-dried AOM samples were performed to indentify the composition of particulate and colloidal AOM (>3.5 kDa) from 3 species of algae. In this technique, an infrared spectrum is generated after passing a beam of infrared (IR) light through the AOM sample. Absorption is recorded when the frequency of the IR light coincide with the vibrational frequency of covalent bonds in certain molecules present in AOM. Seven characteristic IR peaks were recorded, six of which were observed in the 3 AOM samples (Figure 4-9). The functional groups identified based on the IR spectra and the typical organic compounds associated with them are summarised in Table 4-6.

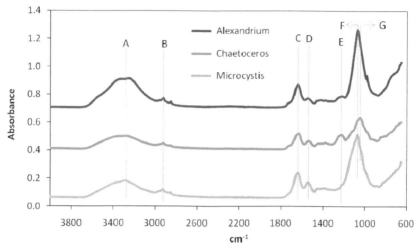

Figure 4-9: FTIR spectra of freeze dried AOM samples from 3 species of algae. Offsets in absorbance spectra were intentionally made for clarity.

High IR absorptions were observed in the following wavelengths: 3400-3300 (Peak A), 1660-1600 (Peak C), 1080-1070 (Peak F) and 1050 (Peak G). Peak A is a broad and intense band corresponding to the O-H stretching vibration of alcohol components usually associated with polysaccharides. Another indication of the presence of polysaccharides is the weak band between 1280 and 1200 cm⁻¹ (Peak E) which is due to C-O stretching and OH deformation of COOH.

Table 4-6: Absorption bands indentified in the infrared (IR) spectra of freeze-dried AOM samples.

Peak*	Wavelength (cm⁻¹)	Functional group	Compound
A	3400-3300 [i]	Stretching OH	Polysaccharides
B	2950-2850 [w]	Stretching -CH_2	Lipids
C	1650-1640 [i]	Stretching C-O & C-N (Amide I)	Proteins
D	1545-1540 [w]	Stretching C-N & bending NH (Amide II)	Proteins
E	1280-1200 [w]	Stretching C-O & OH deformation of COOH	Polysaccharides
F	1080-1070 [i]	CH aromatic	Humic substances
G	1050 [i]	Stretching -S=O	Sugar ester sulphates

Interpretation of IR spectra was based on Mecozzi *et al.* (2001) and Lee *et al.* (2006). * Peak codes as indicated in Figure 4-9; (i) - intense band; (w) – weak band.

The low absorption bands between 2950 and 2850 cm⁻¹ (Peak B) may indicate the presence of the aliphatic compounds possibly associated with lipids. Peak C corresponds to C-O and C-N stretching of primary amides associated with proteins. The presence of proteins is confirmed by the presence of weak bands between 1545 and 1540 cm⁻¹ (Peak D) due to C-N stretching and N-H bending of secondary amides.

Peak F indicates the presence of CH aromatic compounds possibly from humic-like substances. This peak is apparent in AOM samples from AT and MSp but not with CA. This conforms to the findings of FEEM and LC-OCD analyses. Peak G which corresponds to -S=O stretching of sugar ester sulphate groups was apparent in the AOM sample from CA. This further confirms that AOM from CA is highly sulphated as observed in Figure 4-4.

In general, the FTIR findings are consistent with what was reported in marine mucilaginous aggregates (Mecozzi et al., 2001) and algal bloom impacted freshwater (Lee et al., 2006). Hence, the AOM investigated in this study may represent the characteristics of AOM in natural waters. Furthermore, the results confirm the abundance of polysaccharides and proteins as well as the presence of humic like substances and sulphated polysaccharides in AOM released by the 3 species of algae.

4.3.2.4 Lectin-binding affinity

Selective staining using 6 fluorochrome-labelled lectins was applied to identify the functional groups associated with polysaccharides in AOM samples from algae. Each lectin is known to bind with specific functional groups present in polysaccharides to form glycoconjugates. By definition, glycoconjugates are carbohydrates covalently-linked with other chemical species. In this study, the selective binding capacities of selected lectins with AOM from the 3 algal species were visually assessed using multi-channel CLSM imaging. The results are summarised in Table 4-7.

Table 4-7: The lectin binding affinities of polysaccharides from 3 species of bloom-forming algae collected during the stationary-death phase.

Lectin	Specificity	Alexandrium	Chaetoceros	Microcystis
AAL	fucose (Fuc)	++	++	++
UEA-1	fucose (Fuc)	-	+	+
Con-A	α-D-mannose (α-Man), α-D-glucose (α-Glc)	-	-	-
HMA	Sialic acid	-	-	+
PA-I	galactose (Gal)	-	-	++
IAA	N-Acetyl-D-Glucosamine (GlcNAc)	++	-	+

Legend: (-) no or very few glycoconjugates detected; (+) glycoconjugates detected but not ubiquitous; (++) ubiquitous presence of glycoconjugates detected.

AT-derived polysaccharides bound preferentially to AAL and IAA lectins, which have known affinities to fucose and N-Acetylglucosamine sugars, respectively. AOM from CA bonded only with fucose-specific lectins (AAL and UEA-1) while AOM from MSp boded with all but one of the 6 lectins applied. The latter was preferentially bonded with fucose and galactose specific lectins and slightly by sialic acid and N-Acetylglucosamine specific lectins, but no significant binding with α-manose and α-glucose specific lectins was observed. The molecular structures of the above-mentioned functional groups are presented in Annex 4-2.

The results of lectin staining illustrated the chemical diversity of polysaccharides produced by the three species of algae. MSp produced polysaccharides comprised several functional groups (e.g., fucose, sialic acid, galactose and N-Acetylglucosamine). One out of six lectins (i.e., AAL) showed strong affinity to all three AOM samples. Although both AAL and UEA-1 were both identified as fucose-specific lectins, only the former showed consistent binding with all 3 algal polysaccharides. This may indicate that the structures of algal polysaccharides were different in terms of fucose location and were specifically linked to polysaccharide structures in a way that allow good binding with AAL and not with UEA-1.

Images generated from maximum intensity projections of CLSM data sets of AOM stained with fluorophore-labelled lectins provided some insight on how AOM may have been released by the three species of algae (Figure 4-10). AT cells were observed to release large polysaccharide materials (>10µm) after shedding a layer of their cell wall while smaller flocs (<1µm) of similar materials were also released by compromised and deteriorating cells (Figure 4-10a,b). Live cells of CA were observed embedded in large flocs of fibrous and cloud-like polysaccharides (>50 µm). The thin outer layer of the cell wall of MSp comprised glycoconjugates and when the cells have deteriorated, several aggregates linked by glycoconjugates were observed.

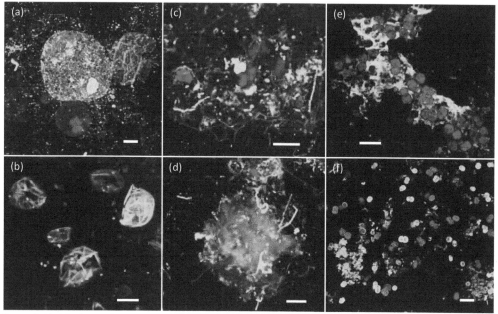

Figure 4-10: Maximum intensity projections of CLSM data sets showing algal aggregates from AT (a, b), CA (c, d) and MSp (e, f) cultures after staining with AAL (a, c, d, f), PA-I (e) and IAA (b) lectins. Color legend: green fluorescence for glycoconjugates; red fluorescence for chlorophyll-a of living algal cells. The scale bar is 10µm.

In general, CLSM images of lectin stained AOM showed similarities with Alcian blue stained AOM in Figure 4-3. This is an indication that TEPs contain substantial amounts of fucose sugars. This was also in agreement with what was found by Zhou *et al.* (1998) in marine algal bloom samples where surface-active TEPs were found to be mainly enriched with deoxysugars (e.g., fucose and rhamnose). However, not all types of saccharides (including rhamnose) were investigated in this study. Further investigations are therefore needed to study the presence of more saccharides in AOM by applying lectins with specific affinity to other functional groups.

4.3.2.5 Cohesion/adhesion potential

Some AOM, specifically TEPs, were reported to be highly sticky (Passow, 2002). To investigate the stickiness (cohesion/adhesion potential) of AOM from 3 algal species, interaction force measurements were performed using AFM. Two types of membranes of polyamide (PA) and polyethersulfone (PES) were selected as reference surfaces to illustrate the potential role of AOM in the fouling of RO and UF membranes, respectively. The force distance curves generated between clean or AOM coated polystyrene microsphere probes and clean or AOM-fouled membrane surfaces are presented in Figures 4-11 and 4-12.

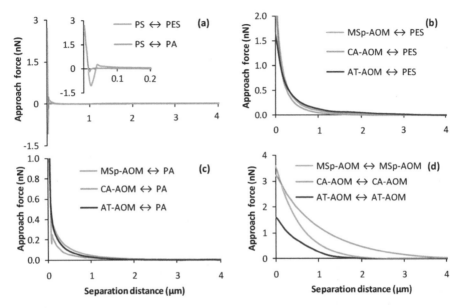

Figure 4-11: AFM approach force-distance curves: (a) clean PS probes approaching clean PES and PA membranes, (b) AOM-coated probes approaching clean PES membrane, (c) AOM-coated probes approaching clean PA membrane, (d) AOM-coated probes approaching AOM-fouled PES membrane.

The generated approach force curves (Figure 4-11) illustrate that short-range attraction forces were observed between clean PS probes and clean membranes. After incubating the probes in AOM solution for 5 days, interaction shifted to rather long range repulsion. This change indicates that indeed a layer of AOM was adsorbed to the surface of the polystyrene microspheres. The attraction between clean polystyrene and clean membranes may be due to van der Waals interaction, hydrogen bonding and/or hydrophobic interaction. The PS particle is hydrophobic and probably prefers contact with the membrane surface over contact with water. Although both polystyrene particle (reported by supplier), the membranes (Hurwitz et al., 2010; Childress and Elimelech, 1996; Ricq et al., 1997) and AOM (Henderson et al., 2008b) are (slightly) negatively charged, electrostatic repulsion does not play a significant role in the surface interaction between clean or AOM-coated polysterene probe and clean membrane because the high ion concentration in the matrix (seawater) screens the charges and limits the range of electrostatic interactions to distances of less than 1 nm (Mosley et al., 2003).

Adhesion/cohesion forces were investigated based on the retract force curves between AOM-coated PS probes and clean or AOM-fouled membranes after 30 seconds of contact time (Figure 4-12). Average retract forces were higher between the AOM-coated PS probe and the AOM-fouled PES membrane than the AOM-coated PS probe and clean membranes (PES and PA). It was also observed that the retract force curves were much broader when AOM was present on both the PS probe and membrane surface (Figure 4-12d). This implies that a high retract force as well as high energy is needed to totally retract (pull-off) bonds between two AOM coated surfaces. In addition, it indicates that the AOM layers can deform and stretch when pulled apart – and only at distances of the order of several microns, do the AOM layers come apart again. The softness and deformability of the AOM layers is corroborated by the approach curves in Figure 4-11d showing a gradual increase in force (needed for compression of the AOM layer) over distances of several microns.

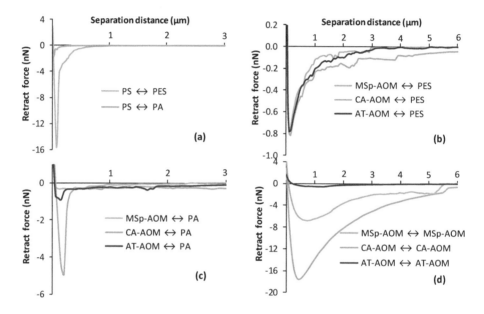

Figure 4-12: AFM retract force-distance curves after 30 seconds of contact time: (a) clean PS probes on clean PES and PA membranes, (b) AOM-coated probes on clean PES membrane, (c) AOM-coated probes on clean PA membrane, (d) AOM-coated probes on AOM-fouled PES membrane.

Hydrogen bonding may have played a major role in the adhesion/cohesion between investigated surfaces. Since hydrogen bonding is a result of electron transfer between electronegative moieties and hydrogen atoms in AOM and on membrane surfaces, the bond strength might be influenced by the membrane surface roughness. The reported surface roughness of PA membranes is generally higher than PES membrane (Pieracci et al., 1999; Hurwitz et al., 2010). This might explain the higher adhesion force of AOM on PA membranes than on PES membrane as the actual surface interaction area is higher in rougher membranes. This is crucial considering that AOMs, particularly TEPs, are deformable and may likely follow the topography of the membrane surface. On the other hand, the cohesion forces between AOMs are influenced not only by hydrogen bonding but also by polymer entanglement during contact between AOM-coated surfaces (Flemming et al, 1997). AOM polymers are generally flexible and elastic, so disentangling them may have occurred in a stepwise manner and therefore requires higher energy to totally detach them.

The average maximum adhesion/cohesion force and total energy needed to pull-off two investigated surfaces from contact are presented in Figure 4-13. AOMs from 3 species of algae showed comparable magnitude in terms of adhesion force and energy on clean PES membrane. However, AOM adhesions on clean PA membranes vary substantially with source algal species. Adhesion force and energy on clean PA was highest for CA-AOM followed by AT-AOM and MSp-AOM, respectively. Cohesion between the AOM-coated probe and AOM-fouled PES membrane was higher (up 15 times) than the adhesion between the AOM-coated probe and clean PES membrane. Average cohesion force and energy was highest for CA-AOM followed by MSp-AOM and AT-AOM, respectively. The measured adhesion forces range from 0.6 to 17.7 nN. These values are 2 to 3 orders of magnitude higher than the typical lift forces occurring in cross-flow membrane reported by Kang *et al.* (2004).

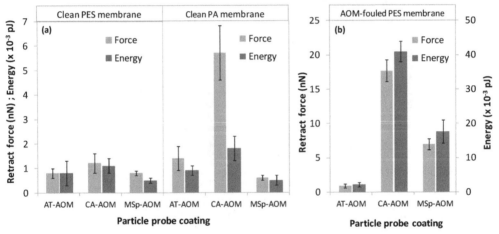

Figure 4-13: Maximum retract (adhesion/cohesion) forces and total energies needed to completely pull-off AOM-coated probes from (a) clean or (b) AOM-fouled membranes after 30 seconds of contact time.

The stickiness of AOM (in terms of adhesive/cohesive strength) can be a crucial indicator of the tendency of AOM to deposit in membrane systems and their role in biofilm formation in RO. Moreover, the sticky and elastic nature of AOM, especially those from CA, means that they can strongly adhere to the pores and surfaces of a UF membrane and that backwashing may not be able to remove them completely. Consequently, non-backwashable fouling occurs, whereby the initial permeability of the membrane is not recovered after the hydraulic backwashing.

To summarise, the stickiness of AOM in terms of adhesive and cohesive strength varies between algal sources. Cohesion between AOM is generally stronger than adhesion of AOM on clean membranes. Thus, more AOM are expected to attached/accumulate on membranes already fouled with AOM than on clean membranes.

4.3.3 AOM rejection by MF/UF membranes

AOM samples from 3 species of algae prepared according to the protocol described in Section 4.2.2 were filtered through MF/UF membranes with various pore sizes. The organic carbon concentrations of the feedwater and permeate were analysed using LC-OCD. AOM rejections in terms of total organic carbon smaller than ~2 μm varied widely with membrane pore sizes and source algae species (Figure 4-14). As expected, UF membranes (100 and 10 kDa MWCO) showed substantially higher removal than MF membranes (0.4 and 0.1μm pore size). AOM from CA were better removed by MF/UF membranes than AOM from MSp and AT, respectively. The membranes substantially removed the biopolymer fraction of AOM while the low molecular weight organic fractions (e.g., building blocks, humic-like substances and organic acids) - a substantial portion of which originated from the medium (Figure 4-6) - were poorly rejected during the filtration experiment.

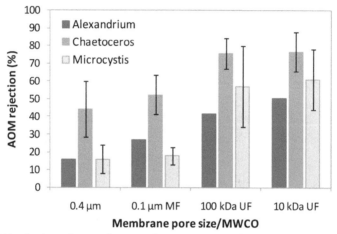

Figure 4-14: AOM rejection of MF and UF membranes in terms of organic carbon (<2μm) measured by LC-OCD.

Since UF/MF membranes are not designed to remove the low molecular weight components of AOM, measuring rejection in terms of biopolymers is a better approach to understand the rejection of these target AOMs. The OC signal responses of the biopolymer fraction, as illustrated in Figure 4-15, decreased with reduction in membrane pore size. The biopolymer peaks of AT-AOM and CA-AOM was almost totally eliminated after 10 kDa UF. However for MSp, the biopolymer peak was still apparent even after 10kDa UF. The differences in removal might be due to differences in the size distribution of biopolymers as shown in Figure 4-7. However, the lower biopolymer removal of MSp-AOM might have been affected by the AOM preparation procedure where filtration through 1 μm pore size filter was performed to remove algal cells from MSp-AOM solution while larger pore size filters (i.e., 5 and 10μm) were used for the preparation of CA-AOM and AT-AOM solutions. Nevertheless, biopolymers which are larger than 2 μm are not quantified during the LC-OCD analysis because the chromatogram columns only allow materials smaller than 2 μm to pass through. Hence, the effect of differences in AOM preparation on biopolymer rejection is assumed to be not substantial.

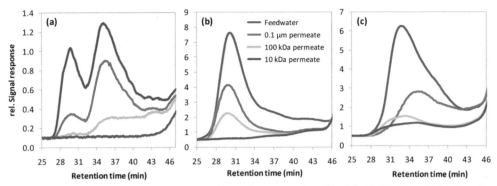

Figure 4-15: OCD chromatogram variations of biopolymers produced by (a) AT, (b) CA and (c) MSp, after passing through various MF and UF membranes.

A quantitative comparison of the rejection of biopolymers by various MF/UF membranes is shown in Figure 4-16. About 14% of biopolymers from AT were rejected by 0.4 μm membrane while 42% were rejected by 0.1 μm MF, 72% by 100kDa UF and 97% by 10kDa UF. The rejections of biopolymers from CA were 47%, 56%, 83% and 95% by 0.4μm, 0.1μm, 100 kDa and 10 kDa membranes, respectively. For biopolymers from MSp, membrane rejections were 42%, 52%, 65% and 83% by 0.4 μm, 0.1 μm, 100 kDa and 10 kDa membranes, respectively.

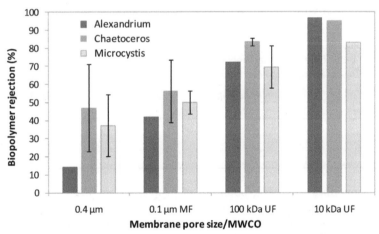

Figure 4-16: Biopolymer rejection of various MF and UF membranes in terms of organic carbon (<2μm) measured by LC-OCD.

Although biopolymers were reported to be larger than 20 kDa (Huber *et al*, 2011), about 3% of biopolymers from AT, 5% from CA and 17% from MSp were not rejected by 10 kDa membranes. This may be due to the wide pore size distribution of the RC membranes. It might also be due to the fact that some algal biopolymers originate from nanogels as small as 3 kDa (Passow, 2000) which means that these biopolymers might have been much smaller than 10 kDa.

The actual pore size of the membrane is expected to decrease over the filtration period as more rejected biopolymer materials accumulate on or inside its pores. As the cake/gel layer build-up and/or compressed on the surface of the membrane, it becomes the main barrier layer and may reject more biopolymers than the membrane itself. On the other hand, TEPs and other biopolymers are fibrillar and highly flexible that it might be capable of squeezing through the membrane pores smaller than their apparent size (Passow, 2000). Hence, rejections of biopolymers by MF/UF membranes are governed by several factors which include concentration of biopolymers in the bulk solution as well as the characteristics of both the membrane and the biopolymers themselves.

In summary, MF/UF membranes can substantially remove biopolymer fraction of AOM but not the low molecular weight fraction (e.g., humic-like substances). Biopolymer removals by conventional UF membranes (100kDa MWCO) can be up to 83% while tighter UF membranes (10 kDA MWCO) can remove up to 97%.

4.3.4 Specific membrane fouling potential of biopolymers

The specific fouling potential of biopolymers in two different size ranges were measured based on serial filtration of AOM samples (stationary-death phase) through 100 kDa and 10 kDa membranes, respectively. MFI-UF was measured based on the hydraulic resistance of the rejected biopolymers in each filtration steps. To determine the amount of biopolymers retained on each membrane, LC-OCD analysis was performed on the feedwater and the UF permeate samples. The recorded biopolymer concentrations and MFI-UF values are presented in Table 4-8.

Table 4-8: Biopolymer concentration in the feed water and UF permeates of 100 kDa and 10 kDa membranes in serial filtration experiments. MFI-UF results in the right columns correspond to the membrane fouling potential of retained biopolymers on 100 kDa and 10 kDa membranes.

Sample	Biopolymers (mg C/L)			MFI-UF (s/L^2)		
	AT	CA	MSp	AT	CA	MSp
Feed	4.10	6.66	2.47			
100 kDa UF	1.15	1.20	0.55	61,000	102,167	33,350
10 kDa UF	0.13	0.33	0.42	25,000	51,000	5,170

The specific membrane fouling potential of each fraction was calculated by dividing the measured MFI-UF with the concentration of biopolymers retained on the membrane. As illustrated in Figure 4-17, the specific membrane fouling potential of larger biopolymers (>100 kDa) were lower than the smaller size fractions (10-100kDa). This can be explained based on the Carman-Kozeny model where the pressure drop through a cake layer is inversely proportional to the square of the size of its components. Larger biopolymers (>100 kDa) from 3 algal sources showed comparable fouling potential while fouling potential of smaller fractions (10-100kDa) vary significantly among source species (CA>MSp>AT).

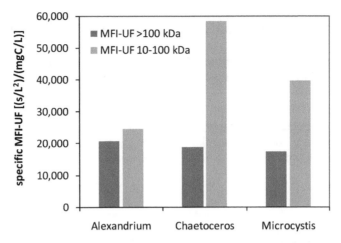

Figure 4-17: Specific membrane fouling potential (MFI-UF) and cake resistance (α) of two size fractions of biopolymers from 3 species of algae.

The higher specific membrane fouling potential of biopolymers not rejected by 100 kDa MWCO UF membranes may have some practical implications to the operation of RO plants. Granular media filters (~150 μm pore size) or UF (>100 kDa MWCO) membranes are usually installed in RO plants to pre-treat raw water. If the pre-treatment system is not effective in removing the smaller colloidal biopolymers (10-100 kDa) from the raw water, these may cause fouling in the RO system downstream. However, installing low MWCO UF membranes (e.g., <10kDa) can potentially remove these materials (Figure 4-16) and minimise possible fouling in the RO system, especially during algal bloom situations.

4.4   Summary and conclusions

1. Batch cultures of 3 species of bloom-forming algae (*Alexandrium tamarense, Chaetoceros affinis, Microcystis sp.*) showed different characteristics in terms of growth and TEP production.

2. Algal-derived organic material (AOM) extracted from algal cultures composed mainly of biopolymers (e.g., polysaccharides and proteins) and some refractory organic matter (e.g., humic-like substances) and/or biogenic substances (e.g., acids, neutrals). Polysaccharides containing fucose and sulphated functional groups were found ubiquitous in all AOM samples.

3. The stickiness of AOM in terms of its adhesive and cohesive strength varies between algal sources. Cohesion between AOM is generally stronger than adhesion of AOM on clean membranes.

4. A linear relationship ($R^2 > 0.88$) was observed between membrane fouling potential (in terms of MFI-UF) and TEP concentration but no apparent relationship was observed between MFI-UF and algal cell concentration in batch cultures.

5. Based on LC-OCD analysis, MF membranes with pore sizes of 0.4 µm and 0.1µm rejected 14-47% and 42-56% of biopolymers, respectively while UF membranes with molecular weight cut-offs of 100 kDa and 10 kDa rejected 65-83% and 83-97% of these material, respectively.

6. The membrane fouling potential of biopolymers rejected by UF membranes differ between algal species. The biopolymer fraction between 10-100kDa showed higher specific membrane fouling potential (MFI per mg C/L biopolymers) than the larger fractions (>100kDa).

References

Alldredge A. L., Passow U., and Logan B. E. (1993) The abundance and significance of a class of large, transparent organic particles in the ocean. Deep Sea Research 40(6), 1131-1140.

Alldredge, A. L., & Silver, M. W. (1988). Characteristics, dynamics and significance of marine snow. Progress in Oceanography, 20(1), 41-82.

Anderson, D. M., Alpermann, T. J., Cembella, A. D., Collos, Y., Masseret, E., & Montresor, M. (2012). The globally distributed genus alexandrium: Multifaceted roles in marine ecosystems and impacts on human health. Harmful Algae, 14, 10-35.

Baghoth, S. A., Sharma, S. K., Guitard, M., Heim, V., Croué, J.P., & Amy, G. L. (2011). Removal of NOM-constituents as characterized by LC-OCD and F-EEM during drinking water treatment. Journal of Water Supply: Research and Technology - AQUA, 60(7), 412-424

Berktay, A. (2011). Environmental approach and influence of red tide to desalination process in the middle-east region. International Journal of Chemical and Environmental Engineering 2 (3), 183-188.

Berman T., and Holenberg M. (2005) Don't fall foul of biofilm through high TEP levels. Filtration & Separation, 42(4), 30-32.

Berman T., Mizrahi R. and Dosoretz C.G. (2011) Transparent exopolymer particles (TEP): A critical factor in aquatic biofilm initiation and fouling on filtration membranes. Desalination, 276, 184-190.

Bhaskar, P. V., & Bhosle, N. B. (2005). Microbial extracellular polymeric substances in marine biogeochemical processes. Current Science, 88(1), 45-53.

Boerlage S.F.E., Kennedy M.D., Tarawneh Z., De Faber R., Schippers J.C. (2004) Development of the MFI-UF in constant flux filtration. Desalination 161 (2):103–113.

Caron , D.A., Garneau, M.E., Seubert, E., Howard M.D.A.,, Darjany L., Schnetzer A., Cetinic, I., Filteau, G., Lauri, P., Jones, B. and Trussell, S. (2010) Harmful algae and their potential impacts on desalination operations off southern California. Water Research 44, 385–416.

Childress, A. E., & Elimelech, M. (1996). Effect of solution chemistry on the surface charge of polymeric reverse osmosis and nanofiltration membranes. Journal of Membrane Science, 119(2), 253-268.

Coble, P. (1996) Characterisation of marine and terrestrial DOM in seawater using excitation-emission matrix spectroscopy. Marine Chemistry 51, 325-346.

Decho, A.W. (1990). Microbial exopolymer secretions in ocean environments: their role(s) in food webs and marine processes Oceanogr. Mar. Biol. Ann. Rev. 28, 73-153.

Flemming, H.C., Schaule, G., Griebe, T., Schmitt, J., & Tamachkiarowa, A. (1997) Biofouling - the achilles heel of membrane processes. Desalination, 113(2-3), 215-225.

Fogg, G. (1983). The Ecological Significance of Extracellular Products of Phytoplankton Photosynthesis. Botanica Marina, 26(1), 1-43.

Frank, B. P., & Belfort, G. (2003). Polysaccharides and sticky membrane surfaces: Critical ionic effects. Journal of Membrane Science, 212(1-2), 205-212.

Henderson, R., Chips, M., Cornwell, N., Hitchins, P., Holden, B., Hurley, S., Parsons, S. A., Wetherill, A. and Jefferson, B. (2008a) Experiences of algae in UK waters: a treatment perspective. Water and Environment Journal 22, 184–192.

Henderson, R.K., Baker, A., Parsons, S.A. and Jefferson, B. (2008b) Characterisation of algogenic organic matter extracted from cyanobacteria, green algae and diatoms. Water Research 42, 3435-3445.

Her, N., Amy, G., McKnight, D., Sohn, J., & Yoon, Y. (2003). Characterization of DOM as a function of MW by fluorescence EEM and HPLC-SEC using UVA, DOC, and fluorescence detection. *Water Research*, 37(17), 4295-4303.

Higgins, M. J., Crawford, S. A., Mulvaney, P., & Wetherbee, R. (2002). Characterization of the adhesive mucilages secreted by live diatom cells using atomic force microscopy. Protist, 153(1), 25-38.

Huber S. A., Balz A., Abert M. and Pronk W. (2011) Characterisation of aquatic humic and non-humic matter with size-exclusion chromatography – organic carbon detection – organic nitrogen detection (LC-OCD-OND). *Water Research* 45 (2), 879-885.

Hurwitz, G., Guillen, G. R., & Hoek, E. M. V. (2010). Probing polyamide membrane surface charge, zeta potential, wettability, and hydrophilicity with contact angle measurements. Journal of Membrane Science, 349(1-2), 349-357.

IOC-UNESCO (2013) What are harmful algae? Retrieved 30 September 2013, Online access http://hab.ioc-unesco.org/

Kang S.T., Subramani A., Hoek E.M.V., Deshusses M.A., Matsumoto M.R. (2004) Direct observation of biofouling in cross-flow microfiltration: mechanisms of deposition and release. Journal of Membrane Science 244 (1–2), 151–165.

Kennedy, M. D., Kamanyi, J., Heijman, B. G. J., & Amy, G. (2008). Colloidal organic matter fouling of UF membranes: Role of NOM composition & size. Desalination, 220(1-3), 200-213.

Kennedy, M. D., Muñoz-Tobar, F. P., Amy, G. L. and Schippers, J. C. (2009) Transparent exopolymer particle (TEP) fouling of ultrafiltration membrane systems. Desalination & Water Treatment 6(1-3) 169-176.

Kirchman D.L., Suzuki Y., Garside C. and Ducklow H.W. (1991) High turnover rates of dissolved organic carbon during a spring phytoplankton bloom. Nature 352, 612 – 614.

Lee, N., Amy, G., Croue, J.P. (2006) Low pressure membrane (MF/UF) fouling associated with allochthonous versus autochthonous natural organic matter. *Water Research* 40, 2357-2368.

Leenheer J.A. and Croué J.P. (2003) Characterizing Aquatic Dissolved Organic Matter: Understanding the unknown structures is key to better treatment of drinking water. Environmental Science & Technology 37 (1), 18A-26A.

Leppard, G.G., 1993. In: Rao, S.S. (Ed.), Particulate Matter and Aquatic Contaminants. Lewis, Chelsea, MI, pp. 169– 195.

Li, L., Gao, N., Deng, Y., Yao, J. and Zhang, K. (2012) Characterisation of intracellular & extracellular algae organic matters (AOM) of Microcystis aeruginosa and formation of AOM-associated disinfection byproducts and odor & taste compounds. *Water Research* 46, 1233–1240.

Martin-Jezequel, V., Hildebrand, M. and Brzezinski, M. (2000) Silicon metabolism in diatoms: Implications for growth. *J. Phycol.* 36, 821-840.

McKnight D.M., Boyer, E.W., Westerhoff, P.K., Doran, P.T., Kulbe, T. and Andersen, D.T. (2001) Spectroflourometric characterization of dissolved organic matter for the indication of precursor organic material and aromaticity. Limnology and Oceanography 46(1), 38-48.

Mecozzi, M., Acquistucci, R., Di Noto, V., Pietrantonio, E., Amici, M., & Cardarilli, D. (2001). Characterization of mucilage aggregates in adriatic and tyrrhenian sea: Structure similarities between mucilage samples and the insoluble fractions of marine humic substance. Chemosphere, 44(4), 709-720.

Mopper, K., Zhou, J., Sri Ramana, K., Passow, U., Dam, H. G., & Drapeau, D. T. (1995). The role of surface-active carbohydrates in the flocculation of a diatom bloom in a mesocosm. Deep-Sea Research Part II, 42(1), 47-73.

Mosley, L. M., Hunter, K. A., & Ducker, W. A. (2003). Forces between colloid particles in natural waters. Environmental Science and Technology, 37(15), 3303-3308.

Myklestad, S. M. (1995). Release of extracellular products by phytoplankton with special emphasis on polysaccharides. Science of the Total Environment, 165, 155-164.

Myklestad, S. M. (1999). Phytoplankton extracellular production and leakage with considerations on the polysaccharide accumulation. *Annali Dell'Istituto Superiore Di Sanita*, 35(3), 401-404.

Myklestad, S., Haug, A., & Larsen, B. (1972). Production of carbohydrates by the marine diatom chaetoceros affinis var. willei (gran) hustedt. II. preliminary investigation of the extracellular polysaccharide. Journal of Experimental Marine Biology and Ecology, 9(2), 137-144.

Neu TR, Woelfl S, Lawrence JR (2004) 3-dimensional differentiation of photo-autotrophic biofilm constituents by multi-channel laser scanning microscopy (single-photon and two-photon excitation). Journal of Microbiological Methods 56: 161-172.

Neu, T. R. (2000). In situ cell and glycoconjugate distribution in river snow studied by confocal laser scanning microscopy. *Aquatic Microbial Ecology, 21*(1), 85-95.

Pankratz, T., 2008. Red tides close desal plants. *Water Desalination Report* 44, 1.

Passow, U. (2000) Formation of transparent exopolymer particles (TEP) from dissolved precursor material. *Marine ecology, Progress series* (Halstenbek) 192, 1-11.

Passow, U. (2002) Transparent exopolymer particles (TEP) in aquatic environments. *Progress in Oceanography* 55(3), 287-333.

Passow, U., and Alldredge, A. L. (1995) A Dye-Binding Assay for the Spectrophotometric Measurement of Transparent Exopolymer Particles (TEP). *Limnology and Oceanography* 40(7), 1326-1335.

Petry, M., Sanz M. A., Langlais C., Bonnelye V., Durand J.P., Guevara D., Nardes W. M., Saemi, C. H. (2007) The El Coloso (Chile) reverse osmosis plant. *Desalination* 203, 141–152.

Pieracci, J., Crivello, J. V., & Belfort, G. (1999). Photochemical modification of 10kDa polyethersulfone ultrafiltration membranes for reduction of biofouling. Journal of Membrane Science, 156(2), 223-240.

Qu, F., Liang, H., Wang, Z., Wang, H., Yu, H. and Li, G. (2012) Ultrafiltration membrane fouling by extracellular organic matters (EOM) of Microcystis aeroginusa in stationary phase: Influences of interfacial characteristics of foulants and fouling mechanisms. *Water Research* 46, 1490-1500.

Ralston J., Larson I., Rutland M., Feiler A., Kleijn M. (2005) Atomic force microscopy and direct surface force measurements (IUPAC Technical Report), Pure and Applied Chemistry, 77, 2149-2170.

Richlen M. L., Morton S. L., Jamali E. A., Rajan A., Anderson D. M. (2010), "The Catastrophic 2008–2009 Red Tide in the Arabian Gulf Region, with Observations on the Identification and Phylogeny of the Fish-killing Dinoflagellate Cochlodinium Polykrikoides", Harmful Algae, 9(2), pp. 163-172.

Ricq, L., Pierre, A., Bayle, S., & Reggiani, J.C. (1997). Electrokinetic characterization of polyethersulfone UF membranes. Desalination 109(3), 253-261.

Rouwenhorst, R.J., Jzn, J.F., Scheffers, A. and van Dijken, J.P. (1991) Determination of protein concentration by total organic carbon analysis. Journal of Biochemical and Biophysical Methods 22, 119-128.

Salinas-Rodríguez S.G., Kennedy, M.D., Amy, G.L., & Schippers, J.C. (2012) Flux dependency of particulate/colloidal fouling in seawater reverse osmosis systems. Desalination and Water Treatment, 42(1-3), 155-162.

Salinas-Rodriguez, S.G., Kennedy, M.D., Schippers J.C. and Amy, G.L. (2009) Organic foulants in estuarine and bay sources for seawater reverse osmosis - Comparing pre-treatment processes with respect to foulant reductions. Desalination & Water Treatment 9, 155-164.

Schippers, J.C. and Verdouw, J. (1980) The modified fouling index, a method of determining the fouling characteristics of of water. Desalination 32, 137-148.

Schurer R., Tabatabai A., Villacorte L., Schippers J.C., Kennedy M.D. (2013) Three years operational experience with ultrafiltration as SWRO pre-treatment during algal bloom. Desalination & Water Treatmen 51 (4-6), 1034-1042.

Schurer, R., Janssen, A., Villacorte, L., Kennedy, M.D. (2012) Performance of ultrafiltration & coagulation in an UF-RO seawater desalination demonstration plant. Desalination & Water Treatment 42(1-3), 57-64.

Spruijt E., van den Berg S.A., Cohen-Stuart M.A., and van der Gucht J. (2012) Direct Measurement of the Strength of Single Ionic Bonds between Hydrated Charges. ACS Nano 6 (6), 5297–5303.

Villacorte, L. O., Kennedy, M. D., Amy, G. L., and Schippers, J. C. (2009) The fate of Transparent Exopolymer Particles (TEP) in integrated membrane systems: Removal through pretreatment processes and deposition on reverse osmosis membranes. Water Research 43 (20), 5039-5052.

WHOI (2013) Harmful algae species. Retrieved 30 September 2013; Available at: http://www.whoi.edu/redtide/species/by-name.

Yamamura, H., Kimura, K., Okajima, T., Tokumoto, H., & Watanabe, Y. (2008). Affinity of functional groups for membrane surfaces: Implications for physically irreversible fouling. Environmental Science and Technology, 42(14), 5310-5315.

Zhou, J., Mopper, K., & Passow, U. (1998). The role of surface-active carbohydrates in the formation of transparent exopolymer particles by bubble adsorption of seawater. *Limnology and Oceanography, 43*(8), 1860-1871.

Zippel, B., & Neu, T. R. (2011). Characterization of glycoconjugates of extracellular polymeric substances in tufa-associated biofilms by using fluorescence lectin-binding analysis. Applied and Environmental Microbiology, 77(2), 505-516.

Annex 4-1: Synthetic seawater and culture media

Table 4-9: Synthetic seawater (SSW) composition.

Chemical ion	Concentration (g/L)
Chlorine, Cl⁻	19.2
Sodium, Na⁺	10.7
Sulfate, SO_4^{2-}	2.7
Magnesium, Mg^{2+}	1.3
Calcium, Ca^{2+}	0.42
Potassium, K⁺	0.39
Hydrogen Carbonate HCO^{3-}	0.16
Carbonate, $CO3^{2-}$	0.001
Bromine, Br⁻	0.05
Total	34.3

Table 4-10: L1 medium for marine dinoflagellates (www.ccap.ac.uk)

Stocks	per liter	stock solution
(1) Trace elements (from CCMP)		
$FeCl_3 \cdot 6H_2O$	3.15 g	-
$Na_2EDTA \cdot 2H_2O$	4.36 g	-
$CuSO_4 \cdot 5H_2O$	0.25 mL	2.45 g/L
$Na_2MoO_4 \cdot 2H_2O$	3 mL	19.9 g/L
$ZnSO_4 \cdot 7H_2O$	1 mL	22 g/L
$CoCl_2 \cdot 6H_2O$	1 mL	10 g/L
$MnCl_2 \cdot 4H_2O$	1 mL	180 g/L
H_2SeO_3	1 mL	1.3 g/L
$NiSO_4 \cdot 6H_2O$	1 mL	2.7 g/L
Na_3VO_4	1 mL	1.84 g/L
K_2CrO_4	1 mL	1.94 g/L
(2) Vitamin mix		
Cyanocobalamin (Vitamin B12)	0.0005 g	-
Thiamine HCl (Vitamin B1)	0.1 g	-
Biotin	0.0005 g	-

Medium	per liter
NaNo3	0.075 g
$NaH_2PO_4.2H_2O$	0.00565 g
Trace elements stock solution (1)	1. 0 ml
Vitamin mix stock solution (2)	1. 0 ml

Table 4-11: Guillard's medium (f/2 + Si) for marine diatoms (www.ccap.ac.uk).

Stocks	per litre
(1) Trace elements (chelated)	
NA_2EDTA	4.16 g
$FeCl_3.6H_2O$	3.15 g
$CuSO_4.5H_2O$	0.01 g
$ZnSO_4.7H_2O$	0.022 g
$CoCl_2.6H_2O$	0.01 g
$MnCl_2.4H_2O$	0.18 g
$Na_2Mo_4.2H_2O$	0.006 g
(2) Vitamin mix	
Cyanocobalamin (Vitamin B12)	0.0005 g
Thiamine HCl (Vitamin B1)	0.1 g
Biotin	0.0005 g
(3) Sodium metasilicate	
$Na_2SiO_3.9H_2O$	30.0g
Medium	
$NaNO_3$	0.075 g
$NaH_2PO_4.2H_2O$	0.00565 g
Trace elements stock solution (1)	1.0 ml
Vitamin mix stock solution (2)	1.0 ml
Sodium metasilicate stock solution (3)	1.0 ml

Table 4-12: BG11 medium for freshwater blue-green algae and protozoa (www.ccap.ac.uk).

Stocks	per litre
(1) $NaNO_3$	15.0 g
(2) K_2HPO_4	4.0 g
(3) $MgSO_4.7H_2O$	7.5 g
(4) $CaCl_2.2H_2O$	3.6 g
(5) Citric acid	0.6 g
(6) Ammonium ferric citrate green	0.6 g
(7) Na_2 EDTA	0.1 g
(8) Na_2CO_3	2 g
(9) Trace metal solution:	
H_3BO_3	2.86 g
$MnCl_2.4H_2O$	1.81 g
$ZnSO_4.7H_2O$	0.22 g
$Na_2MoO_4.2H_2O$	0.39 g
$CuSO_4.5H_2O$	0.08 g
$Co(NO_3)2.6H_2O$	0.05 g
Medium	**per litre**
Stock solution 1	100.0 ml
Stock solutions 2 - 8	10.0 ml each
Stock solution 9	1 ml

Annex 4-2: Molecular structure of selected carbohydrate functional groups

(a) fucose

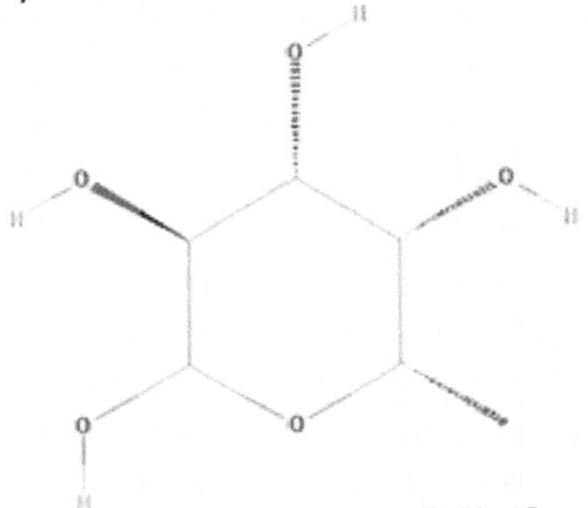

$C_6H_{12}O_5$

(b) α-D-mannose

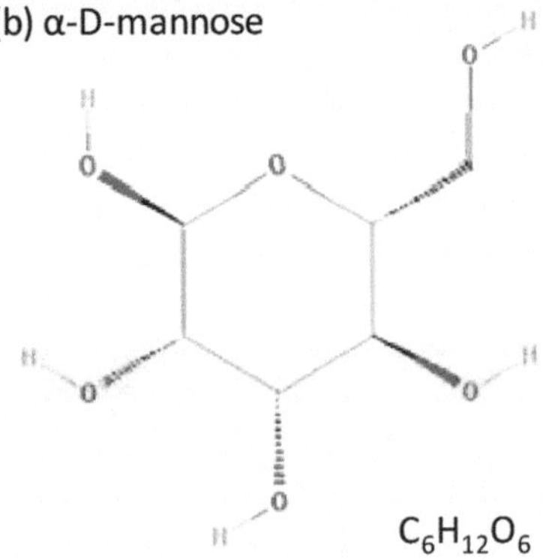

$C_6H_{12}O_6$

(c) α-D-glucose

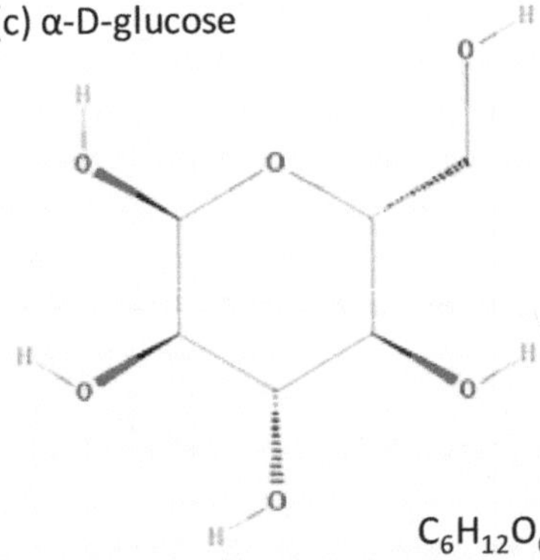

$C_6H_{12}O_6$

(d) galactose

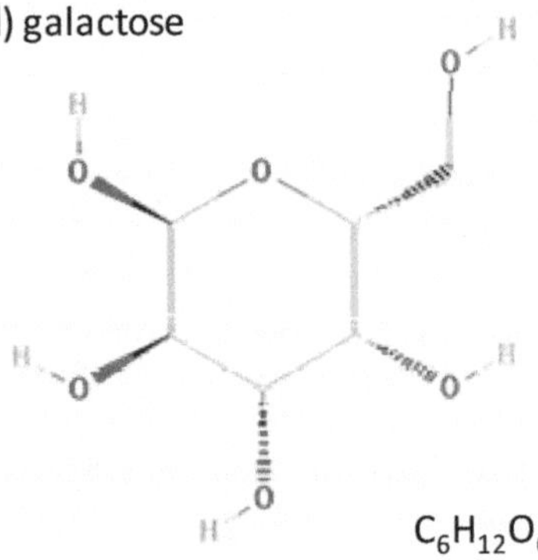

$C_6H_{12}O_6$

(e) sialic acid

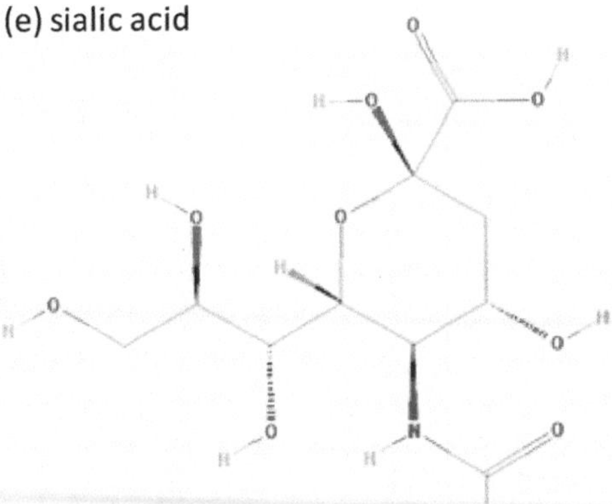

$C_{11}H_{19}NO_9$

(f) N-Acetyl-D-Glucosamine

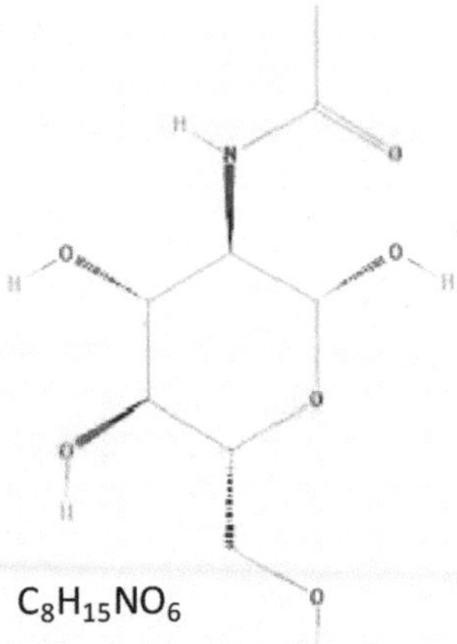

$C_8H_{15}NO_6$

Source: http://www.ncbi.nlm.nih.gov/

5

Measuring transparent exopolymer particles in freshwater and seawater

Contents

Abstract 110
5.1 Introduction 111
    5.1.1 Background 112
    5.1.2 Scope and objectives 115
5.2 Materials and methods 116
    5.2.1 Water samples and synthetic solutions 116
    5.2.2 Alcian blue dye 117
    5.2.3 TEP measurement methods 118
    5.2.4 Filters and filtration set-up 123
    5.2.5 Limit of detection calculation 123
    5.2.6 Dynamic light scattering measurements 124
    5.2.7 Liquid chromatography organic carbon detection (LC-OCD) 124
5.3 Results and discussion 124
    5.3.1 Pre-treatment of Alcian blue solution 124
    5.3.2 Absorption spectra of Alcian blue 125
    5.3.3 Measurement of $TEP_{0.4\mu m}$ and $TEP_{0.1\mu m}$ 126
    5.3.4 Measurement of TEP_{10kDa} 131
    5.3.5 Comparison and application of the two methods 139
5.4 Summary and conclusions 144
References 145

This chapter is an extended version of:

Villacorte, L. O., Ekowati, Y., Calix-Ponce H.N., Schippers, J.C., Amy, G. and Kennedy, M. (2014) Measuring transparent exopolymer particles in freshwater and seawater: limitations, method improvements and applications. *In preparation for Water Research.*

Abstract

Transparent exopolymer particles (TEPs) in aquatic environments are increasingly mentioned as a potential foulant in reverse osmosis, nanofiltration, microfiltration and ultra-filtration membrane systems treating surface water. A number of methods are currently available to measure these substances in freshwater and seawater based on Alcian blue staining and spectrophotometric techniques. However, a critical analysis of these methods identified several questions regarding their reproducibility, calibration, limit of detection and range of TEP sizes taken into account. The main objectives of this study are (1) to improve the reliability of TEP measurements for seawater and freshwater applications, (2) to develop an extended method to measure both particulate (>0.4µm) and colloidal (<0.4 µm) TEPs, (3) to apply and verify the developed methods (with modifications/improvements) in monitoring TEP accumulation in algal cultures (in seawater) and TEP removal in processes of full-scale water treatment plants. The investigations were focused on improving the methods developed by Passow and Alldredge (1995) and Thornton et al. (2007), as these are, in principle, applicable in both saline and freshwater sources. Various modifications were introduced mainly to minimise the interference of dissolved salts in saline samples, to decrease the lower limit of detection, and to measure smaller colloidal TEPs (down to 10kDa). An integrated calibration method was developed for the two modified methods using Xanthan gum as standard TEPs.

5.1 Introduction

The presence of transparent exopolymer particles (TEP) in aquatic environments has been extensively studied over the past two decades, especially in the field of oceanography and limnology (Passow, 2002a). As the name suggests, TEPs are transparent and amorphous organic substances seasonally abundant (e.g., algal blooms) in marine and fresh water environments (Passow, 2002a). TEPs have been observed in various shapes (e.g., strings, disks, sheets or fibers) and sizes, ranging from 3 nm in diameter up to 100s of µm long (Berman and Passow, 2007; Passow, 2000). These materials are a major and essential component of marine mucilaginous aggregates (e.g., marine snow, sea foam) which usually forms during the senescent stage of an algal bloom (Alldredge and Silver, 1988; Alldredge et al., 1993).

In the ocean, TEPs generally originate from exudates and detritus of phytoplankton (micro-algae) and bacterioplankton but they may also originate from some species of macro-algae, oysters, mussels, scallops and sea snails (Passow, 2002b; McKee et al., 2005; Heinonen et al., 2007). TEPs are sticky and gel-like as they comprise mainly hydrophilic, negatively-charged and surface-active acidic polysaccharides (Mopper et al., 1995). Moreover, they are often associated with or they tend to absorb proteins, lipids, trace elements and heavy metals from water (Passow, 2002a), making them a nutritious platform and hotspots for bacterial activity (Bar-Zeev et al., 2012). Such characteristics led some researchers to suspect that they may have an important role in the formation of aquatic biofilms (Alldredge et al., 1993; Passow, 2002a; Bar-Zeev et al., 2012).

Berman and Holenberg were the first to propose the potential role of TEP on biofouling in reverse osmosis (RO) systems (Berman and Holenberg, 2005). Consequently, various studies were conducted to investigate the possible link between TEP and fouling in membrane filtration systems, including RO, micro-/ultra-filtration membranes and membrane bioreactors (Kennedy et al., 2009, Bar-Zeev et al., 2009; Villacorte et al., 2009a,b, 2010a,b; de la Torre et al., 2008). However, most of these studies applied different methods to quantify TEP and vary in terms of TEP size range being considered, making comparison rather difficult.

Marine biologists operationally defined TEPs as particles larger than 0.4 µm considering that they were first discovered by retention on 0.4 µm pore size membrane filters (Alldredge et al., 1993). However, TEPs are not solid particles, but rather agglomeration of particulate and colloidal hydrogels which can vary in size from a few nanometres to hundreds of micrometres (Passow, 2000; Verdugo et al., 2004). Hydrogels are highly hydrated and may comprise more than 99% water, which means they can bulk-up to more than 100 times their solid volume (Azetsu-Scott and Passow, 2004; Verdugo et al., 2004). The majority of these materials are formed abiotically through spontaneous assembly of colloidal polymers in aquatic environment (Chin et al., 1998, Passow, 2000). A substantial fraction of the colloidal precursors of TEP (<0.4µm) are not covered by the current established measurement methods.

5.1.1 Background

5.1.1.1 Alcian blue

There are four methods currently available to measure TEP, all of which are based on selective staining with Alcian Blue dye. Why Alcian blue? The dye is known to be highly selective and forms stable insoluble complexes with target compounds that cannot be easily distained by subsequent treatments. The dye is also widely available and has been routinely used in medical and biological research for many years. However, despite being one of the most widely-used biological stains, the mechanisms involved during reaction of the dye with a specific substrate is still not well understood (Horobin, 1988).

An Alcian Blue (AB) molecule is a tetravalent cation with a copper atom at the centre of its core (Scott *et al.*, 1964; Figure 5-1). The copper core gives the characteristic bluish colour of the dye. In aqueous solutions without extra electrolytes, AB can specifically bind with anionic carboxyl, phosphate and half-ester sulfate groups of acidic polysaccharides, resulting in the formation of neutral precipitates (Ramus, 1977; Parker and Diboll, 1966). AB can also react with carbohydrate-conjugated proteins such as acidic glycoproteins (Wardi and Allen, 1972) and proteoglycans (Bartold and Page, 1985) but not nucleic acids and neutral biopolymers (Karlsson and Björnsson, 2001).

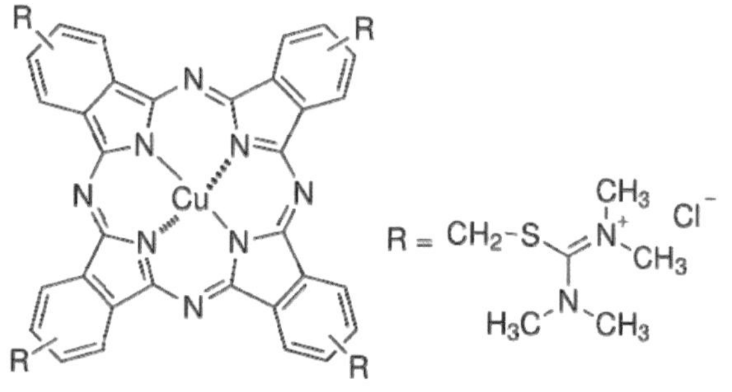

Figure 5-1: Molecular structure of Alcian blue 8GX.

The staining ability of AB depends on the type and density of anionic functional groups associated with the biopolymer materials in the sample. The staining selectivity largely depends on the pH and ionic strength of the AB and sample solution (Horobin, 1988). The staining ability of AB was reported to decrease when pH is lower or higher than 2.5 (Passow and Alldredge, 1995). In high ionic strength solutions, better interactions between anionic polymers and cationic AB can be expected due to compression of the electrical double layer surrounding the AB molecule. AB molecules can spontaneously aggregate in saline solutions resulting in the formation of precipitates not associated with TEP. This is considered as the main drawback of the successful application of AB staining for TEP measurements in seawater. To minimise measurement artefacts due to coagulation, AB staining solutions should always be pre-filtered and should not be directly applied to solutions with high salinity.

5.1.1.2 Microscopic enumeration

The first method to identify and quantify TEPs was introduced by Alldredge *et al.* (1993). The procedure involves filtering water samples through 0.4 µm pore size polycarbonate membranes by vacuum filtration to retain TEPs. AB solution (pH 2.5) is then added to the surface of the filter and the excess stained were flushed through the membrane. The stained TEPs on the membrane were then transferred to a glass slide using the filter-transfer-freeze (FTF) technique. The TEPs on the slides are viewed through a light microscope (200x magnification), counted and sized either manually or semi-automatically with an image analysis system (Mari and Kiørboe, 1996). Concentrations of TEPs are expressed in terms of number of stained particles per ml or in terms of total volume per ml, assuming each particle represents a sphere of equivalent diameter. The FTF, counting and sizing procedures are quite laborious, complicated, time consuming and are not always feasible. The precision of manual sizing and counting particles with light microscope is often poor and would likely result in highly variable results. Furthermore, stained particles often do not show high enough contrast with the background for an image analysis system to be applicable.

5.1.1.3 Spectrophotometric method

To overcome the difficulties of microscopic enumeration, Passow and Alldredge (1995) introduced a less laborious spectrophotometric technique to quantify TEP. This technique is currently the most widely used method to quantify TEP. It applies the same filtration and staining procedure as the method of Alldredge *et al.* (1993) but instead of counting the stained TEP, concentrations are measured based on the absorbance of AB bound to TEPs and calibration with a standard polysaccharide. After filtration and staining, the stained polycarbonate membrane is soaked in 80% sulfuric acid solution to decompose the retained TEP and release the bound AB to the acid solution. The absorbance of the acid solution is then measured at 787 nm wavelength. The absorbance of AB bound to TEP is the difference between total absorbance minus the absorbance of a blank sample. Net absorbance results are converted to TEP concentrations after calibrating the absorbance signals in terms of equivalent concentrations of Xanthan gum per liter (mg X_{eq}/L). The fact that TEP is indirectly measured in terms of absorbance of AB bound to TEP, it does not necessary reflect the actual mass concentration of TEP but rather measures the density of the target functional groups (e.g., carboxyl, sulfate groups) associated with TEP.

The calibration procedure involves dry weight measurement of filters to determine the mass of Xanthan gum retained on polycarbonate filters. This procedure is prone to several inaccuracies during drying (e.g., dust contamination) and weighing (e.g., electrostatic forces interference) of very low quantities (5-50µg) of Xanthan. Moreover, it is rather difficult to prepare a homogeneous and contaminant-free solution of Xanthan for the calibration experiment. An alternative calibration procedure was recently introduced by Kennedy *et al.* (2009) where TOC measurement was used to determine the mass of Xanthan retained on the filters. However, this procedure was later found to be sensitive to organic carbon contamination during standard preparation and filtration and a very accurate TOC analyser is required to perform it. One proposed option to avoid this problem is not to perform a calibration at all, whereby concentrations of TEP are expressed relatively in terms of absorbance per cm per litre of filtered sample (abs/cm/L).

5.1.1.4 Rapid spectrophotometric method

In 2004, Arruda-Fatibello and co-workers developed a rapid spectrophotometric technique to measure TEP (Arruda-Fatibello et al., 2004). Unlike its predecessor, AB stain is not applied to a filter but it is directly applied to the water sample. The water sample is first concentrated 5-10 fold using 0.45 µm pore size tangential membrane tubing. AB solution is then added to the concentrated sample with acetate buffer (pH 4) and stirred. The resulting solution is centrifuged and the absorbance of the supernatant is measured at 602 nm wavelength. The basic assumption is that TEPs form insoluble precipitates and settle during centrifugation. The supernatant absorbance is therefore inversely proportional to the TEP concentration. Calibration is performed following the same procedure but using solutions with different concentrations of Xanthan gum.

Considering that dissolved salts can also form insoluble complexes with AB, the rapid spectrophotometric method is only applicable in fresh water with low salinity (e.g., river and lake water). Moreover, AB staining is performed at pH 4 instead of pH 2.5, which is a remarkable change considering that the staining specificity of AB can significantly vary with pH. At pH 2.5, carboxyl and sulphated polysaccharides can be stained but at pH 1, sulfated polysaccharides are fully stained while carboxyl groups are poorly stained (Passow and Alldredge, 1995). It is still unclear which functional groups are stained at pH 4.0 and perhaps other compounds other than TEPs can be stained at this pH. Therefore, it is likely that the substances measured as TEP with this method are not entirely the same as what was measured in other established TEP methods.

5.1.1.5 Acid polysaccharide method

The latest method which can be used to measure TEP was developed by Thornton et al. (2007). This method was intended to quantify acidic polysaccharides in surface water but follows the same principle as the other established TEP measurement methods. The staining protocol is partly similar to Arruda-Fatibello et al. (2004) but differs in terms of how TEP-AB precipitates are separated from the excess (un-reacted) stain.

Water samples are first adjusted to pH 2.5 (using acetic acid) before the AB stain is added and then mixed. The resulting solution is filtered through a 0.2 µm pore size membrane. The absorbance of the filtrate is measured at 610nm wavelength. It is then assumed that all TEP-AB precipitates were removed during the filtration process. Consequently, the absorbance of the filtrate is inversely proportional to the amount of TEP in the sample. Calibration is performed using the same protocol with either Xanthan gum or Alginic acid as the model TEP. As in the method by Arruda-Fatibello et al. (2004), the above procedure is only applicable for freshwater samples. To overcome this limitation, Thornton *et al.* (2007) introduced a dialysis treatment (1 kDa MWCO cellulose acetate membrane) of the sample prior to staining. With this additional step, the method can also be applied for saline water samples.

The advantage of APS method over other established TEP methods is its potential to measure TEPs smaller than 0.4 μm, which are commonly known as TEP precursors. These colloidal TEPs have been demonstrated to form particulate TEPs by applying bubble adsorption and low shear forces (Passow, 2000). Such conditions often exist in water treatment processes. However, it is unclear if all TEPs and their colloidal precursors actually form precipitates with AB which are large enough to be removed by 0.2 μm filters. Moreover, the application of this method for seawater samples can be rather laborious and prone to contamination. The dialysis pre-treatment itself is performed for 2 days but a longer time may be necessary to remove most of the dissolved salts in seawater samples. During dialysis, some biopolymer materials from the interior walls of the cellulose acetate dialysis bag may be released to the sample which can then cause substantial increase in sample AP concentration (Thornton *et al.*, 2007).

5.1.2 Scope and objectives

A number of TEP quantification methods have been introduced in the last two decades, but their application to water treatment monitoring has not been extensively applied and studied. As TEP is gaining more attention in water research and the membrane-based desalination industry, there is an increasing need for a reliable method to measure these substances in fresh and marine waters. However, a critical analysis of currently available methods identified several questions regarding their reproducibility, calibration and limit of detection. In addition, there is a need to measure colloidal TEP (<0.4 μm) as well since these particles cannot be ignored in studying the potential role of TEP in membrane fouling.

The aim of this study is to improve the reliability of the current TEP methods for better application in water treatment process monitoring and membrane fouling studies. The investigations are focused on improving the methods developed by Passow and Alldredge (1995) and Thornton *et al.* (2007), because these are in principle applicable in both saline and freshwater sources. The specific objectives are:

1) to assess/define the limitations of the existing spectrophotometric methods;
2) to improve their reliability in terms of reproducibility, calibration and limit of detection for seawater and freshwater applications;
3) to develop an extended method to measure both particulate (>0.4 μm) and colloidal TEPs (>0.1 μm and > 10 kDa); and
4) to verify/apply the modified and extended methods in monitoring TEP accumulation in algal cultures (seawater) and TEP removal in various processes of full-scale water treatment plants.

5.2 Materials and methods

5.2.1 Water samples and synthetic solutions

Experiments were performed using natural surface water samples (seawater and fresh water), pre-treated TEP solutions from an algal culture and prepared synthetic seawater solutions.

5.2.1.1 Ultra-pure water and artificial seawater

Ultrapure water (UPW) was used as the blank and base solution of all prepared synthetic water samples. UPW is purified tap water through a series of treatment steps, namely: 1µm filtration > softening > RO filtration > ion exchange > 1µm filtration > GAC filtration > UV > 0.22 µm filtration.

Artificial seawater (ASW) used in blank measurements was prepared based on the typical ion concentration of coastal North Sea water. Minor constituents (e.g., Br⁻, B, Sr^{2+}) which comprise less than 1% of inorganic ions in seawater were not considered. Analytical grade salts were sequentially dissolved in ultrapure water to make up the ASW solution (Table 5-1).

Table 5-1: Inorganic ion composition of artificial seawater (ASW)

Inorganic ions	Concentration (g/L)	Percentage (%)
Chlorine, Cl^-	18.85	55.0
Sodium, Na^+	10.75	31.3
Sulfate, SO_4^-	2.69	7.9
Magnesium, Mg^{2+}	1.17	3.4
Calcium, Ca^{2+}	0.30	0.9
Potassium, K^+	0.38	1.1
Hydrogen Carbonate, HCO_3^-	0.15	0.4
Total dissolved solids (TDS)	34.30	100

5.2.1.2 Standard TEP solutions

Standard solutions were prepared from purified acidic polysaccharides: sodium alginate (Fluka 71240) and Xanthan gum (Sigma 1253). The solutions were prepared by adding 50 mg of dried polysaccharides to 500 mL of UPW while rapidly stirring with a magnetic stirrer. Rapid stirring was maintained for at least 30 minutes until no flocs were visible. The resulting solution is further homogenised 3 times with a tissue grinder (Dounce, Sigma-Aldrich).

5.2.1.3 Algal organic matter (AOM)

A solution of algal organic matter (AOM) collected from a batch culture of *Chaetoceros affinis* was used in the investigations to simulate seawater impacted with severe algal bloom. *C. affinis* is a common species of diatom commonly found in the North Atlantic coastal seawaters and has been reported to produce significant concentrations of TEP (Myklestad *et al.*, 1989). A strain of *C. affinis* (CCAP 1010/27) was inoculated in 2 L glass flasks containing 1.5 L of sterilised f/2+Si medium (Guillard's medium for marine diatoms) in artificial seawater. An artificial white light (40-50 µmol/m².s) was provided to the culture while being continuously mixed on a shaker. After 12 days, the algal cells were allowed to settle in the flask for 24 hours and the supernatant solution containing algal-derived organic matter (AOM) was extracted for the experiments.

5.2.1.4 Water sample collection and storage

Seawater samples were collected from an RO plant seawater intake in the south-western coast of the Netherlands during a spring bloom period (May, 2009). Fresh surface water samples were collected near the surface of the Oude Delft canal during the summer period (August 2009). Water samples were also collected from the raw water and over the treatment processes in two drinking water treatment plants in the Netherlands in May (Plant A) and June 2012 (Plant B). Batches of the collected water samples were either analysed immediately or stored at 4°C and then analysed within 3 weeks after collection.

Two algal cultures (i.e., CA: Chaetoceros affinis and AT: Alexandrium Tamarense) were monitored by collecting samples for TEP analysis every 2-3 days within a 30-60 day incubation period. The detailed protocol for maintaining CA and AT cultures was described in Chapter 4.

5.2.2 Alcian blue dye

Stock solutions of the dye were prepared in acetic buffer solution containing 250 mg/L of Alcian blue (Standard Fluka, Sigma-Aldrich). The buffer solution was prepared by adding drops of acetic acid until the solution pH of 2.5 is reached. The prepared stock solutions were then stirred for 12-18 hours. The final staining solution is taken from the stock solution and pre-filtered through 0.05 µm polycarbonate membrane (Nuclepore, Whatman) before staining. The remaining stock solution is stored in the dark at 4°C. A new stock solution is prepared after 4 weeks.

To measure the apparent concentration of Alcian blue solutions, the copper content of the dye was measured using atomic absorption spectrometry (Perkin Elmer AAnalyst 200). Alcian blue concentration (C_{AB}) was calculated based on the mass proportion of copper for each molecule of Alcian blue ($C_{56}H_{68}Cl_4CuN_{16}S_4$).

$$C_{AB} = \frac{c_{Cu}}{\omega}$$
Eq. 1

where: C_{Cu} is the mass concentration of copper in the staining solution (mg Cu L⁻¹) and ω is the molecular weight fraction of copper in the AB molecule (0.0489).

Absorbance measurements were performed using a UV-Vis recording spectrophotometer (Shimadzu UV-2501PC). Samples were placed in a 1-cm cuvette and then the absorbance was measured at a specific wavelength.

5.2.3 TEP measurement methods

This study investigated the TEP methods developed by Passow and Alldredge (1995) and Thornton et al. (2007) because they are – in principle - applicable in both saline and freshwater water samples. The two protocols with the proposed improvements or amendments are introduced in the following sections. For simplicity, the new methods modified/expanded from Passow and Alldredge (1995) and Thornton et al. (2007) are denoted as $TEP_{0.4\mu m}$ and $TEP_{0.1\mu m}$ (Protocol 1) and TEP_{10kDa} (Protocol 2), respectively from henceforth.

5.2.3.1 Methods for $TEP_{0.4\mu m}$ and $TEP_{0.1\mu m}$

Protocol 1 is a modified and extended version of the method by Passow and Alldredge (1995). In addition to using 0.4 μm pore size membrane (used in the original method), a smaller pore size membrane (0.1 μm) is also utilised in the new method in order to extend the measurement to smaller TEP size range.

Figure 5-2 illustrates Protocol 1 for measuring $TEP_{0.4\mu m}$ and $TEP_{0.1\mu m}$. Water samples are filtered through polycarbonate filters (0.4μm and 0.1μm pore size) by applying a constant vacuum of 0.2 bar. Two ml of UPW is then filtered through the filter-retained TEP by applying <0.2 bar vacuum to wash-out the remaining sample moisture (including inorganic salts) through the filter. Pre-filtered (through 0.05 μm PC filter) AB dye solution (1 mL) is subsequently applied over the filter, allowed to react with the retained TEP for 10 seconds, and then the un-reacted dye is flushed through the filter by vacuum filtration (<0.2 bar). To remove further the remaining un-reacted dye, a rinsing step is performed by filtering 2 ml of ultra-pure water through the filter. The rinsed filter is transferred to a 50 ml glass beaker where 6 ml of 80% sulfuric acid solution is added to decompose TEP and re-dissolve the bound dye. The beaker is gently agitated on a shaker to minimize bubble formation in the acid solution. After 2 hours on the shaker, the acid solution is transferred to 1 cm cuvette and absorbance (A_T) is measured at 787 nm wavelength. The concentration of TEP in the water sample is calculated using Eq. 2.

$$TEP_{0.4\mu m/0.1\mu m} = \frac{f_{787nm}}{V_f}(A_T - A_f - A_s) \qquad \text{Eq. 2}$$

where the total absorbance (A_T) is the absorbance of the dye which reacted with TEP and those adsorbed to the filter (abs/cm); filter blank absorbance (A_f) is the absorbance of the dye adsorbed to the filter (abs/cm); sample correction (A_s) is the absorbance of unstained sample (abs/cm); V_f is the volume of sample filtered (L) and f_{787nm} is the calibration factor (mg Xanthan/abs/cm).

Figure 5-2: Flow diagram of Protocol 1 for measuring $TEP_{0.4\mu m}$ and $TEP_{0.1\mu m}$.

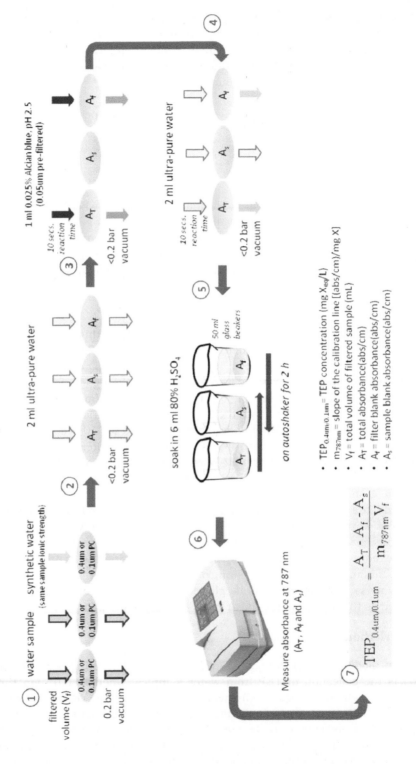

Filter blank (A_f) samples are prepared in the same way as total absorbance but filtering TEP-free blank samples (e.g., synthetic water with similar dissolved salt concentration as the water sample) instead of actual water samples. For sample correction (A_s), water samples were filtered in the same way as determining the total absorbance but skipping the AB staining procedure.

The calibration factor f_{787nm} is the inverse of the best fit linear slope of the standard calibration line.

$$f_{787nm} = \frac{1}{m_{787nm}}$$ Eq. 3

The slope, m_{787nm} is derived from calibration experiments where the dry mass of the standard (Xanthan gum) is plotted against the corresponding absorbance of AB which reacted to it. Passow and Alldredge (1995) used dry weight measurement to determine the mass of Xanthan gum retained on polycarbonate filters. However, this procedure is prone to several inaccuracies and reproducible results could not be obtained. When no calibration is performed, TEP concentrations can be expressed relatively in terms of absorbance per cm per litre (abs/cm/L).

An alternative calibration procedure was developed in this study as a proposed replacement to the calibration method of Passow and Alldredge (1995). The new calibration procedure is described as follows. Firstly, standard solutions (4 ml) containing different concentrations (0-5 mg/L) of Xanthan gum are prepared from the stock solution described in Section 5.2.1.2. For pH adjustment, 0.05 ml acetic acid was added to each solution and then briefly agitated. The solution is stained by adding 1 ml of pre-filtered AB staining solution, mixed for 10 seconds and incubated for 10 mins. Four ml of the resulting solution is then filtered through 0.1 µm PC membrane by vacuum filtration (0.2 bars). The fouled PC membrane was carefully transferred to a 50 ml beaker. Six ml of 80% sulphuric acid solution was added and mixed on a shaker for 2 hours to fully decompose the TEP and dissolve the dye. Afterwards, the absorbance of the acid solution was measured in a spectrophotometer at 787 nm wavelength using a 1 cm cuvette. Assuming all Xanthan in the solution formed precipitates with AB and are eventually retained by a 0.1 µm PC membrane filter, the mass on the filter was calculated by multiplying the Xanthan concentration with the filtered volume of the stained Xanthan solution. Finally, the calculated retained mass and absorbance is plotted and the calibration factor (f_{787nm}) was calculated as the inverse of the linear slope (m_{787nm}) of the plot.

5.2.3.2 Method for TEP$_{10kDa}$

The main modification of Protocol 2 from the method introduced by Thornton *et al.* (2007) is the collection of TEP from water samples, whereby filtration through low molecular weight cut-off (MWCO) membranes is used for TEP retention and then followed by TEP re-suspension in ultrapure water by sonication instead of dialysis treatment. The objective of these steps is primarily to concentrate TEP materials to within detectable levels and to minimise the effect of salinity during the subsequent AB staining. The maximum volume of samples that can be filtered is theoretically unlimited while the retained TEP is re-suspended in 10 ml solution. Hence, the TEP level in the sample can be concentrated prior to AB staining.

Figure 5-2 illustrates Protocol 2 for measuring TEP_{10kDa}. Water samples are filtered through 10 kDa membranes (Millipore RC membrane, 25mm diameter) at constant flux (60 $L/m^2/h$) using a syringe pump (Harvard syringe pump 33). After filtering a specific volume (10-100ml) of sample, the syringe is carefully replaced with a clean syringe containing about 10 ml of air. Air is then injected to the filter holder (60 $L/m^2/h$) until all the remaining water in the feed side of the membrane holder has passed through the membrane. The total filtered sample volume is then measured after collecting all the filtrate. To rinse out saline moistures remaining on the membrane and the retained TEP, 5 ml of UPW is injected to the filter holder at 60 $L/m^2/h$. Air is then again injected until all the rinse water on the feed side of the membrane holder has passed through the membrane. The membrane is then carefully removed from the filter holder and placed feed side down in a clean disposable plastic container (40 ml Unipot with screw cap) containing 10 ml of UPW. The sample is tightly covered, vortexed (Heidolph REAX 2000) for 10 seconds and sonicated (Branson 2510E-MT) for 60 mins. A four ml volume of the re-suspended TEP solution is transferred to a clean 20ml disposable plastic container. To adjust the sample pH to 2.5, 0.05 ml of acetic acid solution was added to the solution (tested with a pH meter). One ml of pre-filtered (through 0.05μm PC membrane) AB solution was added to the sample, which was then mixed vigorously and left to react for 10 minutes. Four ml sample of the TEP-AB solution is then filtered through a 0.1 μm PC filter by vacuum filtration (<0.2 bars). The filtrate was collected in a disposable plastic container (10 ml Unipot with screw cap), transferred to a 1 cm cuvette and absorbance (A_e) was measured at 610 nm wavelength. The TEP_{10kDa} concentration in mg Xanthan equivalent per litre (mg X_{eq}/L) is calculated as follows:

$$TEP_{10kDa} = f_{610nm} \frac{V_r}{V_f} (A_e - A_b) \qquad \text{Eq. 4}$$

$$f_{610nm} = \frac{1}{m_{610nm}} \qquad \text{Eq. 5}$$

where f_{610nm} is the calibration factor [(mg Xanthan/L)/ (abs/cm)], V_r is total volume of re-suspended TEP solution (10 mL), V_f is the volume of filtered sample (mL), A_e is the absorbance of the excess or un-reacted dye (abs/cm), A_b is the absorbance of filtered blank (abs/cm), m_{610nm} is the slope of the calibration line [(abs/cm)/(mg X_{eq}/L)].

The blank absorbance (A_b) is measured to correct for the amount of stain adsorbed by the polycarbonate filter. This was performed following the above-mentioned procedure but replacing the sample with UPW.

Figure 5-3: Flow diagram of Protocol 2 for measuring TEP$_{10kDa}$.

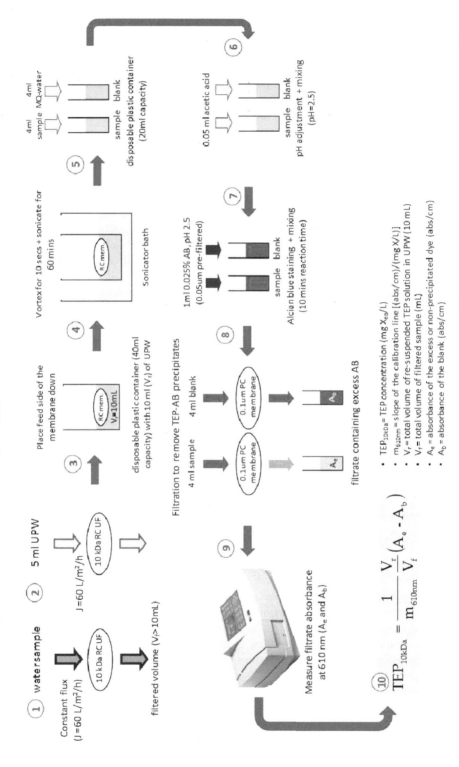

$$TEP_{10kDa} = \frac{1}{m_{610nm}} \frac{V_r}{V_f} (A_e - A_b)$$

- TEP$_{10kDa}$ = TEP concentration (mg X$_{eq}$/L)
- m$_{610nm}$ = slope of the calibration line [(abs/cm)/(mg X/L)]
- V$_r$ = total volume of re-suspended TEP solution in UPW (10 mL)
- V$_f$ = total volume of filtered sample (mL)
- A$_e$ = absorbance of the excess or non-precipitated dye (abs/cm)
- A$_b$ = absorbance of the blank (abs/cm)

To determine the calibration factor (f_{610nm}), different dilutions of a prepared Xanthan gum solution (100 mg/L; Section 5.2.1.2) are stained with Alcian blue and the excess stain absorbance is measured the same way as the sample measurement procedure. The excess dye absorbance versus Xanthan concentration is plotted to determine the slope (m_{610nm}) of the calibration curve. Since concentration is inversely proportional to the excess dye absorbance, the calibration line has a negative slope and the calibration factor (f_{610nm}) has a negative value.

5.2.4 Filters and filtration set-up

Various types of membranes were used for TEP measurement and dye pre-treatment. Track-etched polycarbonate membranes (Nuclepore, Whatman) with nominal pore sizes of 0.4, 0.2, 0.05, 0.03, 0.015 µm were used for vacuum filtration. Polymeric membranes which include 10kDa and 5kDa RC membranes (Millipore) were used for constant flux syringe filtration. The diameter of the membranes used for Protocol 1 and 2 were 47 mm and 25 mm, respectively. To remove possible contaminants, the PC filters were rinsed by flushing >200 ml of UPW through it while RC membranes were soaked for at least 24 hours in UPW and then flushing 5-10 ml UPW prior to sample filtration.

Different vacuum filtration set-ups were used for the two TEP protocols. For Protocol 1, the set-up comprised of a 50 mm glass filter holder (Sartorius) with fiber glass porous support and a vacuum pump (Millipore WP612205) with pressure controller. For Protocol 2, a set-up comprising a 25mm glass filter holder (Sartorius) with PTFE coated stainless steel mesh filter support and a vacuum pump with pressure controller was used. To minimise sample contamination, the filter holder was thoroughly cleaned (after every filtration) by rinsing with ultrapure water. Constant flux filtration (60 L/m^2.h) through RC membranes were performed using a syringe pump (Harvard Pump 33), 60 ml disposable syringe (BD Plastipak™) and 25 mm filter holder (Schleicher & Schuell).

5.2.5 Limit of detection calculation

The lower limit of detection (LOD_{min}) of the TEP methods depends on the variability of the blank absorbance. This was calculated as follows:

Protocol 1: $LOD_{min} = 3\sigma_b\, f_{787nm}\, (1/V_f)$ Eq. 6
Protocol 2: $LOD_{min} = 3\sigma_b\, f_{610nm}\, (V_r\,/V_f)$ Eq. 7

where σ_b is the standard deviation of 10 independently measured blank absorbance (abs/cm). The factor 3 corresponds to a significance level of 0.00135, which means that only 0.135% of blank measurements will statistically yield results that fall above the computed detection limit (Harvey, 2000).

The upper limit of detection (LOD_{max}) is the threshold of the absorbance range which can yield reliable concentration results. For Protocol 1, this limit is determined based on the maximum absorbance at which a linear correlation between absorbance and filtered volume can be observed. For Protocol 2, this limit is the minimum absorbance at which the excess stain absorbance and the standard concentration has a significant linear correlation.

5.2.6 Dynamic light scattering measurements

The size distribution of 2 mg/L Xanthan, prefiltered AB and a mixture of the two solutions were estimated based on dynamic light scattering (DLS) measurements using a Malvern Zetasizer Nano ZS. The DLS technique measures the diffusion of particles moving under Brownian motion and converted to size based on the Stokes-Einstein relationship. The obtained size is the diameter of a sphere with equivalent translational diffusion coefficient as the measured particle, called the hydrodynamic diameter. Due to the complexity of the shape of Xanthan gum (e.g., fibrillar), this technique was only used to compare relative changes in hydrodynamic size distribution rather than the actual size of the material itself. All measurements were performed at 25°C.

5.2.7 Liquid chromatography organic carbon detection (LC-OCD)

Water samples collected from water treatment plants were analysed using liquid chromatography - organic carbon detection (LC-OCD) at DOC-Labor (Karlsruhe, Germany). In this technique, concentrations of high (biopolymers) and low molecular weight (e.g., humics, building blocks, acids, neutrals) organic substances were measured in terms of organic carbon based on size exclusion chromatography as described by Huber et al. (2011). LC-OCD analyses were performed without the inline 0.45 µm pre-filtration to cover part of the larger TEPs (>0.4 µm). Based on the pore size of the sinter filters in the chromatogram column, the theoretical maximum chromatographable size of organic substances without sample pre-filtration is 2 µm (S. Huber, per. com.). The TEPs in the water samples contain an unknown concentration of carbon as biopolymers. Hence, biopolymer concentration is not considered as a direct measure of TEP but was used in this study as a relative indicator of the amount of TEP and their colloidal precursors in the water sample.

5.3 Results and discussion

5.3.1 Pre-treatment of Alcian blue solution

The apparent size distribution of the AB stock solution (250 mg/L) at pH 2.5 was determined by serial filtration through different pore size membranes (i.e., 5, 1, 0.4, 0.2, 0.05, 0.03 and 0 0.015 µm PC filters). About 50% of AB was smaller than 0.2 µm and about 44% were smaller than 0.05 µm (Figure 5-4). Further filtration of the dye through 0.03 µm and 0.015 pore size filters only allowed passage of about 30% and 2% of the initial dye concentration, respectively. When pre-filtering through 0.03 µm filter, it may be necessary to increase the initial dye concentration to avoid possible under-staining of TEP. Moreover, dye filtration through 0.03 µm PC membranes was time consuming (average flux of ~20 L/m²/h at -0.5 bar) and were prone to membrane damage and leakage at vacuum pressures higher than 0.2 bar. For these reasons, pre-filtration of Alcian blue dye through 0.05 µm filters was adopted for the rest of this study.

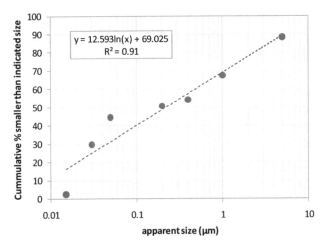

Figure 5-4: Apparent size distribution of 0.025% Alcian blue dye in acetic acid buffer solution (pH 2.5).

Considering that AB solutions are pre-filtered through 0.05µm filters, the dye pigments used in staining TEPs can be as large as 0.05 µm. To minimise the possibility of retaining these materials during TEP measurements, membrane filters used for separating TEP-AB precipitates and excess (unreacted) stain should have pore size larger than 0.05 µm. Hence, the minimum pore size of membranes which can be used for retaining TEP-AB precipitates is 0.1 µm.

5.3.2 Absorption spectra of Alcian blue

According to Ramus (1977), the maximum absorption of AB when dissolved in water is 610 nm wavelength. However when AB is dissolved in sulfuric acid solution, the maximum absorption shifts to 787 nm wavelength (Passow and Alldredge, 1995). The two spectrophotometric methods investigated in this study are based on the absorbance of AB dissolved in different matrix, namely: 80% sulphuric acid solution for Protocol 1 and acetic buffer solution (pH 2.5) for Protocol 2. A spectrum scan of AB in these matrix were performed and the results are presented in Figure 5-5a. The results confirm that the maximum absorbance of alcian blue in 80% sulfuric acid solution lies at 787 nm wavelength. The maximum absorbance of AB in acetic buffer solution (pH 2.5) lies at 610 nm wavelength within the visible light spectrum. For comparison, the absorbance of AB in sulphuric acid at 787 nm is about twice that dissolved in acetic acid buffer solution at 610 nm (Figure 5-5b).

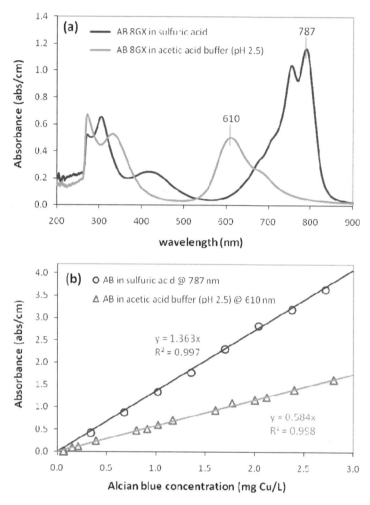

Figure 5-5: (a) Absorption spectra of Alcian blue 8GX (16 mg/L) dissolved in sulphuric acid and acetic acid buffer solution and (b) peak absorbance values of Alcian blue solutions at various concentrations.

5.3.3 Measurement of $TEP_{0.4\mu m}$ and $TEP_{0.1\mu m}$

5.3.3.1 Effect of sample salinity and pre-rinsing

When AB solution is directly applied on membranes with adsorbed moisture from water samples with high salinity (e.g., seawater), instantaneous coagulation of the dye may occur. This can cause retention of unreacted AB on the membrane which eventually translates to higher total absorbance. If not corrected accordingly, this can cause overestimation of the TEP concentration. To minimise this potential effect, it is proposed to first rinse the sample moisture out by filtering UPW through the membrane prior to application of AB solution.

To demonstrate the effect of salinity and the proposed rinsing step, tests were performed using surface water samples with different salinities: Delft canal water (TDS~3 g/L) and coastal North Sea water (TDS~34 g/L). TEP concentrations were measured for both samples with and without the pre-rinsing step. To check the linearity of the results, different volumes of the samples were filtered and total absorbance were plotted against the filtered volume. TEP measurements without pre-rinsing were also performed with ASW and UPW to represent the blank of saline and non-saline samples, respectively. Results are presented in Figure 5-6.

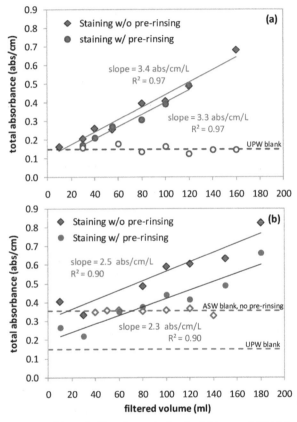

Figure 5-6: TEP measurement with and without pre-rinsing for (a) low and (2) high salinity surface water samples.

The average ASW blank was more than twice the average UPW blank (Figure 5-6). This apparent difference in filter blanks at varying sample salinities was overlooked in the original method where filter blanks were measured by directly staining clean membranes. For the freshwater sample (Figure 5-6a), total absorbance values were all above the average UPW blank absorbance but mostly below the ASW blank even without pre-rinsing. For the seawater sample, absorbance for samples that were not pre-rinsed were mostly above the average ASW blank while the results of the pre-rinsed samples were substantially lower.

In general, samples that were pre-rinsed showed lower absorbance results than those that were not pre-rinsed and the decrease in absorbance was only significant for the seawater sample. Although the total absorbance was observed to substantially decrease, the linear slope of the plot, which represents the relative TEP concentration of the sample, was not significantly affected by the pre-rinsing step. Hence, the effect of sample salinity is mainly on the increase of filter blank and that introducing a pre-rinsing step prior to AB staining can minimise this effect.

The rinsing test results indicate that previously reported TEP concentrations in seawater which were measured using the original protocol by Passow and Alldredge (1995) are likely overestimated due to interference of salinity. However, using the results of the rinsing test, these results can still be recalculated to estimate the actual concentration by replacing the original blank absorbance with the seawater blank absorbance measured in this study (Figure 5-6). This proposed salinity correction was implemented elsewhere in this thesis (Chapter 6; Figure 6-4).

5.3.3.2 Calibration with Gum Xanthan

Calibration using Xanthan gum as TEP standard were performed following the new protocol described in Section 5.2.3.1. Figure 5-7 shows the results of two calibration experiments.

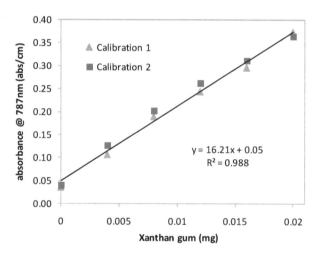

Figure 5-7: Calibration curves with standard Gum Xathan.

The results of the two calibration experiments showed acceptable linearity ($R^2 > 0.98$) and consistent slopes (15.9 and 16.5 abs/cm/mgX), an indication that the new calibration method is reproducible. The average calibration factor which is the inverse of the average slope of the two calibration graphs is 0.062 mg X_{eq}/(abs/cm). This is lower than the factors (0.088 and 0.139) reported by Passow and Alldredge (1995). The difference in calibration factors can be due to variations in the staining capacity of AB or due to differences in the calibration method.

The new calibration procedure does not involve the irreproducible dry weight measurement proposed by Passow and Alldredge (1995) but rather based on the principle that during Alcian blue staining of Xanthan suspensions in UPW, large Xathan-AB precipitates are formed which can be totally retained on 0.1 µm PC filter. It might be possible that some colloidal Xanthan-AB precipitates can still pass through the 0.1 µm filter which may result in underestimation of the staining capacity of the stain. Since the dye solution is pre-filtered through 0.05µm filter, using a lower pore size membrane (<0.1µm pore size) is not considered due to higher risk of retaining AB pigments not bound to TEP. However, further investigation using dynamic light scattering techniques show that most (if not all) Xanthan-AB precipitates can be retained by 0.1 µm PC membrane. The results of this investigation are presented and discussed in detail in Section 5.3.4.5. Overall, the current calibration method is promising because it demonstrated to be reproducible and it eliminates some of the issues in the calibration procedure proposed by Passow and Alldredge (1995).

5.3.3.3 Limit of detection: $TEP_{0.4µm}$ and $TEP_{0.1µm}$

The lower limit of detection (LOD_{min}) of the methods based on Protocol 1 was calculated based on the variability of 10 filter blank measurements using the same batch of staining solution. Blank measurements were performed using 0.4µm ($TEP_{0.4µm}$) and 0.1µm ($TEP_{0.1µm}$) filters and absorptions measured at 787nm wavelength. The measured blanks and the calculated LOD_{min} for 2 blank solutions with different salinities are presented in Table 5-2 and Figure 5-8.

Table 5-2: Blank measurement results of $TEP_{0.4µm}$ and $TEP_{0.1µm}$ methods with fresh and saline samples.

Method	$TEP_{0.4µm}$		$TEP_{0.1µm}$	
TDS (g/L)	0	34	0	34
Blanks measured	10	10	10	10
Ave. blanks (abs/cm)	0.132	0.197	0.099	0.138
Std. deviation (abs/cm)	0.010	0.016	0.008	0.028
3 x std. deviation (abs/cm)	0.029	0.049	0.025	0.083

The average blank and standard deviation were observed to increase with higher salinity (Figure 5-8a and 5-8c). The average of 10 blank measurements using UPW (TDS=0 g/L) are 0.132 (±0.010) and 0.099 (±0.008) abs/cm for $TEP_{0.4µm}$ and $TEP_{0.1µm}$, respectively. Average blanks measured using ASW (TDS=34 g/L) are 0.197 (±0.016) for $TEP_{0.1µm}$ and 0.138 (±0.028) abs/cm for $TEP_{0.4µm}$.

The LOD_{min} is defined as three times the standard deviation of the 10 measured blanks divided by the filtered volume. Consequently, the higher the filtered sample volume, the lower is the expected LOD_{min}. The calculated LOD_{min} of $TEP_{0.4µm}$ and $TEP_{0.1µm}$ for a wide range of filtered volume are presented in Figure 5-8a and Figure 5-8b, respectively. The figures show that in seawater, the LOD_{min} when filtering 100 ml of sample is 0.03 mg X_{eq}/L for $TEP_{0.4µm}$ and 0.05 mg X_{eq}/L for $TEP_{0.1µm}$. Increasing the filtered volume to 300 ml will decrease the LOD_{min} down to 0.01 mg X_{eq}/L and 0.02 mg X_{eq}/L for $TEP_{0.4µm}$ and $TEP_{0.1µm}$, respectively.

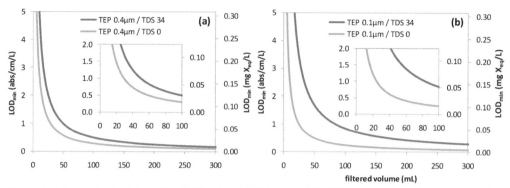

Figure 5-8: Lower limit of detection (LOD$_{min}$) of (a) TEP$_{0.4\mu m}$ and (b) TEP$_{0.1\mu m}$ for fresh (TDS 0) and saline water samples (TDS 34).

In principle, filtering a higher volume of samples will deliver more reliable results up to the point when the retained TEPs exhaust the applied AB dye. This limit (LOD$_{max}$) can be estimated based on the maximum absorbance at which a linear correlation between TEP absorbance and filtered volume can be observed. TEP$_{0.4\mu m}$ and TEP$_{0.1\mu m}$ measurements in AOM samples were performed by filtering different sample volumes. In addition, the filtrate of TEP$_{0.4\mu m}$ was measured using 0.1 µm filters (TEP$_{0.1-0.4\mu m}$). Figure 5-9 shows the total absorbance results with respect to filtered volume.

The LOD$_{max}$ for both TEP$_{0.4\mu m}$ and TEP$_{0.1\mu m}$ was estimated to be at 0.8 abs/cm. Absorbance higher than 0.8 abs/cm is not reliable because most of the dye has already reacted with the TEPs at such level. If all the applied AB dye (1 ml pre-filtered solution) have reacted with TEPs on the membrane, the expected absorbance is about 1.2 abs/cm. The LOD$_{max}$ can be theoretically increased further by increasing the concentration and/or volume of the applied stain. However, the current LOD$_{max}$ is rarely reached when measuring TEP in natural waters. Moreover, increasing the amount of AB dye may increase blank variability and the LOD$_{min}$. Hence, 0.8 abs/cm is considered as the ideal upper threshold of Protocol 1.

Based on the linear slope of absorbance versus filtered volume (between the LOD$_{min}$ and the LOD$_{max}$), the average TEP$_{0.1\mu m}$ concentration (1.97 mg X$_{eq}$/L) is slightly higher than TEP$_{0.4\mu m}$ (1.91 mg X$_{eq}$/L). Apparently for this sample, there is no significant difference when measuring TEP with 0.4 µm or 0.1 µm filters, an indication that the concentration of TEPs not retained by 0.4µm filter but were retained on 0.1 µm filters were relatively marginal. This was supported by the results of TEP$_{0.1-0.4\mu m}$, whereby absorbance results were below the LOD$_{min}$ when samples filtered were less than 40 ml, but absorbance were observed to increase exponentially at higher filtered volumes (Figure 5-9b). The exponential increase was likely due to adsorption of TEP < 0.1 µm on the walls of membrane pores, which may have caused constriction or blocking on the pores and eventually resulting in increased retention of smaller TEPs as more sample is filtered through the membrane.

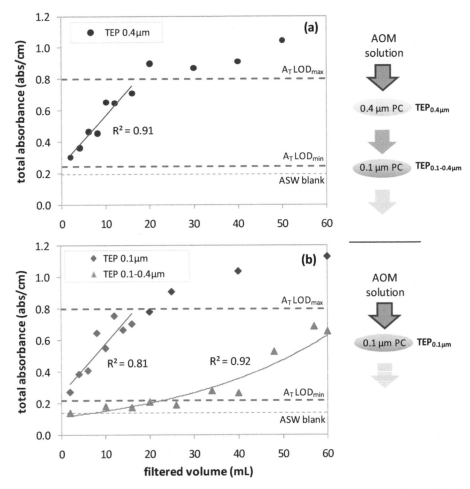

Figure 5-9: TEP concentrations of AOM measured with (a) 0.4 μm and (b) 0.1 μm filters plotted with average filter blanks and limits of detection.

5.3.4 Measurement of TEP_{10kDa}

A key advantage of TEP measurement based on Protocol 2 (TEP_{10kDa}) over Protocol 1 ($TEP_{0.4μm}$ and $TEP_{0.1μm}$) is the lower size range of TEP it can analyse (i.e., down to 10kDa). Various experiments were performed to test and assess the different aspects of the new method. The results and findings are discussed in the succeeding sections.

5.3.4.1 Effect of sample salinity and rinsing

In Protocol 2, AB staining is applied after extracting TEP from 10 kDa membranes in ultra-pure water by sonication in order to prevent interference by dissolved salts in the water sample. To demonstrate the effect of salinity, pre-filtered AB solution (1 ml) was added to solutions (4 ml) prepared from different dilutions of ASW (see Table 5-1). The dye is strongly cationic and can form large flocs instantaneously by reacting with anions in the saline solution. The formed flocs were then removed by filtering the solution through 0.1 µm PC filter. The residual AB was measured based on absorbance of the filtrate solution at 610 nm wavelength.

As presented in Figure 5-10, the residual absorbance of AB reduced substantially with increased ion concentration. For salinities higher than 2 g/L, the measured residual absorbance reduced by at least 50%. These substantial reductions, even at lower salinities, may suggest that AB flocs can form when moisture from saline samples remains on the RC membrane after retention of TEP.

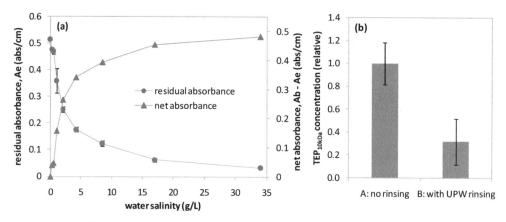

Figure 5-10: (a) Effect on the residual absorbance of Alcian blue when added in UPW solution with different ionic strength. (b) Relative comparison of TEP$_{10kDa}$ concentration in saline AOM solution with and without UPW rinsing in the measurement protocol.

To remove retained saline moisture from the RC membrane, it is proposed to filter UPW through the membrane before extracting the TEP for subsequent AB staining. A comparison of TEP results with and without rinsing during measurement of AOM samples is shown in Figure 5-10b. TEP concentration was overestimated by a factor of 3 when rinsing was not performed. This illustrates that the rinsing procedure is an essential step to minimise interference of dissolved salts, especially in seawater samples.

5.3.4.2 Membrane pore size selection for TEP retention

Selecting the optimal membrane pore size to collect TEP from water samples is an important aspect in the development of Protocol 2 as it indicates which size fraction of TEP can be measured. To determine this, measurements were performed using membranes with different pore sizes (0.4 µm, 0.1 µm, 10 kDa and 5 kDa) to retain TEP in AOM samples. A comparison of the results is presented in Figure 5-11a. As expected, TEP concentrations increased when using smaller pore size of membranes for TEP retention. As a comparison, the TEP retention of a 0.1 µm PC membrane was about 32% higher than 0.4µm membrane while 10 and 5 kDa membranes showed about 100% higher retention.

LC-OCD analyses were also performed for selected membranes to measure retention of biopolymers from AOM and Xanthan gum solutions. As shown in Figure 5-11b, 10kDa membranes retained about 95% of biopolymers from CA, which was about twice the retention of a 0.4 μm membrane. Moreover, the retention of Xanthan on a 10kDa membrane was about 4 times higher than 0.4 μm membrane.

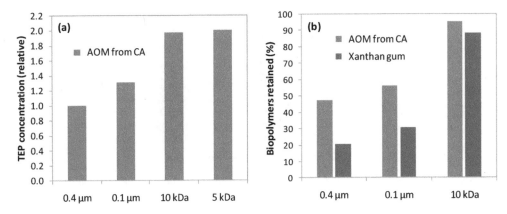

Figure 5-11: (a) TEP measurement from CA culture with different pore size membranes; (b) rejection of biopolymers by 0.4μm, 0.1μm and 10kDa membranes. For (a), the TEP result with 0.4 μm membrane was set as the baseline for relative comparison.

The results of the retention experiments demonstrated that >50% and >35% of TEPs were not retained by 0.4 μm and 0.1 μm pore size membranes, respectively. This suggests that a substantial fraction of colloidal TEPs are not covered by $TEP_{0.4\mu m}$ and $TEP_{0.1\mu m}$ methods. Considering that no substantial difference in TEP concentration between 10 kDa and 5 kDa membranes and the high feed pressure required to filter samples through 5 kDa membranes, 10kDa was eventually selected as the standard membrane pore size for TEP measurements based on Protocol 2.

5.3.4.3 TEP extraction from the membrane

In Protocol 2, TEP extraction from the membrane was performed by vortexing (10 secs) and sonicating (60 mins) the membrane samples in ultra-pure water solution. To illustrate that the current extraction procedure is sufficient to extract TEP from the membrane, tests were performed by filtering 20 ml of AOM solution through 10 kDa RC membranes and then TEP were extracted from the membrane with and without pre-vortexing followed immediately by 30 or 60 mins of sonication. The measured net absorbance of AB which reacted to the extracted TEP materials using the different extraction procedures are presented in Figure 5-12.

TEP retained on 10k Da RC membrane and sonicated for 60 mins showed slightly higher (3-10%) net absorbance than those sonicated for 30 mins. Pre-vortexing the sample may contribute to the extraction of TEP from the membrane as it can aid detachment of loosely attached TEP and enhance further extraction during the succeeding sonication step. Filter blank measurements of RC membranes vortexed and sonicated (60 mins) did not detect apparent release of TEP during extraction.

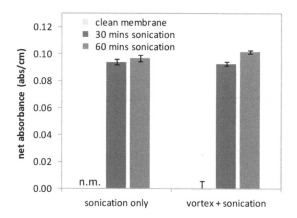

Figure 5-12: Effect of vortexing and sonication time on the extraction of TEP retained on 10kDa RC membranes.

5.3.4.4 Staining time

Tests were performed to investigate if 10 minutes of reaction time was sufficient to fully stain TEP in Xanthan and AOM solutions. The net AB absorbance (A_b-A_e) taken up by TEP was compared for samples stained for different reaction times (Figure 5-13). For both samples, there was no substantial difference in TEP concentration for reaction times between 1 and 60 mins. Hence, a reaction time of 10 mins was assumed to be sufficient to fully stain TEP in UPW. This also supports previous findings that reaction between TEP and AB in a low electrolyte matrix is instantaneous and that a longer staining time is generally not necessary (Passow and Alldredge, 1995; Thornton et al., 2007).

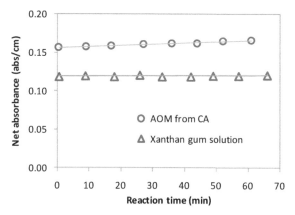

Figure 5-13: Effect of staining time on the net absorbance of TEP from AOM and Xanthan gum.

5.3.4.5 Retention of TEP-AB precipitates

The staining procedure of Protocol 2 involves application of AB on TEPs in UPW solution at pH 2.5. The TEP in the solution are expected to complex with AB molecules and form precipitates which can be removed by the following filtration step. Thornton *et al.* (2007) employed manual syringe filtration with a 0.2 μm pore size cellulose acetate (CA) membranes to removed TEP-AB precipitates. In the current method, vacuum filtration through 0.1 μm pore size track-etched PC membranes are applied. Track-etched PC membranes have a more defined pore size structure and narrower size distribution than other polymeric membranes. Hence, retention of TEP-AB precipitates on PC membranes is expected to be more consistent than with CA membranes. In addition, the filtration pressure in Protocol 2 is standardised by using a pressure-controlled pump to avoid the inconsistencies of manual syringe filtration.

To demonstrate that using a 0.1 μm pore size membrane is more reliable in removing TEP-AB precipitates than a larger pore size membrane, tests were performed using solutions of Xanthan gum (5 mg/L) and AOM. The measured absorptions of stained samples after filtration through 0.4 μm, 0.2 μm and 0.1 μm membranes were deducted from the filtrate absorption of stained blank solutions to compare their net absorbance. As shown in Figure 5-14, the net absorbance results with the 0.1 μm membrane was up to 60% higher than with a 0.2 μm membrane and up to 150% higher than with a 0.4 μm membrane. This may indicate that some TEP-AB precipitates were able to pass through 0.4 μm and 0.2 μm filters. It is also possible that some colloidal TEP-AB precipitates can still pass through the 0.1 μm membrane. However, using a lower pore size membrane (e.g., 0.05 μm pore size filters) is no longer feasible considering that there is a risk of overestimating the results due to retention of un-reacted AB (see Figure 5-4).

To check if additional un-reacted AB was retained as it passed through a layer of retained AB-TEP precipitates, a series of measurements were performed for stained samples of different concentrations of Xanthan (0-5 mg X/L). The stained samples were first filtered through 0.2 μm and 0.1 μm membranes and the filtrate were collected for absorbance measurements. Stained blank samples were then filtered through the same membrane and the filtrate absorbance was measured. The results of the test are presented in Figure 5-14b.

The reduction in absorbance with increasing Xanthan concentration was linear when using 0.1 μm membranes while a non-linear trend was observed when 0.2 μm membranes were used. The 0.2 μm membranes did not retain substantial amounts of precipitates at lower concentrations (<2 mg Xathan/L) but at higher concentrations, accumulation of precipitates probably clogged or constricted the pores of the membrane which then eventually resulted in retention of more precipitates. On the other hand, 0.1 μm membranes consistently retained precipitates as demonstrated by its linear trend. Further retention of un-reacted AB on fouled membrane for concentrations <4 mg X/L was rather marginal for both 0.2 μm and 0.1 μm membranes (Figure 5-14b). This means that the higher retention of 0.1 μm at this range was likely not due to further retention of un-reacted AB by the layer of retained TEP-AB precipitates but rather retention of smaller TEP-AB precipitates.

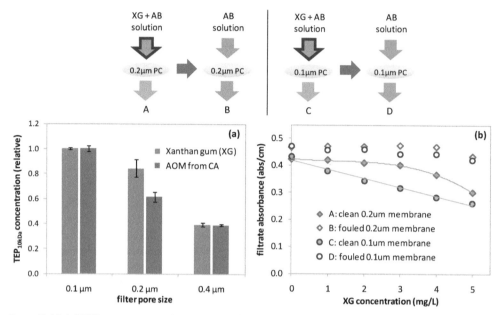

Figure 5-14: (a) TEP_{10kDa} concentrations measured by retaining TEP-AB precipitates on PC membranes of various pore sizes; (b) Residual AB absorbance for different concentrations of Xanthan gum after filtration through 0.2 μm and 0.1 μm membranes. Un-shaded data points correspond to the absorbance of blank samples filtered through the membrane after retention of TEP-AB precipitates.

The removal of TEP-AB precipitates can be further explained by comparing the hydrodynamic size distribution of Xanthan, AB dye and the precipitates formed after reaction of the two. Figure 5-15 shows that the hydrodynamic size range (200-440 nm) of 2 mg/L Xanthan does not overlap with those of pre-filtred AB (Peak 1: 2.5-4 nm; Peak 2: 22-38 nm). The apparent hydrodynamic size range of Xanthan was larger than 100 nm which make them likely to be retained by 0.1 μm PC membranes.

The hydrodynamic size distribution of AB-stained Xanthan showed three characteristic peaks with average sizes of 220, 65 and 12 nm, respectively. The first peak (110-400 nm) likely comprises Xanthan-AB precipitates considering that it overlaps with the size range of unstained Xanthan peak (200-450 nm). The average hydrodynamic size of AB-stained Xanthan (220 nm) was smaller than those of unstained Xanthan (310 nm). The shift in size to lower size range can be an effect of coiling of Xanthan fibrils into more compact structures after reaction with AB molecules, leading to reduction of the overall hydrodynamic diameter. Theoretically, these materials are more compact or rigid and are easier to remove by filtration than flexible unstained Xanthan with higher hydrodynamic diameter. The two smaller size peaks of AB-Xanthan solution may be attributed to the two peaks of AB solution. The two AB peaks appear to have shifted to larger hydrodynamic size ranges (Peak 1: 7.2-17 nm; Peak: 32-110 nm). The reason for the apparent shift might be due to coagulation of the dye considering that the solution was agitated after addition of AB to Xanthan solution.

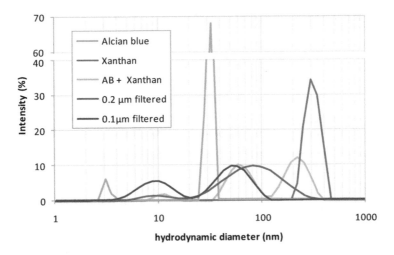

Figure 5-15: Hydrodynamic size distribution of 2 mg/L Xanthan, prefiltered AB dye and after reaction between AB and Xanthan.

After filtration of Xanthan-AB solution through 0.2 μm PC membrane, a broader peak appeared between 20 and 330 nm (Figure 5-15). At such range, it is likely that this peak comprises both excess AB and stained Xanthan not removed by the 0.2 μm PC membranes. Filtration of AB-stained Xanthan solution through 0.1 μm PC membrane resulted in total removal of the first Xanthan-AB peak (110-400nm). This is an indication that the 0.1 μm PC membrane is sufficient in removing TEP-AB precipitates from stained samples.

5.3.4.6 Calibration with Xathan Gum

Absorbance results from TEP measurements were calibrated based on the staining capacity of Xanthan gum of known concentration. Calibrations were performed for two ranges of Xanthan concentrations (Figure 5-16). The calibration factor (f_{610nm}) of Xanthan measured for concentration range of 0-5 mg/L was about -30 (mg Xanthan/L)/(abs/cm). The calibration results from 3 experiments performed with different batches of AB and Xanthan solution showed similar slopes (Figure 5-16a). Although the 3 calibration experiments showed variations in initial absorbance at 0 mg/L, the calibration factors were similar. When the concentration range of Xanthan used in the calibration was increased to 0-10 mg/L, the calibration factor was observed to increase slightly increased from -30.4 to -34 (mg X_{eq}/L)/(abs/cm). Linear regression between absorbance and standard concentration showed coefficient of determination (R^2) higher than 98, an indication of the high precision of the calibration method.

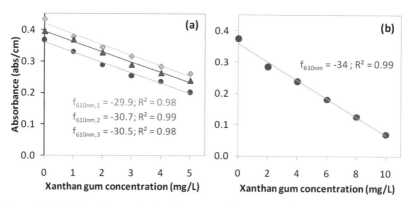

Figure 5-16: Calibration graphs for 2 concentration ranges of Xanthan gum standard.

The calibration factors (with Xanthan standard) determined in this study were 5-34% lower than what was observed by Thornton *et al.* (2007) and 25% higher than what was recorded by Arruda-Fatibello *et al.* (2004). The substantial differences might be due to differences in procedure for separating TEP-AB precipitates or the membranes used to retain them.

In this study, calibration was performed using 0.1 µm PC membrane instead of 0.2 µm cellulose acetate used by Thornton *et al.* (2007). As shown in Figure 5-15, using a 0.1 µm pore size membrane can retain up to 40% more TEP-AB precipitates than using a 0.2 µm pore size membrane; hence, a decrease in calibration factor is expected when shifting to lower pore size membrane. On the other hand, Arruda-Fatibello *et al.* (2004) used centrifugation (3000 rpm) instead of filtration to separate TEP-AB precipitates from excess stain. Although filtration through 0.1 µm membranes is expected to yield higher removal of precipitates than centrifugation at 3000 rpm, calibration results by Arruda-Fatibello *et al.* (2004) indicate higher removal by centrifugation. This can be attributed the higher solution pH (pH 4.0) and the procedure (no pre-filtration) used by the author in preparing the stain. AB dye can be more prone to coagulation at higher pH and not pre-filtering the stain before applying it to the standard solution may result in overestimation of the calibration factor because some un-reacted but coagulated AB dye might settle together with the TEP-AB precipitates during centrifugation.

5.3.4.7 Limit of detection: TEP_{10kDa}

The lower limit of detection of TEP_{10kDa} was calculated based on the standard deviation of 10 independently measured blanks using the same batch of staining solution. The average absorbance of 10 blank measurements was 0.413 abs/cm and the standard deviation (σ_b) is 0.01 abs/cm (Figure 5-17a).

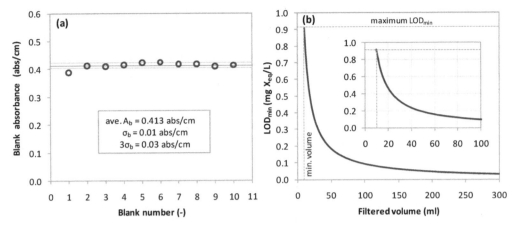

Figure 5-17: (a) Variation of measured blanks for protocol 2 and (b) calculated LOD_{min} as a function of filtered sample volume.

Similar to Protocol 1, the LOD_{min} for Protocol 2 can be lowered by increasing the volume of sample filtered. A typical filtered volume used for surface water samples is 50-100 ml, but a much smaller volume is sufficient (10-30 ml) for algal culture samples where TEP concentration is usually much higher. The LOD_{min} for a wide range of filtered sample volume was calculated according to Eq. 7 in Section 5.2.5 and the results are presented in Figure 5-17b. The LOD_{min} for un-concentrated samples (10 ml filtered volume) is 0.91 mg X_{eq}/L. For concentrated samples (>10ml filtered volume), the LOD_{min} can go down to 0.09 mg X_{eq}/L for 100 ml filtered volume or 0.05 mg X_{eq}/L for 200 ml filtered volume (Figure 5-17b). For waters with very low concentrations of TEP, filtering large volume of samples might be necessary to make sure TEP concentration is above the LOD_{min}.

5.3.5 Comparison and application of the two methods

The methods developed in this study are not redundant but rather complimentary as they measure different but overlapping fractions of TEP. Protocol 1 can measure particulate TEP and a fraction of colloidal TEP ($TEP_{0.4\mu m}$ and $TEP_{0.1\mu m}$). Protocol 2 can measure both particulate and a substantial fraction of colloidal TEP (TEP_{10kDa}). Therefore, both methods can be applied in monitoring the presence of TEP in the raw water of treatment plants. $TEP_{0.4\mu m}$ and $TEP_{0.1\mu m}$ measurement is simpler and faster to perform; which is convenient for routine monitoring of TEP in the raw water. However, to measure the removal of TEP over the treatment processes, TEP_{10kDa} measurement is more appropriate because it covers a wider size range of TEP.

5.3.5.1 Integrated calibration

The calibration method used in Protocol 1 ($TEP_{0.4\mu m}$ and $TEP_{0.1\mu m}$) can be integrated with the calibration method used in Protocol 2 (TEP_{kDa}). The integrated calibration method follows the same procedure as in Protocol 2 but with some additional steps to incorporate calibration for Protocol 1. The additional step is to soak the 0.1 μm PC membrane used to retain TEP-AB precipitates in 6 ml of 80% sulfuric acid solution.

The beaker with the soaked membrane is incubated on a shaker for 2 hours and then the absorbance of the acid solution is measured at 787 nm wavelength.

The net absorbance (total absorbance minus the filter blank) results of the two protocols are plotted in relation to the concentration of Xanthan (mg/L) in the solution as shown in Figure 5-18a. The calibration factor at 610 nm wavelength is about twice the calibration factor at 787 nm wavelength. This was expected considering that the absorbance signal for the same concentration of AB at 787 nm is about twice that measured at 610nm (see Figure 5-5b). For Protocol 1, the calibration factor is calculated as the inverse of the slope of retained mass (instead of concentration used in Protocol 2) of Xanthan (mg) and absorbance at 787nm. The concentration values in the x-axis in Figure 5-18a was converted to mass of retained Xanthan by multiplying concentration with the amount of standard solution filtered (i.e., 4 ml). The converted results is plotted in Figure 5-18b.

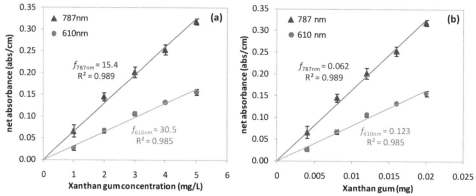

Figure 5-18: Comparison of calibration graphs for Protocol 1 and 2 showing absorbance in relation to (a) concentration and (b) mass of stained Xanthan.

The integrated calibration protocol can be useful when measuring both TEP_{10kDa} and $TEP_{0.4\mu m}$ or $TEP_{0.1\mu m}$ in the same water sample. Consequently, the analyses time can be minimised for both parameters.

5.3.5.2 Effect of sample storage time

Storing samples for a period of time before analysis may lead to variation between the measured and actual TEP concentration. The effect of storage time was investigated by monitoring concentrations of TEP_{10kDa} and $TEP_{0.4\mu m}$ in AOM samples for a period of 17 days. Laboratory prepared AOM samples were stored in 2 L glass bottle at 4°C temperature. The results of the monitoring are shown in Figure 5-19.

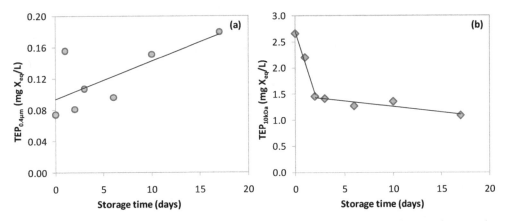

Figure 5-19: TEP$_{0.4\mu m}$ and TEP$_{10kDa}$ monitoring in laboratory prepared AOM sample stored in 2 L glass bottle at 4°C for 18 days.

TEP$_{0.4\mu m}$ concentrations vary over time with a relatively increasing trend. At day 17, TEP$_{0.4\mu m}$ was more than twice the initial concentration. The increase in concentration might be due to coagulation of colloidal TEPs smaller than 0.4 µm over time.

TEP$_{10kDa}$ concentration rapidly decreased (by ~45%) for the first 3 days in storage and then a slow decrease was observed until the end of monitoring. At day 17, TEP$_{10kDa}$ was about 60% lower than the initial concentration. The decrease might be due to adsorption of TEP materials on the inner walls of glass sample bottle and/or microbial degradation. Further investigations are necessary to further understand the mechanisms involved in the TEP loss or increase as well as on how to develop better measures to preserve samples (e.g., sample bottle, freezing, preservative addition).

A major consequence of storing samples, for instance those collected in water treatment processes, is underestimation or overestimation of the treatment removal efficiencies. Lower removal efficiency can be recorded when a proportional decrease in concentration occurred in the influent and effluent samples. The removal percentage can be also underestimated if the percentage of TEP lost in the influent sample is higher than in the effluent sample while removal can be overestimated if the percentage TEP lost in the influent sample is lower than in the effluent sample. Overall, it is important that samples be analysed immediately after sampling to obtain reliable TEP concentrations in both influent and effluent samples.

5.3.5.3 TEP monitoring in marine algal cultures

The two protocols were applied to monitor up to 60 days the TEP produced in 2 batch cultures of common species of bloom-forming marine algae: *Alexandrium tamarense* (AT) and *Chaetoceros affinis* (CA). Samples were collected within the incubation period to measure algal cell concentration (direct cell counting), TEP$_{0.4\mu m}$ (Protocol 1) and TEP$_{10kDa}$ (Protocol 2). TEP concentrations vary substantially for the two algal species at different growth phases (Figure 5-20a,b). CA produced up to about 3 times more TEP than AT. In both cultures, the abundance of TEP$_{10kDa}$ was substantially higher than $TEP_{0.4\mu m}$. On average, CA and AT produced at least 5 times more TEP_{10kDa} than $TEP_{0.4\mu m}$ (Figure 5-20c). In general, this results indicate that the colloidal fraction (<0.4μm) is likely the dominant fraction of TEP comprising algal-derived organic matter.

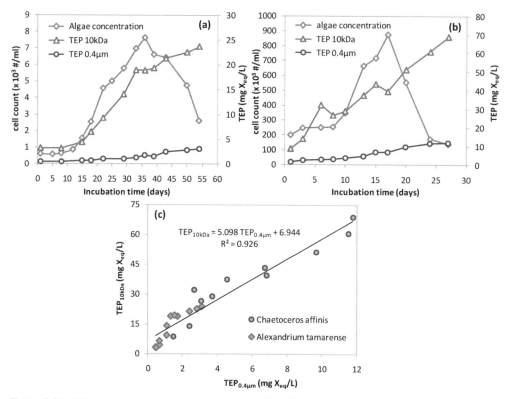

Figure 5-20: TEP measurements in batch cultures of (a) *Alexandrium tamarense* and (b) *Chaetoceros affinis* and (c) relationship between measured TEP_{10kDa} and $TEP_{0.4\mu m}$.

5.3.5.4 TEP monitoring in water treatment processes

$TEP_{0.4\mu m}$ and TEP_{10kDa} were monitored over the treatment processes of two drinking water treatment plants. For comparison, biopolymer concentrations based on LC-OCD analysis were also measured.

Plant A treats raw water from a lake source by various treatment processes such as microstraining, coagulation-flocculation-sedimentation, rapid sand filtration (RSF), granular activated carbon (GAC) filtration, ultrafiltration (UF) and reverse osmosis (RO) followed by post-treatment. Samples were collected during the summer period (July, 2012) when the raw water contained 0.05 mg X_{eq}/L of $TEP_{0.4\mu m}$ and 1.55 mg X_{eq}/L of TEP_{10kDa}. High removal of TEP was observed after coagulation-sedimentation-sand filtration. Further TEP removal was also recorded in the subsequent UF treatment while RO completely removes all remaining TEPs (Figure 5-21a).

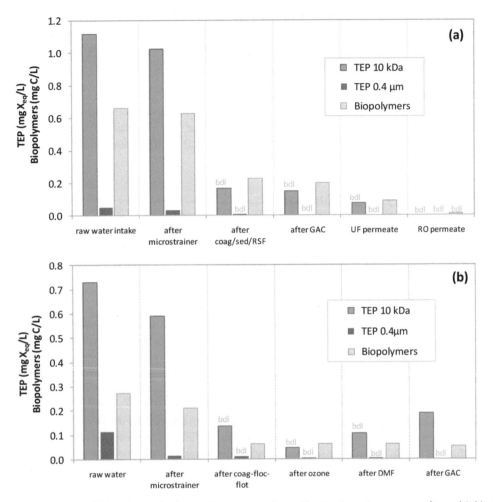

Figure 5-21: TEP and biopolymer concentration measured over the treatment processes of two drinking water treatment plants.

Plant B treats water from a reservoir by microstraining, coagulation-flocculation-flotation, ozonation, dual media filtration (DMF) and GAC filtration. Samples were collected during the spring algal bloom period (May, 2012) when the raw water comprises 0.11 mg X_{eq}/L of $TEP_{0.4\mu m}$ and 0.73 mg X_{eq}/L of TEP_{10kDa}. A significant drop in TEP_{10kDa} concentration (down to below detection limit) was observed after coagulation-flocculation-flotation. However, concentration increased to above detection limit after water pass through granular activated carbon (GAC) filter, possibly due to sloughing of biofilm that were likely present in the media.

In both plants, $TEP_{0.4\mu m}$ removal over the treatment process was higher than both TEP_{10kDa} and biopolymers. This can be attributed mainly to the size of $TEP_{0.4\mu m}$ which covers mainly the particulate size range while TEP_{10kDa} and biopolymers covers both the particulate and colloidal size ranges. UF and conventional coagulation followed by sedimentation-RSF or flotation showed substantial removal (>50%) of both TEP_{10kDa} and biopolymers. The similarities of the removal of TEP_{10kDa} and biopolymers may indicate that TEPs are a significant fraction of the biopolymer materials in the raw water of the two treatment plants.

The plant monitoring results further demonstrates that TEPs in surface water can be mainly colloidal in nature at least during the algal bloom period and that measuring this fraction might be essential to better evaluate the removal efficiency of water treatment processes. Investigations on the removal of TEP over various treatment processes will be further discussed in Chapter 6.

5.4 Summary and conclusions

This study focused on improving the methods developed by Passow and Alldredge (1995) and Thornton *et al.* (2007) to measure particulate and colloidal TEPs in seawater and freshwater samples. The following are the highlights of the study:

1. The two existing TEP methods showed a tendency to overestimate TEP concentration when measuring brackish and saline water samples due to coagulation of the Alcian blue dye in the presence of dissolved salts. To minimise this problem, a membrane rinsing procedure was introduced to remove saline moisture from the membrane after retention of TEP. The rinsing step was effective in lowering the analytical blank and the detection limit of both methods.

2. A modified methods for measuring TEPs larger than 0.4 μm ($TEP_{0.4\mu m}$) and larger than 0.1 μm ($TEP_{0.1\mu m}$) based on the method introduced by Passow and Alldredge (1995) were proposed, with respect to minimising interference of salinity and improving reproducibility and the level of detection.

3. A method which is able to measure smaller colloidal TEPs down to 10 kDa was developed and tested. This method is partly based on the principles used in the method of Thornton *et al.* (2007). The new method enables the measurement of different fractions of TEP by making use of membranes with different pore sizes.

4. An integrated calibration protocol was developed using Xanthan gum as the standard for TEP for the two modified methods. Calibration results were more reproducible than the existing calibration methods.

5. $TEP_{0.4\mu m}$ and TEP_{10kDa} measurements were successfully applied to monitor TEP production in marine algal cultures and TEP removal over the treatment processes of two full-scale water treatment plants. Colloidal TEPs (<0.4μm) were more abundant than particulate TEPs (>0.4μm) in both algal cultures and natural water samples.

6. Storing water samples for several days before analysis can lead to underestimation of the actual TEP concentration. Hence, TEP measurements should be performed immediately after sample collection.

References

Alldredge, A. L., Passow, U. and Logan, B. E. (1993) The abundance and significance of a class of large, transparent organic particles in the ocean. Deep Sea Research Part I: Oceanographic Research Papers 40(6), 1131-1140.

Alldredge, A. L., Passow, U., & Logan, B. E. (1993). The abundance and significance of a class of large, transparent organic particles in the ocean. Deep-Sea Research I, 40, 1131–1140.

Arruda-Fatibello, S. H. S., Henriques-Vieira, A. A. and Fatibello-Filho, O. (2004) A rapid spectrophotometric method for the determination of transparent exopolymer particles (TEP) in freshwater. Talanta 62(1), 81-85.

Azetsu-Scott, K., and Passow, U. (2004) Ascending marine particles: Significance of transparent exopolymer particles (TEP) in the upper ocean. Limnol. Oceanogr, 49(3), 741-748.

Bartold P.M. and Page R.C. (1985) A microdetermination method for assaying glycosaminoglycans and proteoglycans. Analytical Biochemistry 150 (2), 320–324.

Bar-Zeev E., Berman-Frank I., Girshevitz O., and Berman T. (2012) Revised paradigm of aquatic biofilm formation facilitated by microgel transparent exopolymer particles. PNAS 109 (23), 9119-9124.

Bar-Zeev, E., Berman-Frank, I., Liberman, B., Rahav, E., Passow, U. and Berman, T. (2009) Transparent exopolymer particles: Potential agents for organic fouling and biofilm formation in desalination and water treatment plants. Desalination & Water Treatment 3, 136–142.

Berman, T., and Holenberg, M. (2005) Don't fall foul of biofilm through high TEP levels. Filtration & Separation 42(4), 30-32.

Berman, T., and Passow, U. (2007) Transparent Exopolymer Particles (TEP): an overlooked factor in the process of biofilm formation in aquatic environments. Available from Nature Precedings <http://dx.doi.org/10.1038/npre.2007.1182.1>

Chin W.-C., Orellana M.V., and Verdugo P. (1998) Spontaneous assembly of marine dissolved organic matter into polymer gels. Nature 391, 568–572.

de la Torre, T., Lesjean, B., Drews, A., and Kraume, M. (2008) Monitoring of transparent exopolymer particles (TEP) in a membrane bioreactor (MBR) and correlation with other fouling indicators. Water Science and Technology 58 (10), 1903–1909.

Harvey D. (2000) Modern analytical chemistry. McGraw-Hill, USA.

Heinonen, K.B., Ward, J.E., Holohan, B.A. (2007) Production of transparent exopolymer particles (TEP) by benthic suspension feeders in coastal systems. Journal of Experimental Marine Biology and Ecology 341 (2007) 184–195

Horobin, R. W. (1988) Understanding histochemistry: Selection, evaluation and design of biological stains. Wiley, New York.

Huber S. A., Balz A., Abert M. and Pronk W. (2011) Characterisation of aquatic humic and non-humic matter with size-exclusion chromatography – organic carbon detection – organic nitrogen detection (LC-OCD-OND), Water Research 45 (2), 879-885.

Karlsson M. and Björnsson S. (2001) Quantitation of Proteoglycans in Biological Fluids Using Alcian Blue. Proteoglycan Protocols, Methods in Molecular Biology 171, 159-173.

Kennedy, M. D., Muñoz-Tobar, F. P., Amy, G. L. and Schippers, J. C. (2009) Transparent exopolymer particle (TEP) fouling of ultrafiltration membrane systems. Desalination & Water Treatment 6(1-3) 169-176.

Mari X., & Kiørboe T. (1996). Abundance, size distribution and bacterial colonization of transparent exopolymer particles (TEP) in the Kattegat. Journal of Plankton Research 18, 969–986.

McKee, M. P., Ward, J. E., MacDonald, B. A. and Holohan, B. A. (2005) Production of transparent exopolymer particles by the eastern oyster Crassostrea virginica, Marine Ecology Progress Series 183, 59-71.

Mopper, K., Zhou, J., Sri Ramana, K., Passow, U., Dam, H. G., and Drapeau, D. T. (1995) The role of surface-active carbohydrates in the flocculation of a diatom bloom in a mesocosm. Deep-Sea Research Part II, 42(1), 47-73.

Myklestad, S., Holm-Hansen, O. Varum, K.M. and Volcani, B.E. (1989) Rate of release of extracellular amino acids and carbohydrates from marine diatom Chaetoceros Affinis. Journal of Plankton Research 11 (4), 763-773.

Parker B.C. and Diboll A.G. (1966) Alcian Stains for Histochemical Localization of Acid and Sulfated Polysaccharides in Algae. Phycologia 6 (1), 37-46.

Passow and Alldredge, 1995 Passow, U. and Alldredge, A. L. (1995) A Dye-Binding Assay for the Spectrophotometric Measurement of Transparent Exopolymer Particles (TEP). Limnology and Oceanography 40(7), 1326-1335.

Passow, U. (2000) Formation of transparent exopolymer particles (TEP) from dissolved precursor material. Marine Ecology Progress Series 192, 1-11.

Passow, U. (2002a) Transparent exopolymer particles (TEP) in aquatic environments. Progress in Oceanography 55(3), 287-333.

Passow, U. (2002b) Production of transparent exopolymer particles (TEP) by phyto-and bacterioplankton. Marine Ecology Progress Series 236, 1-12.

Ramus, J. (1977) Alcian Blue: A Quantitative Aqueous Assay for Algal Acid and Sulfated Polysaccharides. Journal of Phycology 13, 345-348.

Scott, J. E., Quintarelli, G., & Dellovo, M. C. (1964). The chemical and histochemical properties of alcian blue - I. the mechanism of alcian blue staining. Histochemie, 4(2), 73-85.

Thornton, D. C. O., Fejes, E. M., DiMarco, S. F. and Clancy, K. M. (2007) Measurement of acid polysaccharides (APS) in marine and freshwater samples using alcian blue. Limnology and Oceanography: Methods 5, 73–87.

Verdugo, P., Alldredge, A. L., Azam, F., Kirchman, D. L., Passow, U. and Santschi, P. H. (2004) The oceanic gel phase: a bridge in the DOM–POM continuum. Marine Chemistry 92, 67-85.

Villacorte, L. O., Kennedy, M. D., Amy, G. L., and Schippers, J. C. (2009a) Measuring Transparent Exopolymer Particles (TEP) as indicator of the (bio)fouling potential of RO feed water. Desalination & Water Treatment 5, 207-212.

Villacorte, L. O., Kennedy, M. D., Amy, G. L., and Schippers, J. C. (2009b) The fate of Transparent Exopolymer Particles (TEP) in integrated membrane systems: Removal through pretreatment processes and deposition on reverse osmosis membranes. Water Research 43 (20), 5039-5052.

Villacorte, L. O., Schurer, R., Kennedy, M., Amy, G. and Schippers, J.C. (2010a) Removal and deposition of Transparent Exopolymer Particles (TEP) in seawater UF-RO system. IDA Journal 2, 45-55.

Villacorte, L. O., Schurer, R., Kennedy, M., Amy, G., Schippers, J.C. (2010b) The fate of transparent exopolymer particles in integrated membrane systems: a pilot plant study in Zeeland, The Netherlands. Desalination & Water Treatment 13, 109-119.

Wardi, A. H. and Allen, W. S. (1972) Alcian blue staining of glycoproteins. Analytical Biochemistry, 48 (2) 621-623.

Fate of transparent exopolymer particles in integrated membrane systems

Contents

Abstract 148

6.1   Introduction 149

6.2   Materials and Methods 151
    6.2.1   Water samples, collection and storage 151
    6.2.2   Integrated membrane systems (IMS) 151
    6.2.3   TEP measurements 153
    6.2.4   Total organic carbon (TOC) and chlorophyll-a 153
    6.2.5   Liquid chromatography organic carbon detection (LC-OCD) 153
    6.2.6   Modified fouling index – ultrafiltration (MFI-UF) 153
    6.2.7   RO membrane autopsy 155

6.3   Results and discussion 156
    6.3.1   TEP in different water sources 156
    6.3.2   Removal by various treatment processes 161
    6.3.3   TEP accumulation in RO system 169

6.4   Conclusions 173
    References 173

This chapter is an extended version of:

Villacorte, L. O., Ekowati Y., Calix-Ponce H.N., Schurer R., Lohmann B., Kennedy M.D., Amy G. and Schippers J.C. (2013) Monitoring Transparent Exopolymer Particles (TEP) in integrated membrane systems. *In Preparation for Water Research.*

Abstract

The presence of transparent exopolymer particles (TEP) in surface water sources has been overlooked for many years as a potential foulant in reverse osmosis (RO) systems. This chapter investigates the fate of TEP over the treatment processes of various RO plants using newly developed and improved spectrophotometric methods (Chapter 5) to measure particulate and colloidal TEPs (TEP$_{0.4\mu m}$ and TEP$_{10kDa}$). Variations in TEP concentrations in the raw water in relation to seasonal changes in water temperature and algal concentration were examined and the removal efficiencies of different treatment processes were compared based on reduction of TEP, total organic carbon, biopolymers, humics and membrane fouling potential. Deposition of TEPs and other organic materials in RO systems was investigated by performing membrane autopsies.

TEP monitoring (1-3 years) in seawater and freshwater sources for 2 RO plants revealed that high TEP concentrations occurred mainly during the spring algal bloom (March-May). Ultrafiltration and conventional coagulation followed by flotation or by sedimentation and rapid sand filtration (RSF) were effective in removing TEP$_{0.4\mu m}$ and TEP$_{10kDa}$. Comparable reductions were also observed with biopolymer concentration (measured by LC-OCD) and membrane fouling potential (in terms of MFI-UF) but not with total organic carbon concentration. Data collected from 5 plants showed TEP$_{10kDa}$ concentration showed better correlation with MFI-UF than TEP$_{0.4\mu m}$, biopolymers or humic concentration, an indication that colloidal TEPs may cause more (serious) fouling in membrane systems than other types of organic matter.

Membrane autopsies of RO elements taken from 2 plants revealed substantial accumulation of TEPs and bacteria in RO membranes and spacers. However, some of these TEPs may have been produced locally by biofilm bacteria.

6.1 Introduction

Despite recent advances in pre-treatment technologies, the occurrence of organic and/or biological fouling remains a major challenge in the operation of reverse osmosis (RO) systems treating seawater and fresh (surface) water sources (Flemming et al., 1997; Baker and Dudley, 1998; Kruithof et al., 1998; van Agtmaal et al., 2007). Organic fouling and biofouling in RO membranes are often inter-connected. Organic fouling is caused by substantial accumulation of natural organic matter (NOM), which may result in a substantial decrease of normalized flux in RO membranes. The deposited organic matter can provide good attachment sites and a growth platform for microorganisms which feed on the available (biodegradable) nutrients in the feed water. If nutrients (e.g., C, N, P) are not limited, these microorganisms can multiply rapidly and excrete more NOM, mainly extracellular polymeric substances (EPS), leading to biofilm formation and eventually biofouling. Moreover, the deposited organic substances, which are often sticky, may enhance the deposition of other colloids/particles from the feedwater to the membrane/spacers and further aggravate fouling problems.

In aquatic systems, extracellular polymeric substances (EPS) produced by bacteria and algae have a major role in the cohesion between cells and EPS-coated particles, as well their adhesion to other surfaces. EPS are microbial polymers comprising a wide variety of macromolecules (e.g., polysaccharides, proteins, glycoproteins and glycolipids) which is often dominated by polysaccharides (40-95%; Flemming and Wingender, 2001). Ahimou et al. (2007) demonstrated that biofilm cohesiveness has a strong correlation with polysaccharide rather than protein concentrations supporting the initial theory of Sutherland (2001) of polysaccharides being a "strong and sticky framework of biofilms". In RO membranes, the physical structure of a biofilms is described as either "compact and gel-like" or "slimy and adhesive" (Baker and Dudley, 1998; Schneider et al., 2005), which may indicate abundance of polysaccharides. Whether the accumulation of polysaccharides in RO membranes is mainly due to local production by biofilm microorganisms or by gradual deposition of EPS from the RO feedwater, is still largely unknown.

Over the last two decades, an abundant form of EPS in surface water called transparent exopolymer particles (TEP) has been described as a major agent in the aggregation of colloids and particles in surface water environments. Several studies have reported the presence of TEPs in fresh and marine surface water (Alldredge et al., 1993; Berman and Viner-Mozzini, 2001; Passow, 2002a) and recently in wastewater (de la Torre et al., 2008; Kennedy et al., 2009). TEPs mainly originate from polysaccharides released by phytoplankton and bacterioplankton (Passow, 2002a, 2002b) as well as from biological detritus of higher organisms including macroalgae (Thornton, 2004) and some bivalve molluscs (McKee, 2005). In the ocean, TEPs were reported to form abiotically from colloidal biopolymers of 1-3 nm in diameters and up to hundreds of nanometers long, some of which can even pass through 8 kDa pore size membranes (Passow, 2000; Santschi et al., 1998).

Berman and Holenberg (2005) first proposed a potential link between TEP and RO biofouling. They proposed TEP to be the "major initiator" of biofilm formation in RO systems, which could potentially lead to biofouling. TEP comprise mainly acidic polysaccharides which are rather sticky. Their relative stickiness is reported to be between 100 and 10,000 times more than most suspended particles in aquatic systems (Passow, 2002). TEPs sticky character makes them likely to attach to RO membranes, providing favourable surfaces for bacterial colonization and may thereby initiate biofilm development in the system. Moreover, a number of studies reported that significant percentages (2-68 %) of bacterial population in seawater were found attached to or embedded in TEPs (Alldredge et al., 1993, Passow and Alldredge, 1994). As TEP can be a potential carrier of bacteria to the RO system, it may not only serve as an initiator but may also play a vital role in enhancing microbiological growth in the system.

Passow and Alldredge (1995) operationally defined TEP as Alcian blue stainable material retained on 0.40 µm PC filters, while those not retained are considered as dissolved TEP precursors (Passow, 2000) or hydrogels (Verdugo et al., 2004). However, the International Union of Pure and Applied Chemistry (IUPAC) categorise materials as particulate (suspended) > 1 µm, colloidal 0.001–1 µm and dissolved <0.001 µm. Following this definition means $TEP_{0.4\mu m}$ encompass both particulate and colloidal size ranges while their precursors fall within the colloidal size range. Nevertheless, TEPs do not behave like most colloidal/particulate materials as they are highly flexible and may squeeze through filter pores smaller than their apparent size at high pressure or at high filtration rate. Colloidal TEPs are more likely to pass through media filters including MF/UF membranes. These colloidal TEPs may eventually form particulate TEP under certain conditions that promote coagulation (Passow, 2000; Li and Logan, 1997), which occur in most treatment processes. In general, TEP may undergo a series of changes depending on the treatment conditions (e.g., shear forces, retention times, temperature, etc.) in the plant. Thus, studying and monitoring these substances is potentially important in selecting effective pre-treatment strategies for RO systems.

TEP has been widely studied in the fields of oceanography and limnology but limited information is currently available on its relevance to water treatment. A number of methods are currently available to quantify and monitor TEPs. These methods either involve direct counting (Alldredge et al., 1993) or spectrophotometric measurements (Passow and Alldredge, 1995; Arruda-Fatibello et al., 2004; Thornton et al., 2007). Although these techniques were introduced several years ago, TEP measurement for water treatment applications is still not widely studied (Liberman and Berman, 2006; de la Torre et al., 2008; Bar-Zeev et al. 2009; Kennedy et al., 2009; Villacorte et al., 2009; van Nevel et al., 2012). Essential modifications of two of the spectrophotometric methods (e.g., $TEP_{0.4\mu m}$, TEP_{10kDa}) were introduced in Chapter 5 (this thesis) in order to improve their reliability for such applications.

The main objectives of this study are as follows:
1. to apply the improved spectrophotometric methods ($TEP_{0.4\mu m}$ and TEP_{10kDa}) to measure particulate and colloidal TEP in various water treatment facilities;

2. to compare variations of TEP in different water sources and their removal over the treatment processes;
3. to assess relationships between TEP, membrane fouling potential (MFI-UF) and other NOM parameters (i.e, TOC, biopolymers and humic concentration); and
4. to investigate accumulation of TEP in reverse osmosis systems.

6.2 Materials and Methods

6.2.1 Water samples, collection and storage

Water samples were collected from various water treatment installations. Sample volumes of 0.5 to 1 litre were collected in clean glass or plastic sample bottles. Majority of the samples were analysed batch-wise within 48 hours after sampling. However, it was necessary for some batch of samples to be stored (at ~4°C) for up to 2 weeks before analyses can be performed.

6.2.2 Integrated membrane systems (IMS)

Investigations were conducted for 3 full-scale (Plants A, B and C) and 1 pilot (Plant C) RO installation in the Netherlands and the United Kingdom (UK). For comparison, a conventional drinking water facility (Plant E) in the Netherlands was also investigated. The 5 plants treat water from various sources and employ different pre-treatment processes prior to the RO system (Table 6-1 and 6-2). To assess the presence of TEP, water samples were collected from selected points within each plant including the raw water, and over the treatment processes (Figure 6-1). Treatment efficiencies of each treatment steps were evaluated based on removal of TEP, TOC, biopolymer and humics, and change in membrane fouling potential.

Table 6-1: Overview of 5 water treatment facilities investigated in this study.

Plant	Type	Source water	Product water	Capacity[*] (m^3/h)	Treatment processes
A	Full-scale	River	Industrial	n.o.	coag-floc → poly[a]-sed → RSF → inline-coag → DMF1 → DMF2 → CEx → AEx → RO
B	Full-scale	Lake	Drinking	2300	MS → coag-floc- sed → RSF → GAC→ UF→ RO
C	Full-scale	Reservoir	Industrial	155	MS → inline-coag → UF → RO
D	Pilot	Seawater	Drinking	15	MS → inline-coag → UF → RO
E	Full-scale	Reservoir	Drinking	2000	MS → coag→ flot→ Oz → DMF → GAC → ClO_2

[*] plant water production capacity
n.o. = RO system was still not operational during sampling
sed = sedimentation
RSF = rapid sand filtration
coag-floc = coagulation-flocculation
poly = cationic polymer addition
DMF = dual media filtration
CEx = cation exchange
AEx = anion exchange

RO = reverse osmosis
MS = microstrainer
GAC = granular activated carbon filtration
UF = ultrafiltration
flot = flotation
Oz = ozonation
ClO_2 = chlorination
[a] cationic polymer dose = 5 mg/L polyacrylamide polymers

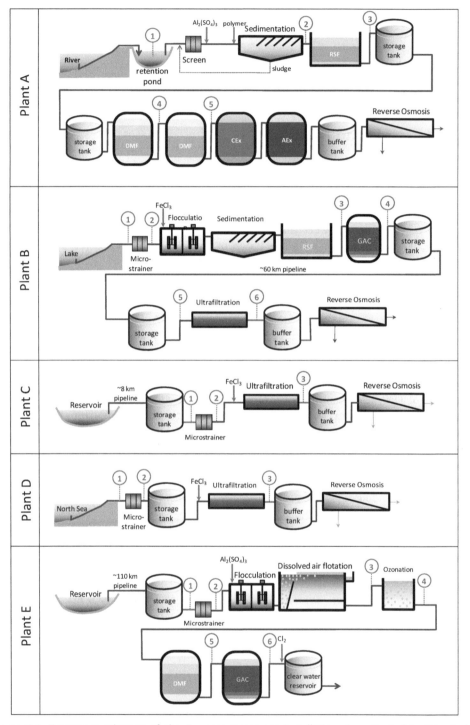

Figure 6-1: Treatment schemes of the 5 water treatment installations investigated in this study. Abbreviations: RSF = rapid sand filter; DMF = dual (granular) media filter; CEx = cation exchange; AEx = anion exchange; GAC = granular activated carbon filter.

Table 6-2: Summary of coagulant application in 5 plants.

Plant	Treatment process	Coagulant	Dose (mg/L)
A	coagulation → flocculation	Al^{3+}	4.7
	inline-coagulation→ DMF	PACl	2
B	coagulation → flocculation	Fe^{3+}	6.9
C	inline-coagulation → UF	Fe^{3+}	0 - 2
D	inline-coagulation → UF	Fe^{3+}	0.6 - 4
E	coagulation → flotation	Al^{3+}	4

6.2.3 TEP measurements

Colloidal and particulate TEPs were measured based on the spectrophometric methods modified from Passow and Alldredge (1995) and Thornton *et al.* (2007). The protocols to measure $TEP_{0.4\mu m}$ and TEP_{10kDa} are described in Chapter 5 (this thesis).

6.2.4 Total organic carbon (TOC) and chlorophyll-a

TOC concentration in water samples was measured using a Shimadzu $TOC\text{-}V_{CPN}$ analyzer based on combustion catalytic oxidation/NDIR method. Chlorophyll-a concentrations were measured by Aqualab B.V. (Netherlands) in accordance with the Dutch standard protocol (NEN 6520).

6.2.5 Liquid chromatography organic carbon detection (LC-OCD)

Water samples collected from water treatment plants were analysed using liquid chromatography - organic carbon detection (LC-OCD) at DOC-Labor (Karlsruhe, Germany). In this technique, concentrations of biopolymers, humic substances and other low molecular weight (LMW) organic substances were measured in terms of organic carbon based on size exclusion chromatography and data processing using a customised software program CHROMCalc (DOC-Labor, Karlsruhe; Huber *et al.*, 2011). LC-OCD analyses were performed without pre-filtration to include the larger TEP fraction (>0.4μm). Based on the pore size of the sinter filters in the SEC column, the theoretical maximum size limit of organic substances that can be analysed without sample pre-filtration is 2 μm (S. Huber, *per. com.*).

6.2.6 Modified fouling index – ultrafiltration (MFI-UF)

The modified fouling index (MFI) was developed by Schippers and Verdouw (1980) to measure the membrane fouling potential of water based on the cake filtration fouling mechanism. Initially, MFI was measured using membranes with 0.45 or 0.05 μm pore sizes and at constant pressure. However, it was later found that particles smaller than the pore size of these membranes most likely play a dominant role in particulate fouling. In addition, it became clear that the predictive value of MFI measured at constant pressure was limited. For these reasons, MFI test measured at constant flux (with ultra-filtration membranes) was eventually developed (Boerlage *et al*, 2004; Salinas-Rodriguez *et al.*, 2012). In this study, the latest MFI-UF constant flux method was applied to measure the membrane fouling potential of water samples from 5 water treatment facilities.

Assuming that cake or gel filtration is the dominant fouling mechanism during filtration of surface waters, the change in total resistance is directly proportional to the increase in total mass of the deposited foulants/cake on the membrane surface. So during constant flux (J_p) filtration at any time (t), cake resistance (R_c) is dependent to the specific cake resistance (α) and the bulk foulant concentration (C_b).

$$R_c = \alpha\, C_b J_P\, t \qquad\qquad\qquad\qquad\qquad\qquad \text{Eq. 1}$$

where α is a function of C_b and the fouling index (I):

$$\alpha = \frac{I}{C_b} \qquad\qquad\qquad\qquad\qquad\qquad\qquad \text{Eq. 2}$$

Substituting Eq. 2 to the resistance in series model:

$$J_P = \frac{\Delta P}{\eta\,(R_m + R_c)} = \frac{\Delta P}{\eta\,[R_m + (I J t)]}$$

$$\Delta P = J_p\, \eta\, R_m + J_P^2\, \eta\, I\, t \qquad\qquad\qquad\qquad \text{Eq. 3}$$

where ΔP is the total pressure drop across the membrane, η is the dynamic water viscosity and R_m is the membrane resistance.

The fouling index (I) of the feed water is calculated based on the minimum slope of the ΔP versus time plot recorded during constant flux filtration. Consequently, the index was normalised to the standard reference conditions set by Schippers and Verdouw (1980) known as the modified fouling index (MFI):

$$MFI = \frac{\eta_{20^\circ C}\, I}{2\, \Delta P_0 A_0^2} \qquad\qquad\qquad\qquad\qquad \text{Eq. 4}$$

where $\eta_{20^\circ C}$ is the water viscosity at 20°C (0.001003 Pa.s), ΔP_0 is the standard feed pressure (2 bar) and A_0 is the standard membrane area (13.8 x10^{-4} m^2).

To better assess the fouling potential of both particulate and colloidal material (including TEP$_{10kDa}$) in water, low MWCO UF membranes (10 kDa RC, Millipore) were used instead of the conventional 100 kDa UF membranes. Sample filtration was set at 60 L/m^2.h flux using a syringe pump (Harvard Pump 33) and the change in trans-membrane pressure (ΔP) is monitored over time for up to 60 mins. Finally, the recorded ΔP is plotted against time and the fouling index (I) is then calculated based on the minimum slope.

6.2.7 RO membrane autopsy

The lead RO element from a first stage pressure vessel (PV) in Plant C and the lead and last elements (1st stage PV) of Plant D were taken from the RO systems. The membrane elements were packed in plastic bags and stored at 4°C. Membrane autopsies were performed within 4 days after the membrane elements were removed from the RO plant. Each spiral wound element was opened lengthwise by cutting the glass fibre casing and then carefully unrolled for sampling. Several sections of the membrane and spacer were cut-off along the length of the leaf of the spiral wound membrane with relatively intact deposits. Some of the samples were prepared in squares of 2 cm width, stained with Alcian blue and then rinsed with demineralised water. The stained samples were placed on glass slides, viewed and photographed using an Olympus BX51 microscope (magnification=200x). More samples (membrane + spacer) were cut into 3 x 3 cm squares, submerged in 50 ml of UPW water and sonicated (Branson 2510E-MT) for 1 hour to extract accumulated organic materials. The extracted solutions were analysed within 24 hours for $TEP_{0.4\mu m}$ and biopolymer concentrations (using LC-OCD). Moreover, additional samples were sent to Het Waterlaboratorium (Haarlem, Netherlands) to measure adenosine triphosphate (ATP) based on the protocol developed by van der Wielen and van der Kooij (2010).

Scanning electron microscopy (SEM) analyses of autopsied RO membranes was performed at the Wetsus Laboratory (Leeuwarden, Netherlands). For this analysis, 3 cm x 3 cm membrane and spacer samples from Plant D were cut from the middle part of a leaf from each element. Samples were then submerged in artificial seawater and stored at 4°C for 2 days before biological preparation was performed. Biological preparation (fixation) of samples was needed before SEM imaging to prevent total dehydration of protein substances and prevent collapse of bacterial cells during the drying process. The fixation process included soaking the samples in glutaraldehyde (Sigma-Aldrich, 3802) solution followed by ethanol dehydration. The glutaraldehyde serves as a cross-link between protein molecules to keep them intact after dehydration. The treated membrane samples were mounted on metal stubs with double sided tape, then coated with gold and analysed by SEM (Jeol/EO JSM 6480). Coating the samples with gold was performed using a low vacuum gold sputter which allows a very thin layer of gold to form on the sample surface. The gold coating prevents the accumulation of static electric charge on the biological sample during electron irradiation.

6.3 Results and discussion

6.3.1 TEP in different water sources

The locations of the 5 water treatment facilities investigated in this study and their corresponding water sources are presented in Figure 6-2. Plant A is located in Scotland (UK) while Plants B, C, D and E are located in the Netherlands.

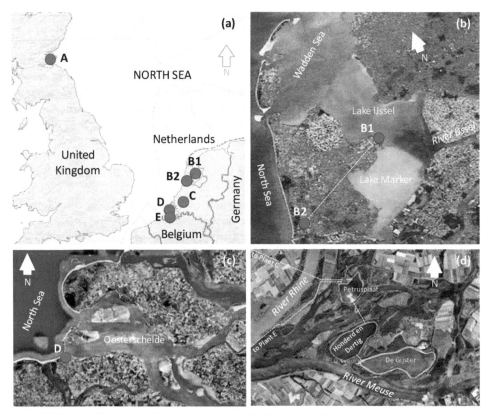

Figure 6-2: (a) Location map of 5 water treatment facilities investigated in this study; (b) Lake Ijssel during an algal bloom in May 30, 2009 and location of Plant B; (c) satellite image of the Oosterschelde area and location of Plant D; (d) satellite image of the 3 source reservoirs of Plant C and E. Map/image sources: (a) U. Dedering/www.wikipedia.org; (b) www.eosnap.com; (c,d) maps.google.com.

The raw water for Plant A is abstracted from the River Leven which flows from Loch Leven, the largest naturally eutrophic freshwater lake in Scotland. The lake has been documented in several papers and reports to have suffered periodic cyanobacterial blooms since the 1960s with a yearly average chlorophyll-a concentration of ~39 µg/L (Carvalho and Kirika, 2005). The blooms were mainly associated with high influx of phosphorus entering the lake and relatively low flushing rate (May *et al.*, 2012). Samples were collected in mid-October. Historical records show that both diatom and cyanobacterial blooms is prevalent in the lake during early autumn

(Carvalho and Kirika, 2005). $TEP_{0.4\mu m}$ and TEP_{10kDa} concentrations of the raw water sample were 0.07 and 0.71 mg X_{eq}/L, respectively.

Plant B represents two separate but interconnected water treatment facilities in North Holland. Plant B1 comprises the conventional water treatment system, the product water of which is transported (via a pipeline) for about 50 km to Plant B2 for further treatment with UF and then RO (Figure 6-2b). The raw water intake for Plant B is located near the west coast of Lake Ijssel. This relatively shallow lake (5.5 m average depth) is the largest freshwater body in the Netherlands and is also known to have periodic algal blooms especially during summer seasons. Intake water sample were collected in June 2012 when chlorophyll-a and cell concentrations were 16 µg/L and 8,200 cells/ml, respectively. This can be considered as a "mild" bloom, considering that peak cell concentrations of 30,000-40,000 cells/ml were observed in the source water (Lake Ijssel) over the previous 2 years (Figure 6-3). $TEP_{0.4\mu m}$ and TEP_{10kDa} concentrations of the raw water sample (June, 2012) were 0.05 and 1.12 mg X_{eq}/L, respectively.

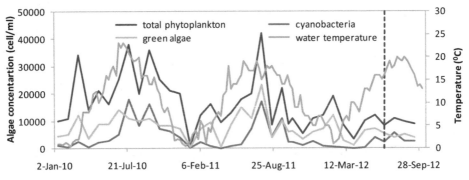

Figure 6-3: Historical data of micro-algae concentration and water temperature in Lake Ijssel (Data provided by PWN). Broken line indicates the date when samples were collected for this study.

The raw water of both Plant C and E are transported via pipeline from the Petrusplaat reservoir. Plant C and E are located about 8 and 110 km from Petrusplaat, respectively. The Petrusplaat is at the downstream end of a series of 3 storage reservoirs for water diverted from the River Meuse (Figure 6-2d). Cholorophyll-a, TOC and $TEP_{0.4\mu m}$ concentrations were monitored at the River Meuse intake and the 3 reservoirs (i.e., De Gijster, Honderd en Dertig and Petrusplaat) between February and December 2009 (Figure 6-4b,c,d,e). Although the 4 samples were collected from 4 interconnected water bodies, chlorophyll-a concentrations were substantially different. These large variations are expected because of the long water retention time of each of these reservoirs, which is estimated to be between 5-13 weeks.

$TEP_{0.4\mu m}$ concentrations also vary among the 3 reservoirs and the River Meuse but consistently showed maximum concentrations in mid-spring (April). The spike in $TEP_{0.4\mu m}$ concentrations coincided with peaks in chlorophyll-a concentration during this period. More peaks in chlorophyll-a were observed during the summer period (especially for Petrusplaat; Figure 6-4e) but $TEP_{0.4\mu m}$ concentrations were significantly lower than during the spring bloom. The difference in $TEP_{0.4\mu m}$ production during the spring and summer period might be attributed to the varying ability of dominant algal species to release TEP and/or the different governing conditions (e.g., temperature, nutrient stress, etc.) that drives such release. The spring algal bloom in the Petrusplaat reservoir was dominated by *Cryptophyceae* (flagellated algae) and some species of diatoms while the summer bloom was dominated by diatoms and colony forming green algae (*Pandorina*). TOC concentrations in the River Meuse and the 3 reservoirs showed comparable trends, all showing a rapid drop in concentrations (1.5-2.3 mg C/L) in early spring just before the onset of spring bloom. TOC concentrations then rapidly recovered during the bloom and fluctuated between 2.6 and 4.5 mg C/L for the rest of the year.

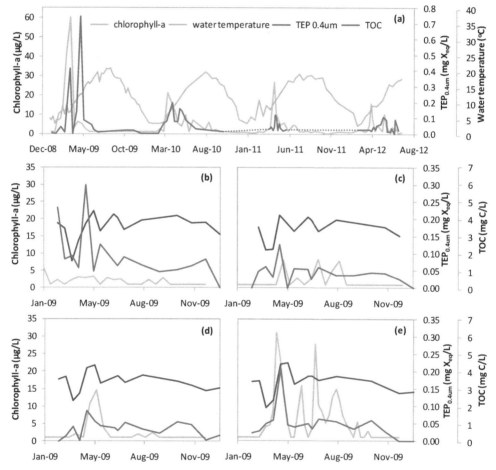

Figure 6-4: Variations of $TEP_{0.4\mu m}$ and other related parameters in the source waters of Plant C, D and E: (a) intake water of Plant D in 2009-2012, (b) River Meuse intake in 2009, (c) De Gijster reservoir in 2009, (d) Honderd en Dertig reservoir in 2009 and (e) Petrusplaat reservoir in 2009. Chlorophyll-a and temperature data provided by Evides. TEP results in year 2009 were measured using the original method of Passow and Alldredge (1995) but were salinity corrected by recalculating the results using blanks with similar salinity with the samples instead of ultra-pure water (see Chapter 5, Figure 5-7).

Plant D is a seawater UF-RO pilot plant. The raw water is pumped to the plant through an open intake of about 4 meters below the seawater surface and about 200 meters from the coastline. The intake point is near the entrance of a shallow bay (Oosterschelde) which is largely affected by the tidal movements in the open North Sea (Figure 6-2C). Water temperature at the intake can be as low as 2.5°C in winter and as high as 21°C in summer.

For more than 3 years (February 2009–July 2012), various samples were collected from the raw water of Plant D to monitor chlorophyll-a and $TEP_{0.4\mu m}$ concentrations. Both chlorophyll-a and $TEP_{0.4\mu m}$ concentrations varied substantially through the years (Figure 6-4a). Both parameters were typically high during the spring season (March-May) only and remain substantially low for the rest of the year, suggesting that the measured $TEP_{0.4\mu m}$ were mainly of algal origin.

The highest chlorophyll-a (60 µg/L) and $TEP_{0.4\mu m}$ (0.74 mg X_{eq}/L) concentrations in Plant D were recorded in spring of 2009. The 2009 bloom was considered severe for this source water as the chlorophyll-a concentrations of the succeeding 3 years were at least 2.3 times lower (<26 µg/L). The 2009 spring bloom was dominated by *Thalassiosira* (diatom) and *Pheaeocystis* (foam-forming algae) while *Pheaeocystis* and *Chaetoceros* (diatoms) dominated in 2010, *Chrysochromulina* (flagellated algae) in 2011 and *Spermatozopsis* (green algae) and *Rhodomonas* (flagellated algae) in 2012. TEP concentrations seem to be higher when diatoms and/or *Pheaeocystis* are the dominant bloom-forming species in the source water. Several diatoms (e.g., *Chaetoceros*) have been considered as important producers of TEP in seawater (Passow, 2002a). *Pheaeocystis* is a unicellular, photosynthetic alga known for their ability to form floating colonies with hundreds of cells embedded in a polysaccharide gel matrix which can multiply massively during blooms. During the decline of the bloom and as the wind sweeps them up to the sea shore, the decaying remains of these colonies eventually formed thick foams (Figure 6-5a). These colonies/flocs were observed to contain substantial amounts of TEP (Figure 6-5b).

Figure 6-5: (a) Foam accumulation near the intake structure of Plant D during the *Pheaeocystis* bloom in May 2010 (Photo: A. Alhadidi); (b) Alcian blue stained algal flocs observed in the raw water sample collected during the bloom period.

Among the 5 plants sampled, Plant D (seawater source) had the highest $TEP_{0.4\mu m}$ concentration (0.74 mg X_{eq}/L) while Plant B (lakewater source) recorded the highest

TEP_{10kDa} (1.12 mg X_{eq}/L) concentration (Table 6-3). However, TEP_{10kDa} were not measured before the year 2012 which means concentrations during the severe 2009 bloom period could not be included in this comparison. TEP concentrations measured in the raw waters of Plants C and D were substantially different (Table 6-3) despite originating from the same source (Figure 6-2d). This may be attributed to variations on the date/time of sampling and/or due to some changes/processes (e.g. TEP attachment/detachment from pipe wall) occurring during pipeline transport of the raw water (8km to Plant C and 110km to Plant E). In general, $TEP_{0.4\mu m}$ concentrations observed in this study are comparable to what was reported in fresh surface water (0.04–9 mg X_{eq}/L; Kennedy et al., 2009; de Vicente et al., 2010; van Nevel et al., 2012) and seawater (0.05-11 mg X_{eq}/L; Passow, 2002a). TEP_{10kDa} concentrations were within the range (0.2-22 mg X_{eq}/L) reported by Arruda-Fatibello et al. (2004) and Thornton et al. (2007). Remarkably, 75-98% of TEP measured in this study were in the colloidal size range based on the difference between $TEP_{0.4\mu m}$ and TEP_{10kDa}. This further suggests the importance of measuring the colloidal fraction as they are likely abundant in surface water sources, and with their smaller size, might be more persistent in terms of removal by pre-treatment and more detrimental to membrane filtration.

Table 6-3: Summary of raw water parameters of water treatment facilities investigated in this study.

Plant code	Sampling period	Chlorophyll-a µg/L	TEP (mg X_{eq}/L)		Biopolymers mg C/L	TOC mg C/L	MFI-UF s/L^2
			$TEP_{0.4\mu m}$	TEP_{10kDa}			
A	Oct 12	n.m.	0.07	0.71	0.90	6.36	13900
B	Jun 12	16	0.05	1.12	0.66	5.82	22600
C	Apr 12-May 12	<2 – 18[a]	0.02-0.03	0.19	0.13	3.65	9900
D	Feb 09-May 13	<2 -60	<0.01-0.74	0.43-1.49[b]	0.06-0.48	1.35-2.0	420-16000[c]
E	Apr 12-May 12	<2 – 18[a]	0.01-0.11	0.73	0.27	3.69	18350

[a] Chlorophyll-a measured at the source water reservoir (Petrusplaat), 8 km and 110 km from Plant C and E, respectively.
[b] TEP_{10kDa} measured is samples collected between May 2012 and May 2013 only.
[c] MFI-UF data are based on measurements of 2012 and 2013 samples only.

The organic carbon concentrations in terms of TOC and biopolymers were in the range of 1.35-6.36 mg C/L and 0.06-0.90 mg C/L, respectively. The biopolymer concentration (measured by LC-OCD) observed in Plant A is one of the highest biopolymer concentration ever reported in a surface water RO plant. Although biopolymers only represent 5-25% of total organic carbon, they have been reported to be major foulant in various membrane-based water treatment processes (Amy, 2008). However, biopolymer measurements using LC-OCD are limited to a size range smaller than 2 µm (when analysed without pre-filtration) or in most cases smaller than 0.45 µm (with pre-filtration). This means that a substantial fraction of TEPs, which can be as large as 100s of µm, were not covered in LC-OCD measurements. Moreover, TEP composition of these biopolymers is still unknown and may vary substantially in different water sources. Hence, a high biopolymer concentration does not necessarily mean high TEP concentration.

The recorded fouling potential of the source waters based on MFI-UF measurements was highest for Plant B followed by Plants E, A, C and D. The MFI-UF was measured with a 10 kDa membrane, which means that it is expected to cover the effect of both $TEP_{0.4\mu m}$ and TEP_{10kDa}. The fouling potential observed in this study are between 420 and 22600 s/L^2, some of which were higher than what was reported (100-12500 s/L^2) by Salinas-Rodriguez (2011) in brackish and seawater when measured at flux rates between 50 and 150 $L/m^2/h$.

6.3.2 Removal by various treatment processes

Water samples were collected in 5 water treatment facilities in 2012 and 2013 to monitor the fate of TEP over treatment processes using the improved measurement method of $TEP_{0.4\mu m}$ and the newly developed TEP_{10kDa}. Other relevant parameters such as MFI-UF, TOC, biopolymer and humics concentrations were also measured for comparison.

6.3.2.1 $TEP_{0.4\mu m}$ and TEP_{10kDa}

To monitor the fate of TEP through different treatment processes, $TEP_{0.4\mu m}$ and TEP_{10kDa} concentrations were measured over the pre-treatment processes of 5 water treatment installations between April 2012 and May 2013. Results are presented in Figure 6-6.

Filtration through microstrainers (30-100µm) in 4 plants resulted in average reductions of 47% (5-87%) of $TEP_{0.4\mu m}$ and 20% (8-48%) of TEP_{10kDa}. The substantial reduction in $TEP_{0.4\mu m}$ by microstrainers indicates the presence of TEP material larger than 30-100 µm in the raw water. This is expected considering TEP larger than 200 µm has been reported during algal bloom (Passow 2002a). It is also possible that the high removal might be due to possible shearing (grinding) of large TEP flocs in microstrainers to such a size that a significant part of the material can pass through membranes (0.4 µm and 10 kDa) used for TEP measurement.

Conventional coagulation (coagulation-flocculation) followed by flotation or sedimentation + RSF was found to be effective in reducing the concentration of TEP to below or close to the detection limit (TEP_{10kDa}<0.15 mg X_{eq}/L; $TEP_{0.4\mu m}$<0.01 mg X_{eq}/L). The mechanism of TEP removal by conventional coagulation can be attributed to charge neutralisation of negatively charged TEP by the cationic coagulant metal species (e.g. Fe^{3+}, Al^{3+}) and/or sweep flocculation of TEP by the large aggregates of hydroxide precipitates of the inorganic coagulant (e.g., $Al(OH)_3$, $Fe(OH)_3$). In the case of Plants A and B, sweep flocculation is likely the main mechanism, considering the rather high coagulant doses (20-30 mg/L). In Plant E, the lower coagulant dosage (4 mg Fe^{3+}/L) might have been compensated by the flotation step which may have further enhanced removal of un-coagulated TEP by bubble scavenging (Mopper et al., 1995; Zhou et al., 1998). Azetsu-Scott and Passow (2004) demonstrated that particle-free TEP gels have the tendency to float and accumulate on the surface of the water column due to its low specific gravity (~0.86). Furthermore, the lower TOC concentration in the raw water of Plant E (~3.5 mg C/L) compared to Plants A and B (~6 mg C/L) might be the reason for the better efficiency of the coagulation step, resulting in high TEP removal despite lower coagulant dosage.

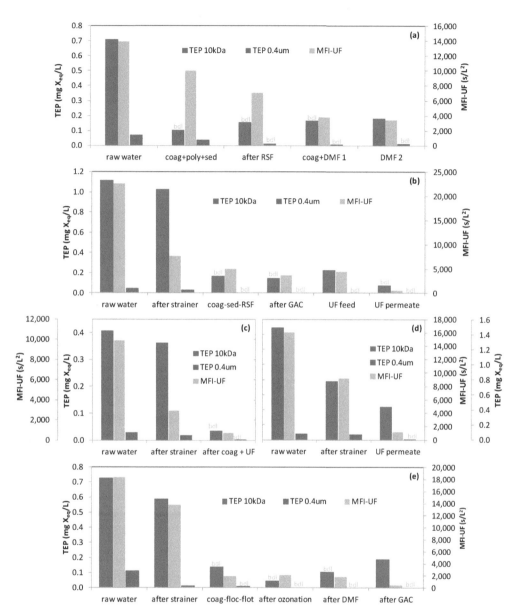

Figure 6-6: TEP concentrations and membrane fouling potentials (MFI-UF) of water samples collected along the treatment processes of 5 plants: (a) Plant A, (b) Plant B, (c) Plant C, (d) Plant D and (e) Plant E. Abbreviations: coag = coagulation; floc = flocculation, sed = sedimentation, RSF = rapid sand filtration, DMF = dual media filtration, GAC = granular activated carbon filtration, UF = ultrafiltration, flot = flotation, bdl = below detection limit.

In general, most of the TEP-coagulant flocs were large enough to be effectively separated from the water phase by flotation or sedimentation + rapid sand filtration. The addition of cationic polymers (5 mg/L polyacrylamide) in Plant A may have an insignificant effect on the TEP flocculation process considering TEP removals were similar in Plant B and E where no cationic polymers were added and coagulant doses were much lower. Nevertheless, the nature of the TEP particles can be different in the different sources, which might have a major impact on the removal efficiency.

The UF systems in Plants B, C and D were also able to reduce the concentrations of $TEP_{0.4\mu m}$ down to below the detection limit (<0.01 mg X_{eq}/L). TEP_{10kDa} concentrations in UF permeate of Plants B and C were also below the detection limit (<0.15 mg X_{eq}/L). However, permeate concentration in Plant D was above the detection limit (~0.4 mg X_{eq}/L). Considering that the UF systems in the three plants are using a similar type of membrane (Pentair X-flow Xiga UF, 150 kDa MWCO), the substantial difference in TEP_{10kDa} removal might be attributed to the difference in size distribution of TEP in the feedwater. TEPs are highly flexible and may pass through pores smaller that their hydrodynamic size, which means that the size of TEPs measured in the UF permeate were not necessarily smaller than the nominal pore size of the UF membrane (~0.03 μm). On the other hand, soon after the start of a filtration cycle, gels start to accumulate narrowing the pore channels and a higher TEP rejection can be expected. The lower rejection in Plant D might be largely due to the effect of high salinity in the water during TEP measurement. The effect of the coagulation step on the TEP removal efficiency of UF could not be properly assessed in the current study and further studies are still needed.

In Plant A and E, the dual (granular) media filters (DMF) with and without inline coagulation did not show any apparent TEP removal. However, such assessment may not be accurate because $TEP_{0.4\mu m}$ and TEP_{10kDa} concentrations in the DMF feedwater were either below or just slightly above the detection limit. Proper assessment on the effect of ozonation on TEP was also not possible for similar reason.

A substantial increase in TEP_{10kDa} concentration was observed after the GAC filter in Plant E. Since GAC can serve as biological filters, the increase may be due to the release of TEPs from biofilm accumulating on the GAC media.

In general, the TEP concentrations in the influent and effluent of the different treatment processes reported in this section might have been underestimated due to TEP loss during storage of samples. Some water samples were stored (at 4°C) between 2 to 15 days before analysis. However, it was discovered later that TEP concentration can reduce substantially during storage (see Chapter 5, Figure 5-20). Consequently, the TEP removal efficiencies illustrated in Figure 6-6 might have been underestimated assuming TEP loss in the influent and effluent samples are proportional. The removal percentage can be also underestimated if the percentage of TEP lost in the influent sample is higher than the effluent sample while removal can be overestimated if the percentage TEP lost in the influent sample is lower than in the effluent sample. Moreover, various effluent samples were detected below the detection limit which means the removal efficiencies cannot be measured precisely (in terms of %). Hence, further investigations are needed to measure TEP removal of different pre-treatment processes without prolonged sample storage and at a much higher TEP concentrations in the feed water, such as in severe algal blooms.

6.3.2.2 Membrane fouling potential (MFI-UF)

Membrane fouling potential of the water in terms of MFI-UF was monitored over the treatment processes of 5 water treatment facilities. The comparison of MFI-UF values for each plant is presented in Figure 5-6. An overview of the change in MFI-UF over the different treatment processes is shown in Table 6-4.

Table 6-4: Change in MFI-UF values over the different treatment processes of 5 plants.

Treatment processes	Range	Average reduction
Micro-straining	25 -71	51
Coagulation - sedimentation - rapid sand filtration	34 - 50	42
Coagulation - flotation	86	86
(Inline coagulation) - UF	76 -87	81
(Coagulation) - dual media filtration	11-46	23
Granular activated carbon (GAC) filtration	28-77	53
Ozonation	-	-

Pre-treatment with microstrainers (30-100 µm) can led to substantial reduction of MFI-UF. Such reductions were between 25% and 71% in the 4 plants applying these strainers. This can be attributed to the removal of large flocs of organic/inorganic materials (including particulate TEPs) from the raw water. However, it cannot be excluded that the microstrainer may have caused grinding of some organic material (e.g., TEP) to smaller particles which could then pass through the pores of 10 kDa UF membrane used in MFI-UF measurements.

Between 34-50% reduction of MFI-UF was observed for coagulation-flocculation followed by sedimentation and RSF (Plants A and B). However, a substantially higher reduction (86%) was observed when coagulation-flocculation was followed by flotation (Plant E). The corresponding MFI-UF values in the product waters of coagulation-sedimentation-RSF processes were much higher for Plant A (7000 L/s^2) and Plant B (5000 L/s^2) than after coagulation-flotation process in Plant E (1900 L/s^2). This is despite the fact that the corresponding TEP$_{10kDa}$ concentrations in the product water were relatively similar for the 3 plants (0.14-0.17 mg X$_{eq}$/L).

The possible reason for the large difference in MFI-UF reduction could be due to differences in organic carbon concentration in the source waters and/or the specific membrane fouling potential of the TEP coming from the source waters or colloids of other nature. The effect of other biopolymers and biopolymer-metal complexes may have contributed as well. For example, formation of TEP-iron or TEP-aluminium complexes during flocculation at high coagulant doses (as in the case of Plant A and B). These TEP complexes might not react fully with Alcian blue during TEP measurement and will not be measured as such. Moreover, these neutral complexes might have lower inter-particle repulsion forces thereby forming a more compact cake with high filtration resistance during MFI-UF measurements. Further investigations are necessary to confirm such hypothesis.

UF removed 88% of membrane fouling potential in Plant B, 76% in Plant C and 87% in Plant D. The MFI-UF of the product waters of the 3 UF systems were between 520 and 1200 s/L². This was comparable to what was reported in a similar study by Salinas-Rodriguez (2011).

The first DMF in Plant A with inline coagulation reduced MFI-UF by 46% while the second DMF without inline coagulation removed only 11%. The latter was comparable to the DMF performance of Plant E. MFI-UF reductions through GAC filtration were 28% in Plant B and 77% in Plant E. No apparent change in MFI-UF was observed after ozone treatment (Plant E). However, it cannot be excluded that the removal efficiency reported above was affected by the preceding treatment step(s).

In general, the reported MFI-UF values in the influent and effluent of the different treatment processes might have been lower than the actual values due to TEP loss during storage of samples (2-15 days). Hence, it is likely that the actual MFI-UF reduction after the treatment process is higher than what is reported in this section.

6.3.2.3 Organic carbon

Algae and bacteria release diverse forms of organic matter including various types of biopolymers (e.g., TEPs, proteins, neutral polysaccharides, etc.) and low molecular weight organic substances. To investigate the removal of these materials, LC-OCD analyses were performed on the same set of water samples used in TEP and MFI-UF monitoring.

The typical variations of LC-OCD-OND chromatograms over the treatment processes are presented in Figure 6-7. In this figure, the raw water is from Lake Ijssel where TOC concentration was about 6 mg C/L and the specific UV absorbance was 2.7 L/mg.m. The organic matter in the raw water comprised biopolymers, humic substances, building blocks, low molecular weight (LMW) neutrals and LMW acids. The organic nitrogen detector (OND) indicates significant concentrations of nitrates, reflected in the OND chromatogram as double peaks at retention times between 55-85 mins (Huber et al., 2011). The organic N to organic C ratio of the biopolymer fraction was rather low (1:20). Considering that protein substances contain organic nitrogen but not polysaccharides, a low C:N ratio indicates that polysaccharides (e.g., TEPs) are the dominant types of biopolymers present in the water.

The chromatogram series in Figure 6-7 also represents the changes in concentration of the different fractions of natural organic matter. A decrease in signal means removal by the corresponding treatment processes. There was no apparent change in chromatograms after the raw water passed through the microstrainer. The coagulation-flocculation-sedimentation-RSF processes removed a substantial amounts of OC signals from both biopolymers and LMW organic substances (e.g., humics), as well as the ON signals from nitrates. GAC filtration did not remove substantial OC or ON signals. After the water was transported via pipeline for 50 km., a slight increase of biopolymers was observed. The latter was largely removed after UF. Finally, the RO system removed almost all of the OC and ON signals.

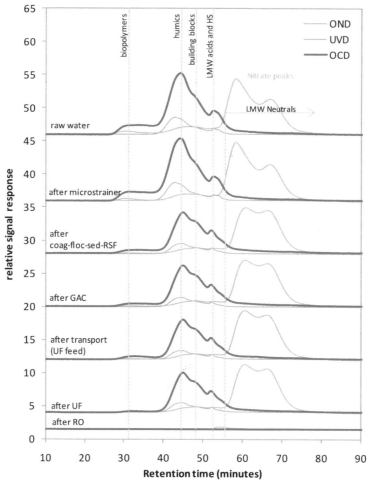

Figure 6-7: LC-OCD-OND chromatograms of water samples collected over the treatment processes of Plant B. Retention time is an indication of the molecular weight (MW) of the organic matter fraction – the higher the retention time, the smaller the MW of the fraction. The chromatograms are presented with offsets from the original signal values for clarity and better comparison.

Figure 6-8 shows concentrations of TOC, biopolymers and humics over the treatment processes of the 5 plants. The microstrainers in the 4 plants investigated recorded less than 6% removals of TOC and humics while 4-22% of biopolymers were removed. Coagulation followed by sedimentation and then RSF recorded average removals of 39%, 46% and 64% of TOC, humics and biopolymers, respectively. On other hand, coagulation followed by flotation removed 31% TOC, 33% humics and >71% biopolymers (Table 6-5).

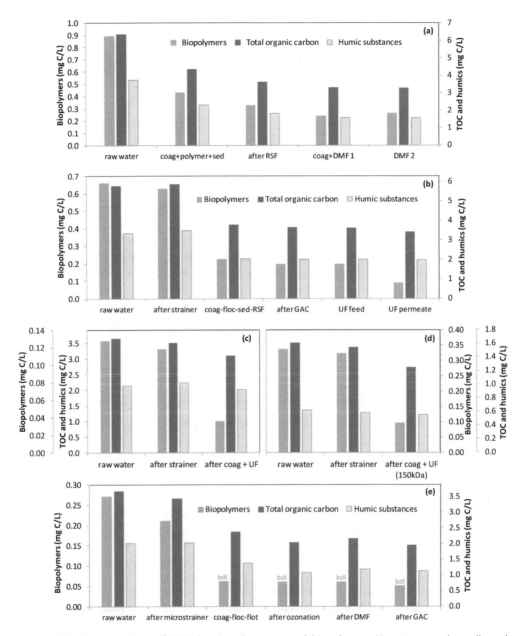

Figure 6-8: Concentrations of TOC, humic substances and biopolymers in water samples collected along the treatment processes of 5 plants: (a) Plant A, (b) Plant B, (c) Plant C, (d) Plant D and (e) Plant E. Abbreviations: coag = coagulation; floc = flocculation, sed = sedimentation, RSF = rapid sand filtration, DMF = dual media filtration, GAC = granular activated carbon filtration, UF = ultrafiltration, flot = flotation, bdl = below detection limit (for biopolymers = 0.075 mg C/L).

Table 6-5: Summary of relative reduction (%) in TOC, humic substances and biopolymer concentrations over the various treatment processes.

Treatment processes	TOC	Humics	Biopolymers
Micro-straining	0 - 6	0 - 5	4 - 22
Coagulation - sedimentation - rapid sand filter	35 - 43	41 - 51	63 - 64
Coagulation - flotation	31	33	>71
(Inline coagulation) - ultrafiltration	11 - 19	2 - 9	55 - 70
(Coagulation) - dual media filtration	(-6) - 2	(-9) - 0	(-10) - 0
Granular activated carbon filtration	4 - 10	1 - 5	12 - 15
Ozonation	14	21	bdl

The average removal of the UF system of 3 plants (Plants B, C and D) was 13%, 5% and 63% of TOC, humics and biopolymers, respectively. DMF treatment in Plants A and E showed positive and negative removals possibly due to release of organic materials from the filter media. Average removals of GAC filters in Plants B and E were 7% of TOC, 3% of humics and >12% of biopolymers. Ozone treatment in Plant E resulted in 21% reduction of humics possibly due to oxidation.

Since some samples were stored (2-15 days) before LC-OCD analyses, the reported organic carbon concentration might be lower than the actual concentration. However, a recent investigation by Tabatabai (2013) showed that TOC and biopolymer concentrations of algal organic matter stored at 4°C were stable for at least 6 months. This may indicate that the effect of storage was not substantial on the LC-OCD results and that the removal efficiencies reported in this section are reliable.

6.3.2.4 Statistical relationship

The results of the 6 parameters measured in water samples collected from the 5 plants were consolidated. Linear regression analyses were performed to evaluate the relationship between the membrane fouling potential (in terms of MFI-UF) and concentrations of $TEP_{0.4\mu m}$, TEP_{10kDa} and biopolymers. Among the 3 parameters, TEP_{10kDa} showed the highest linear correlation ($R^2=0.67$) with MFI-UF followed by biopolymers and $TEP_{0.4\mu m}$ (Figure 6-9). Considering that TEP_{10kDa} has a better relationship with MFI-UF than with biopolymers may indicate that TEP_{10kDa} has a substantial impact on MFI-UF results and can likely cause more (serious) fouling in membrane systems than other forms of biopolymers. Nevertheless, the observed correlation is not that strong possibly due to the influence of the different treatment processes on the samples analysed, such as the effect of coagulant, TEP oxidation during ozonation and released of bacterial-derived TEP materials from DMF and GAC media.

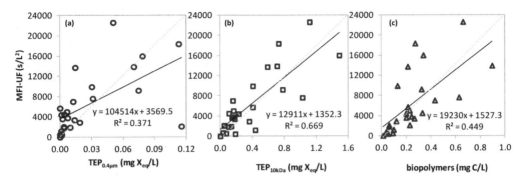

Figure 6-9: Scatter plots between membrane fouling potential in relation to (a) $TEP_{0.4\mu m}$, (b) TEP_{10kDa} and (C) biopolymers. Solid lines are the best fit linear curve of the plot while broken lines represent the 45^0 line (R^2=1).

Further regression analyses were performed for different combinations of the 6 parameters. The calculated regression coefficients are summarised in Table 6-6.

Table 6-6: Linear regression coefficient (R^2) matrix of 6 parameters investigated in this study.

Parameters	$TEP_{0.4\mu m}$	TEP_{10kDa}	Biopolymers	Humics	TOC	MFI-UF
$TEP_{0.4\mu m}$	1.000					
TEP_{10kDa}	0.304	1.000				
Biopolymers	0.207	0.381	1.000			
Humics	0.004	0.018	0.089	1.000		
TOC	0.001	0.026	0.138	**0.989**	1.000	
MFI-UF	0.371	**0.669**	0.449	0.036	0.055	1.000

As expected, TOC and humics showed very poor relationship (R^2<0.06) with MFI-UF. Humic substances are rather dissolved (~1000 Da) and were likely not retained on 10kDa membranes during MFI-UF measurements. The very high regression coefficient (R^2=0.99) between humics and TOC was expected considering that majority of TOC (>60%) in the water samples comprise humic substances (Figure 6-8). TEP_{10kDa} correlates better with biopolymers than $TEP_{0.4\mu m}$ possibly due to the wider size range overlap between TEP_{10kDa} (>10kDa) and biopolymers (<2µm). On the other hand, a substantial fraction of $TEP_{0.4\mu m}$ was not measured as biopolymers due to the upper size limit (~2 µm) of the LC-OCD analysis.

6.3.3 TEP accumulation in RO system

Autopsies were performed on membrane elements taken from Plant C and D. Samples of membrane elements from both plants were stained with Alcian blue and observed by using an optical microscope. Elements taken from Plant D were investigated more in detail by using LC-OCD, $TEP_{0.4\mu m}$, TOC, ATP and SEM analyses.

Figure 6-10 shows the results of Alcian blue staining of membrane samples taken from Plant C and D. TEP-like materials were observed in both plants. Some of these substances were found attached around and at the corners of the membrane feed spacer mesh.

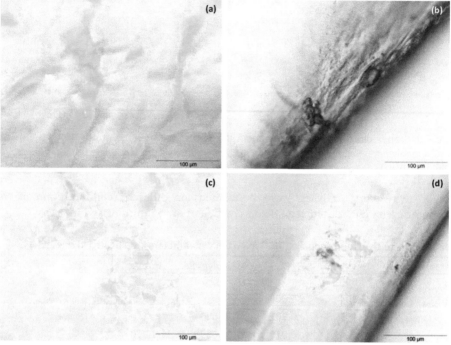

Figure 6-10: Photographs of TEP-like (green to blue) substances observed on RO membrane and spacer samples after staining with Alcian Blue. (a) membrane, first element, Plant C; (b) spacer, first element, Plant C; (c) membrane, first element, Plant D; (d) spacer, first element, Plant D.

Some fibrils and blobs of different shapes and sizes were also found on other fouled membrane samples, mostly exhibited the physical characteristics of TEP. Some of these TEP-like gels appeared to contain other types of particulate matter, resembling a heterogeneous layer of slime. This observation was consistent with what was observed in other RO/NF membrane autopsy studies (Tran *et al.*, 2007; Ning and Troyer, 2007).

Two membrane elements (1st and 6th elements) were taken from Plant D to perform a more detailed autopsy. Both elements were running for about a year without chemical cleaning. Further investigations were conducted on these two RO membrane elements to determine the composition of accumulated organic materials. Based on the LC-OCD chromatogram (Figure 6-11a), the organic carbon composition of extracted organic materials were mainly biopolymers. A low UV signal observed at 27 and 48 minutes retention times might be attributed to the presence of inorganic colloidal materials and traces of aromatic compounds, respectively. Similarly, two organic nitrogen peaks were observed possibly due to the presence of protein compounds (25 mins) and organic nitrogen associated with low molecular weight aromatic compounds. The high C:N ratio (8.2) of biopolymers may indicate the substantial presence of polysaccharides.

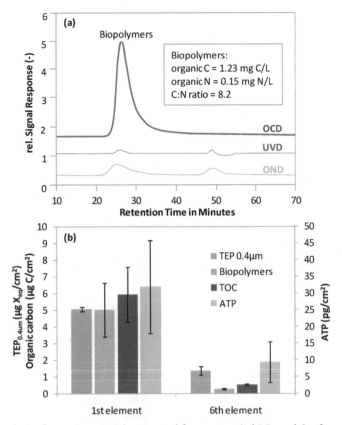

Figure 6-11: Analysis of organic materials extracted from autopsied RO modules from Plant D: (a) LC-OCD-OND chromatogram of samples extracted from the lead RO element; (b) Concentrations of TEP$_{0.4\mu m}$, biopolymers, TOC, and ATP extracted from lead and last RO element.

The accumulation of TEP$_{0.4\mu m}$, TOC, biopolymers and ATP on the 1st and 6th RO elements were measured based on the organic materials extracted from the membrane samples. TEP$_{0.4\mu m}$ accumulation was much higher (3.7 times) in the lead element than in the last element (Figure 6-11b). This was consistent with the accumulation of biopolymers and TOC. LC-OCD analyses indentified biopolymers to comprise 46-89% of TOC in the extracted organic materials. Moreover, the concentration of ATP extracted from the membranes was higher in the lead element (32 pg ATP/cm^2) than in the last element (9 pg ATP/cm^2). The observed ATP accumulations were lower than the estimated biofouling threshold (1000 pg ATP /cm^2) reported by Vrouwenvelder et al. (2008). The RO system before the autopsy did not suffer substantial decline in normalized flux or substantial increase in net pressure drop.

The SEM images (Figure 6-12) of autopsied RO membranes illustrated that the 1st element is more fouled than the 6th element. The 1st element sample was completely covered with foulants, which was relatively homogenous over the membrane area analysed. Several bacterial cells were also found (spheres with arrow in Figure 6-

12a), most of which were embedded in the foulant layer. The foulant accumulation in the 6th element RO membrane was relatively heterogeneous and covered about 50-70% of the membrane surface. In several parts of the sample, the RO membrane surface is still visible. Only sporadic and isolated bacterial cells were found on the 6th element.

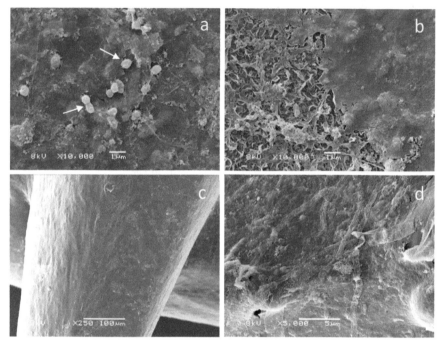

Figure 6-12: SEM images of RO membrane samples from Plant D: (a) foulant accumulation with bacterial cells (spheres) visible on top of the 1st element membrane, (b) 6th element membrane partially covered with foulants, (c) spacer fibre samples from 6th element with white small dots identified as bacterial cells, and (d) high magnification image at the crosslink of the membrane spacer showing accumulation of long fibrous organic materials.

A spacer sample of the 1st element was also analysed using SEM. The spacer was relatively clean compared to the membrane surface. Most of the foulant accumulations were situated in the intersections of the spacer mesh (Figure 6-12c,d). A similar observation was reported by Vrouwenvelder et al. (2009). Moreover, several bacterial cells were also observed attached to the spacer. The SEM images further indicate accumulation of biofilms on RO membrane and spacers, specifically in the 1st element.

Various studies have identified polysaccharide substances as a major component of organic materials found on fouled NF/RO membranes (Park et al., 2006; Karime et al., 2008). However, not all of the polysaccharide substances in fouled RO membranes may have originated from the feedwater as TEPs because biofilm bacteria can produce considerable amount of polysaccharides at different phases of their growth as cell coatings, release or detritus lysed from dying/dead cells (Flemming, 2002; Xu and Chellam, 2005). A recent study by Bar-Zeev et al. (2009)

demonstrated that TEP may accumulate on RO membranes using glass slides. They found that most of the accumulated TEPs during the first few hours of exposure to seawater originated from the RO feed water and were not locally produced by biofilm bacteria. However, on the long term, TEP accumulation may primarily originate from exopolymer release by biofilm bacteria.

The RO membranes autopsied in this study had been exposed to seawater for about a year without chemical cleaning. This means that TEPs from bacterial origin likely contributed considerably to the total accumulation. It might also be possible that TEPs from the RO feedwater may have contributed to the biofilm development by enhancing further attachment of bacteria and other biopolymers in the system gradually over time. The higher biofilm accumulation in the first element is likely driven by the higher membrane flux in the lead elements of an RO pressure vessel.

6.4 Conclusions

1. TEP monitoring in the raw water (seawater and freshwater) of 2 RO plants revealed that high TEP concentrations occurred mainly during the spring algal bloom seasons (March-May).

2. Ultrafiltration (with and without inline coagulation) and conventional coagulation followed by flotation or by sedimentation and rapid sand filtration (RSF) were effective in lowering $TEP_{0.4\mu m}$ and TEP_{10kDa} concentrations down to below the detection limit.

3. Significant reductions in biopolymer concentration (measured by LC-OCD) and membrane fouling potential (in terms of MFI-UF) of the water were also observed after UF, coagulation-sedimentation-RSF and coagulation-flotation treatments.

4. Data collected from 5 plants showed better correlation between MFI-UF and TEP_{10kDa} than between MFI-UF and $TEP_{0.4\mu m}$, biopolymers or humic concentrations, an indication that colloidal TEPs can likely cause more (serious) fouling in membrane systems than other types of organic matter.

5. Membrane autopsies of RO elements taken from 2 plants revealed substantial accumulation of TEPs and bacteria in RO membranes and spacers. Some of these TEPs may have been produced locally by biofilm bacteria.

References

Ahimou, F., Semmens, M. J., Haugstad, G., and Novak, P. J. (2007) Effect of Protein, Polysaccharide, and Oxygen Concentration Profiles on Biofilm Cohesiveness. Applied and Environmental Microbiology 73(9), 2906-2910.

Alldredge, A. L., Passow, U. and Logan, B. E. (1993) The abundance and significance of a class of large, transparent organic particles in the ocean. Deep Sea Research Part I: Oceanographic Research Papers 40(6), 1131-1140.

Amy, G. (2008) Fundamental understanding of organic matter fouling of membranes, Desalination 231, 44-51.

Arruda-Fatibello, S. H. S., Henriques-Vieira, A. A. and Fatibello-Filho, O. (2004) A rapid spectrophotometric method for the determination of transparent exopolymer particles (TEP) in freshwater. Talanta 62(1), 81-85.

Azetsu-Scott, K., and Passow, U. (2004) Ascending marine particles: Significance of transparent exopolymer particles (TEP) in the upper ocean. Limnol. Oceanogr, 49(3), 741-748.

Baker, J. S. and Dudley, L. Y. (1998) Biofouling in membrane systems—A review. Desalination 118(1-3), 81-89.

Bar-Zeev, E., Berman-Frank, I., Liberman, B., Rahav, E., Passow, U. and Berman, T. (2009) Transparent exopolymer particles: Potential agents for organic fouling and biofilm formation in desalination and water treatment plants. Desalination & Water Treatment 3, 136–142.

Berman, T. and Viner-Mozzini, Y. (2001) Abundance and characteristics of polysaccharide and proteinaceous particles in Lake Kinneret. Aquatic Microbial Ecology 24, 256-264.

Berman, T., and Holenberg (2005) M., Don't fall foul of biofilm through high TEP levels. Filtration & Separation 42(4), 30-32.

Boerlage, S. F. E., Kennedy, M., Tarawneh, Z., De Faber, R., & Schippers, J. C. (2004). Development of the MFI-UF in constant flux filtration. Desalination, 161(2), 103-113.

Carvalho L. and Kirika, A. (2005) Loch Leven 2003: physical, chemical and algal aspects of water quality, Report to Scottish Natural Heritage.

de la Torre, T., Lesjean, B., Drews, A., and Kraume, M. (2008) Monitoring of transparent exopolymer particles (TEP) in a membrane bioreactor (MBR) and correlation with other fouling indicators. Water Science and Technology 58 (10), 1903–1909.

de Vicente, I., Ortega-Retuerta, E., Mazuecos, I.P., Pace, M.L., Cole, J.J., Reche, I., 2010. Variation in transparent exopolymer particles in relation to biological and chemical factors in two contrasting lake districts. Aquatic Sciences 72, 443-453.

Flemming, H. C. (2002) Biofouling in water systems–cases, causes and countermeasures. Applied Microbiology and Biotechnology 59, 629-640.

Flemming, H. C., and Wingender, J. (2001) Relevance of microbial extracellular polymeric substances (EPSs)–Part I: Structural and ecological aspects. Water Science and Technology 43(6), 1-8.

Flemming, H. C., Schaule, G., Griebe, T., Schmitt J. and Tamachkiarowa A. (1997) Biofouling--the Achilles heel of membrane processes. Desalination 113(2-3), 216-225.

Huber S. A., Balz A., Abert M. and Pronk W. (2011) Characterisation of aquatic humic and non-humic matter with size-exclusion chromatography – organic carbon detection – organic nitrogen detection (LC-OCD-OND). Water Research 45 (2), 879-885.

Karime, M., Bouguecha, S. and Hamrouni, B. (2008) RO membrane autopsy of Zarzis brackish water desalination plant. Desalination 220(1-3), 258-266.

Kennedy, M. D., Muñoz-Tobar, F. P., Amy, G. L. and Schippers, J. C. (2009) Transparent exopolymer particle (TEP) fouling of ultrafiltration membrane systems. Desalination & Water Treatment 6(1-3) 169-176.

Kruithof, J. C., Schippers, J. C., Kamp, P. C., Folmer, H. C. and Hofman, J. (1998) Integrated multi-objective membrane systems for surface water treatment: pretreatment of reverse osmosis by conventional treatment and ultrafiltration. Desalination 117(1-3), 37-48.

Li, X., and Logan, B. E. (1997) Collision Frequencies between Fractal Aggregates and Small Particles in a Turbulently Sheared Fluid. Environmental Science and Technology 31(4), 1237-1242.

Liberman, B., and Berman, T. (2006) Analysis and monitoring: MSC - a biologically oriented approach. Filtration & Separation 43(4), 39-40.

May, L., Defew, L. H., Bennion, H. and Kirika A. (2012) Historical changes (1905–2005) in external phosphorus loads to Loch Leven, Scotland, UK, Hydrobiologia 681(1), 11-21.

McKee, M. P., Ward, J. E., MacDonald, B. A. and Holohan, B. A. (2005) Production of transparent exopolymer particles by the eastern oyster Crassostrea virginica, Marine Ecology Progress Series 183, 59-71.

Mopper, K., Zhou, J., Sri Ramana, K., Passow, U., Dam, H. G., & Drapeau, D. T. (1995). The role of surface-active carbohydrates in the flocculation of a diatom bloom in a mesocosm. Deep-Sea Research Part II, 42(1), 47-73.

NEN 6520, (1981) Water: Spectrophotometric determination of chlorophyll a content. Nederlandse Norm.

Ning, R. Y. and Troyer, T. L. (2007) Colloidal fouling of RO membranes following MF/UF in the reclamation of municipal wastewater. Desalination 208(1-3), 232-237.

Park, N., Kwon, B., Kim, S. D. and Cho J. (2006) Characterizations of the colloidal and microbial organic matters with respect to membrane foulants. Journal of Membrane Science 275(1-2), 29-36.

Passow, U. (2000) Formation of transparent exopolymer particles (TEP) from dissolved precursor material. Marine Ecology Progress Series 192, 1-11.

Passow, U. (2002a) Transparent exopolymer particles (TEP) in aquatic environments, Progress In Oceanography 55(3-4), 287-333.

Passow, U. (2002b) Production of transparent exopolymer particles (TEP) by phyto-and bacterioplankton. Marine Ecology Progress Series 236, 1-12.

Passow, U. and Alldredge, A. L. (1995) A Dye-Binding Assay for the Spectrophotometric Measurement of Transparent Exopolymer Particles (TEP). Limnology and Oceanography 40(7), 1326-1335.

Passow, U., and Alldredge, A. L. (1994) Distribution, size, and bacterial colonization of transparent exopolymer particles TEP in the ocean. Marine Ecology Progress Series 113(1-2), 185–198.

Salinas-Rodriguez S.G. (2011) Particulate and organic matter fouling of SWRO systems: Characterization, modelling and applications. Ph.D. thesis, UNESCO-IHE/TUDelft, Delft.

Salinas-Rodríguez S.G., Kennedy, M.D., Amy, G.L., & Schippers, J.C. (2012) Flux dependency of particulate/colloidal fouling in seawater reverse osmosis systems. Desalination and Water Treatment, 42(1-3), 156-162.

Santschi, P.H., Balnois, E., Wilkinson, K.J., Zhang, J., Buffle, J. (1998) Fibrillar polysaccharides in marine macromolecular organic matter as imaged by atomic force microscopy and transmission microscopy. Limnology and Oceanography 43(5), 896-908.

Schippers, J.C. and Verdouw, J. (1980) The modified fouling index, a method of determining the fouling characteristics of water. Desalination 32, 137-148.

Schneider, R. P., Ferreira, L. M., Binder, P. and Ramos, J. R. (2005) Analysis of foulant layer in all elements of an RO train. Journal of Membrane Science 261(1-2), 152-162.

Sutherland, I. A. (2001) Biofilm exopolysaccharides: a strong and sticky framework. Microbiology 147, 3-9.

Tabatabai S.A.A. (2013) In-line coagulation for UF membranes. Ph.D. thesis (in preparation), UNESCO-IHE/TUDelft, Delft.

Thornton, D. C. O. (2004) Formation of transparent exopolymeric particles (TEP) from macroalgal detritus. Marine Ecology Progress Series 282, 1-12.

Thornton, D. C. O., Fejes, E. M., DiMarco, S. F. and Clancy, K. M. (2007) Measurement of acid polysaccharides (APS) in marine and freshwater samples using alcian blue. Limnology and Oceanography: Methods 5, 73–87.

Tran, T., Bolto, B., Gray, S., Hoang, M. and Ostarcevic, E. (2007) An autopsy study of a fouled reverse osmosis membrane element used in a brackish water treatment plant. Water Research 41(17), 3916-3923.

van Agtmaal, J., Huiting, H., de Boks, P. A. and Paping, L. (2007) Four years of practical experience with an Integrated Membrane System (IMS) treating estuary water. Desalination 205(1-3) 26-37.

van der Wielen, P. W. J. J., & van der Kooij, D. (2010). Effect of water composition, distance and season on the adenosine triphosphate concentration in unchlorinated drinking water in the netherlands. Water Research, 44(17), 4860-4867.

van Nevel, S., Hennebel, T., De Beuf, K., Du Laing, G., Verstraete, W., & Boon, N. (2012). Transparent exopolymer particle removal in different drinking water production centers. Water Research, 46(11), 3603-3611.

Verdugo, P., Alldredge, A. L., Azam, F., Kirchman, D. L., Passow, U. and Santschi, P. H. (2004) The oceanic gel phase: a bridge in the DOM–POM continuum. Marine Chemistry 92, 67-85.

Villacorte, L. O., Kennedy, M. D., Amy, G. L., and Schippers, J. C. (2009a) The fate of Transparent Exopolymer Particles (TEP) in integrated membrane systems: Removal through pretreatment processes and deposition on reverse osmosis membranes. Water Research 43 (20), 5039-5052.

Villacorte, L. O., Kennedy, M. D., Amy, G. L., and Schippers, J. C. (2009b) Measuring Transparent Exopolymer Particles (TEP) as indicator of the (bio)fouling potential of RO feed water. Desalination & Water Treatment 5 (1-3), 207-212.

Vrouwenvelder, J.S., Graf von der Schulenburg, D.A., Kruithof, J.C., Johns, M.L., van Loosdrecht, M.C.M. (2009). Biofouling of spiral wound nanofiltration and reverse osmosis membranes: A feed spacer problem. Water Research 43, 583-594.

Vrouwenvelder, J.S., Manolarakis, S.A., Van der Hoek, J.P., Van Paassen, J.A.M, Van der Meer, W.G.J., Van Agtmaal, J.M.C., Prummel, H.D.M., Kruithof, J.C., Van Loosdrecht, M.C.M. 2008. Quantitative biofouling diagnosis in full scale nanofiltration and reverse osmosis installations. Water Research 42, 4856-4868.

Xu, W. and Chellam, S. (2005) Initial Stages of Bacterial Fouling during Dead-End Microfiltration. Environmental Science and Technology 39(17), 6470–6476.

Zhou, J., Mopper, K., & Passow, U. (1998). The role of surface-active carbohydrates in the formation of transparent exopolymer particles by bubble adsorption of seawater. Limnology and Oceanography, 43(8), 1860-1871.

Fouling of ultrafiltration membranes by algal biopolymers in seawater

Contents

Abstract 178
7.1 Introduction 179
7.2 Theoretical background 180
    7.2.1 Membrane fouling 180
    7.2.2 Fouling reversibility 181
    7.2.3 Effect of salinity on particle interaction 184
7.3 Experimental methods 184
    7.3.1 Artificial seawater 185
    7.3.2 Algal organic matter 185
    7.3.3 Model biopolymers 186
    7.3.4 Liquid chromatography – organic carbon detection (LC-OCD) 186
    7.3.5 Transparent exopolymer particles (TEP) 186
    7.3.6 Dynamic light scattering measurements 186
    7.3.7 Atomic force microscopy 187
    7.3.8 Filtration experiments 187
7.4 Results and discussion 189
    7.4.1 Characteristics of biopolymers 189
    7.4.2 Membrane fouling potential and effect of solution chemistry 194
    7.4.3 Reversibility of biopolymer fouling by hydraulic backwashing 203
    7.4.4 Calculation of projected chemical cleaning interval 207
7.5 Conclusions 208
References 209

This chapter is a modified version of:

Villacorte L.O., Karna B., Ekowati Y., Kleijn M., Amy, G., Schippers, J.C. and Kennedy, M.D. (2013) Fouling of ultrafiltration membrane by algal biopolymers in seawater. *In preparation for Journal of Membrane Science*.

Abstract

Biopolymers produced by microscopic algae in seawater sources are suspected to be one of the main causes of fouling in ultrafiltration (UF) membranes. This chapter investigates the fouling caused by biopolymers produced by common marine algae (Chaetoceros affinis) at different solution pH and ionic strength. Furthermore, the role of major mono- and di- valent cations in seawater was also investigated by varying the cation composition of the feed water matrix. The UF fouling propensities were measured based on the membrane fouling potential in terms of the modified fouling index (MFI) and the observed non-backwashable fouling rate in an inside-out capillary UF membrane. For comparison, UF fouling caused by model biopolymers (SA: sodium alginate and XG: xanthan gum) were also investigated under similar conditions.

The experimental results show that the fouling potential of biopolymers is largely affected by the ionic strength of the matrix. At high ionic strength (seawater), lowering the solution pH may result in higher fouling potential. It was also demonstrated that Ca²⁺ ions have a major influence on the fouling propensity of biopolymers among the major cations abundant in seawater. In general, the effect of solution chemistry on the fouling behaviour of algal-derived biopolymers is substantially different to what was observed in model polysaccharides (i.e., sodium alginate and gum xanthan) widely used in UF fouling experiments.

7.1 Introduction

Microbial-derived organic material in seawater comprise of heterogeneous mixtures of polysaccharides, proteins, nucleic acids, lipids and various low molecular weight compounds. In seawater, where they are usually associated with an algal bloom event, polysaccharides comprise a major fraction (up to 90%) of this pool (Myklestad, 1995). Most microscopic algae (e.g., diatoms, dinoflagellates, cyanobacteria) produce substantial amounts of acidic polysaccharides under normal conditions but more often in response to low nutrient condition at the height of an algal bloom or due to high metal stress condition (Leppard, 1997). This anionic group of biopolymers has been dubbed by marine scientists as transparent exopolymer particles or TEPs (Alldredge *et al.*, 1993; Passow, 2002).

TEPs have been identified to play a major role in the formation of mucilaginous aggregates in seawater (Alldredge, 1999; Alldredge *et al.*, 1993). They also play a major role in the flocculation of inorganic particles/colloids in aquatic systems through destabilisation and/or complexation with dissolved cations (e.g., Ca^{2+}) and trace metals (Wilkinson *et al.*, 1997; Leppard, 1997; Hung *et al.*, 2003). Due to their high stickiness, TEPs may act as glue for colliding particles, readily forming aggregates with solid particles of various sizes such as bacteria, phytoplankton, mineral clays, or detritus, which may eventually lead to sedimentation (Passow, 2002; Dam and Drapeau 1995; Logan *et al.* 1995; Engel 2000). On the other hand, TEPs are generally lighter than water (700-840 kg/m^3) and thus, it can retard sedimentation of aggregates (Azetsu-Scott and Passow, 2004).

The significance of studying TEPs was recently recognised in the field of membrane technology because of their potential link to the occurrence of biological fouling in reverse osmosis (RO) systems (Berman and Holenberg, 2005; Villacorte *et al.*, 2009a; Bar-Zeev *et al.*, 2012) and organic fouling in ultrafiltration (UF) membranes (de la Torre *et al.*, 2008; Kennedy *et al.*, 2009). Moreover, the presence of TEP has been increasingly reported, over the last few years, in the water sources of various membrane-based water treatment plants (Villacorte *et al.* 2009b; 2010a,b; Gasia-Bruch *et al.*, 2011; Schurer *et al.*, 2012;2013; Van Nevel *et al.*, 2012).

Algal biopolymers like TEPs and their precursors have been reported to cover a wide range of sizes, ranging from 3 nm in diameter up to 100s of μm in length (Passow, 2000; 2002). This basically covers the range of material which can be retained by UF and RO membranes. TEPs are highly hydrated (>99% water by weight) as they tend to retain water while only allowing some to pass through (Verdugo *et al.*, 2004; Azetsu-Scott and Passow, 2004). This means that TEPs can bulk-up to a hundred times their dry volume and may provide substantial resistance to permeate flow during membrane filtration even at low (mass) concentration. Additionally, the relative stickiness of TEP is reported to be between 100 and 10,000 times more than most suspended particles in aquatic systems (Passow, 2002), which make it more likely to adhere to membrane surfaces and/or pores than any other aquatic organic substances. TEPs are also expected to strongly adhere to UF membranes whereby hydraulic cleaning (backwashing) may no longer be effective in adequately restoring the initial membrane permeability. This may result in increase of chemical consumption due to additional membrane cleaning (e.g., CEB, CIP) and/or additional pre-treatment (e.g., in-line coagulation).

Over the years, various studies demonstrated a more important role of biopolymers on the UF fouling propensity of surface waters than other types of organic substances in surface water (Lee N. et al., 2006; Amy et al., 2008). Biopolymers comprise the high molecular weight fraction of natural organic matter (typically >20 kDa), which are more likely to be retained by UF membranes (2-500 kDa MWCO) than dissolved organics such as aquatic humics (typically <1 kDa). Hence, they are also more likely to cause fouling in UF systems.

Over the years, various attempts were made to investigate the fouling propensity of biopolymers in relation to the chemistry of the water. Most of these studies use commercially available substances such as sodium alginate as model biopolymers. Such studies show that the fouling propensity of sodium alginate can be substantially affected by the pH, ionic strength and ionic composition of the water (Katsoufidou et al., 2007; van de Ven et al., 2008; Sioutopoulos et al., 2010). However, sodium alginate may not fully represent the fouling behaviour of algal biopolymers commonly found in surface waters.

The objective of this study is to investigate the UF fouling propensity of biopolymers produced by a common species of marine algae at different solution pH, ionic strength and cation composition. The fouling propensity were measured based on the modified fouling index (MFI-UF) and short term UF filtration tests with periodic backwashing. For comparison, fouling with model polysaccharides such as sodium alginate (macro-algal polysaccharide) and xanthan gum (bacterial polysaccharide) were also investigated under similar conditions.

7.2 Theoretical background

7.2.1 Membrane fouling

Filtration of liquids through a semi-permeable membrane can be explained based on the resistance in series model.

$$J_P = \frac{\Delta P}{\eta\, R} = \frac{\Delta P}{\eta\, (R_m + R_b + R_c)} \qquad\qquad \text{Eq. 1}$$

where J_P is the membrane permeate flux, ΔP is the trans-membrane pressure, η is the dynamic water viscosity, R is the total filtration resistance, R_m is the membrane resistance, R_b is the blocking resistance and R_c is the cake/gel resistance.

Assuming that cake or gel filtration is the dominant fouling mechanism in dead-end UF system, the change of total resistance is proportional to the increase in total mass of the deposited foulants on the membrane surface. For constant flux filtration, the cake resistance is dependent on the specific cake resistance (α) and the bulk foulant concentration (C_b).

$$R_c = \alpha\, C_b J_P\, t \qquad\qquad\qquad \text{Eq. 2}$$

Schippers and Verdouw (1980) proposed measuring the membrane fouling potential of the feedwater by defining the fouling index (I) as:

$$I = \alpha\, C_b \qquad\qquad\qquad \text{Eq. 3}$$

Based on the Carman-Kozeny model, α is inversely proportional to the diameter of the foulant materials and the cube of cake porosity.

$$\alpha = \frac{180(1-\varepsilon)}{\rho_p d_p \varepsilon^3} \qquad\qquad\qquad \text{Eq. 4}$$

where ε is the cake porosity, ρ_p is the density and d_p is diameter of particles.

Substituting Eq. 2 and Eq. 3 to Eq. 1 to derive the constant flux equation:

$$J_P = \frac{\Delta P}{\eta\,[R_m + (I\, J\, t)]}$$

$$\Delta P = J_p\, \eta\, R_m + J_P^2\, \eta\, I\, t \qquad\qquad\qquad \text{Eq. 5}$$

The fouling potential of the feedwater is calculated based on the minimum slope of ΔP versus time plot recorded during constant flux filtration (Boerlage et al., 2004; Salinas-Rodriguez, 2011). To calculate the modified fouling index (MFI-UF), the fouling index is normalised to the standard reference conditions set by Schippers and Verdouw (1980):

$$MFI - UF = \frac{\eta_{20^\circ C}\, I}{2\, \Delta P_o A_o^2} \qquad\qquad\qquad \text{Eq. 6}$$

where $\eta_{20^\circ C}$ is the water viscosity at 20°C, ΔP_o is the standard feed pressure (2 bar) and A_o is standard membrane area (13.8×10^{-4} m²).

7.2.2 Fouling reversibility

Schippers et al. (1981) developed a fouling prediction model for RO systems based on the MFI. In dead-end UF systems, this prediction model can only be applicable for the first filtration cycle since dead-end membranes are cleaned regularly and the increase in hydraulic resistance is mostly diminished after hydraulic cleaning (e.g., backwashing).

Figure 7-1 illustrates the four categories of fouling in dead-end MF/UF systems, namely:

1. Physically reversible or backwashable fouling (BW) - fouling which can be mitigated by hydraulic backwashing and/or other physical cleaning methods.

2. Physically irreversible fouling or non-backwashable fouling (nBW) - fouling which cannot be mitigated by backwashing and/or other physical cleaning methods.

3. Chemically reversible fouling (CR) - fouling which can be diminished by chemical cleaning (e.g., CEB, CIP). Chemically enhanced backwashing (CEB) is performed by adding cleaning chemicals on the filtrate water and then used for backwashing. Cleaning-in-place (CIP) is performed by feeding cleaning chemical solution to the UF system, soak the membrane and/or continuously recycling the solution to the system for several minutes (>30 minutes).

4. Chemically irreversible fouling (CiR) - fouling which cannot be diminished by chemical cleaning (e.g. CEB, CIP). This is considered as a permanent form of fouling.

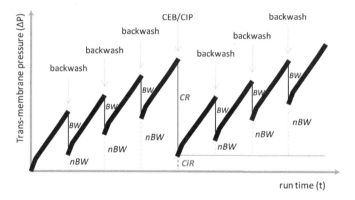

Figure 7-1: Graphical representation of the different categories of fouling in dead-end MF/UF systems.

Large-scale UF membrane systems are generally operating at constant-flux. Fouling in these systems is measured based on the increase of trans-membrane pressure (ΔP) and/or decrease of permeability ($J/\Delta P$) over the filtration period. In an ideal situation, regular physical cleaning by backwashing is effective in recovering the initial permeability of the membrane (Figure 7-2). However in most situations, backwashing is not always effective in totally recovering the initial permeability. The rate of fouling can gradually increase over the succeeding filtration cycles due to the effect of cake compression and/or due to reduction of effective filtration area. This non-ideal situation is expected to occur due to adsorption of sticky organic substances (e.g., biopolymers) onto membrane pores and surfaces, making conventional hydraulic backwashing ineffective in totally removing these foulants (Figure 7-2).

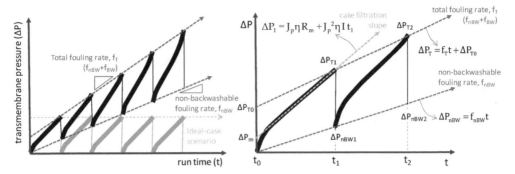

Figure 7-2: Fouling development in constant flux dead-end UF system.

The relationship between backwashable and non-backwashable fouling can be defined as follows:

$$f_T = f_{BW} + f_{nBW}$$ Eq. 7

Where:
f_T = total fouling rate (bar/h);
f_{BW} = backwashable fouling rate (bar/h);
f_{nBW} = non-backwashable fouling rate (bar/h).

In the ΔP versus time plot, f_T is the average slope of the final ΔP recorded in successive filtration cycles while f_{nBW} is the average slope of the initial ΔP recorded in successive filtration cycles. A graphical illustration of this relationship is shown in Figure 7-2. From this figure, two linear equations can be derived:

$$\Delta P_T = \Delta P_{T,0} + f_T\, t$$ Eq. 8

$$\Delta P_{nBW} = f_{nBW}\, t$$ Eq. 9

where $\Delta P_{T,0}$ is the y-intercept of the total fouling rate curve (Figure 7-2).

The cake filtration equation intersects with the total fouling (f_T) equation at $t = t_1$; where t_1 is equivalent to the run time of one filtration cycle. Combining Eq. 8 with Eq. 5 and then substituting Eq. 7 to calculate $\Delta P_{T,0}$:

at $t=t_1$: $\Delta P_{T1} = \Delta P_1$

$$f_T t_1 + \Delta P_{T,0} = J_p\, \eta\, R_m + J_P^2\, \eta\, I\, t_1$$

$$\Delta P_{T,0} = J_p\, \eta\, R_m + J_P^2\, \eta\, I\, t_1 - f_{BW}\, t_1 - f_{nBW}\, t_1$$ Eq. 10

Substituting $\Delta P_{T,0}$ to the total fouling equation (Eq. 8):

$$\Delta P_T = f_T\, t + J_p\, \eta\, R_m + J_P^2\, \eta\, I\, t_1 - f_{BW}\, t_1 - f_{nBW}\, t_1 \qquad\qquad \text{Eq. 11}$$

From Eq. 11, an equation can be derived to calculate the maximum trans-membrane pressure ($\Delta P_{T,N}$) after a specific number of filtration cycles (N).

$$\Delta P_{T,N} = N\, f_T\, t_1 + J_p\, \eta\, R_m + J_P^2\, \eta\, I\, t_1 - f_{BW}\, t_1 - f_{nBW}\, t_1$$

$$\Delta P_{T,N} = J_p\, \eta\, R_m + J_P^2\, \eta\, I\, t_1 + (N-1)\, f_{BW}\, t_1 + (N-1)\, f_{nBW}\, t_1 \qquad \text{Eq. 12}$$

Moreover, the calculated filtration time to reach a threshold of trans-membrane pressure (ΔP_c) to perform chemical cleaning (e.g., CEB or CIP) can be derived from Eq. 11 as follows:

$$t_c = \frac{\Delta P_c - J_p\, \eta\, R_m - (J_P^2\, \eta\, I - f_{BW} - f_{nBW})\, t_1}{f_{BW} + f_{nBW}} \qquad\qquad \text{Eq. 13}$$

This threshold is based on the minimum allowable permeability or maximum allowable pressure drop increase usually defined by the membrane manufacturer or operator of the UF system.

7.2.3 Effect of salinity on particle interaction

In solutions with low ionic strength, natural organic matter are expected to be smaller in size and have a rather loose and flexible structure as a result of high inter-particle charge repulsion by the electrical double layer around it (Braghetta *et al.*, 1997). At higher ionic strength, the electrical double layer is compressed enhancing aggregation and leading to formation of larger material with a more rigid and dense structure (Faibesh *et al.*, 1998). This effect can be further explained by the Debeye screening length (κ^{-1}) relationship which is a measure of the distance of influence (sphere of influence) of a particle.

$$\kappa^{-1} = \left(\frac{\varepsilon_0\, \varepsilon_r\, k_B T}{2 \times 10^3 N_A e^2 I_s} \right)^{0.5} \qquad\qquad \text{Eq. 14}$$

where ε_0 is the vacuum permittivity, ε_r is the relative permittivity of the background solution, k_B is the Boltzmann constant, T is the absolute temperature, N_A is the Avogadro's number, e is the elementary charge and I_s is the ionic strength. In this equation, the inter-particle distance of influence is inversely proportional to the ionic strength, an indication that an increase in salinity will promote aggregation of particles. Similarly at higher ionic strength, membrane-foulant interaction is less repulsive which means increased contact between particles and the membrane and concentration polarisation is lesser.

7.3 Experimental methods

Experiments focused on investigating the UF fouling propensity of biopolymers extracted from a common species of algae in seawater. For comparison, model biopolymers such as algae-derived sodium alginate and bacteria-derived xanthan gum were also investigated.

7.3.1 Artificial seawater

The artificial seawater (ASW) was prepared based on the typical ion concentration of coastal North Sea waters. Minor constituents (e.g., Br-, Sr^{2+}) which comprise less than 1% of inorganic ions in seawater were not considered. The ASW was prepared by sequentially dissolving analytical grade salts (stirred for >2 hours) to make up the artificial seawater solution (Table 7-1). Sodium bicarbonate was added to the solution to provide sufficient buffering capacity. The ambient pH of ASW was between 7.8 and 8.2. For some of the experiments, the pH was adjusted by adding few drops of 1% HCl or 0.5M NaOH, accordingly.

Table 7-1: Inorganic ion composition of model artificial seawater (ASW)

Inorganic ions	Concentration (g/L)	Percentage (%)
Chlorine, Cl-	18.85	55.0
Sodium, Na$^+$	10.75	31.3
Sulfate, SO$_4$-	2.69	7.9
Magnesium, Mg^{2+}	1.17	3.4
Calcium, Ca^{2+}	0.30	0.9
Potassium, K$^+$	0.38	1.1
Hydrogen Carbonate, HCO$_3$-	0.15	0.4
Total dissolved salts (TDS)	34.30	100

7.3.2 Algal organic matter

Chaetoceros affinis is one of the common species of algae in the ocean, specifically in the coastal waters of the North Atlantic Ocean where they are one of the major producers of micro-algal biopolymers (Myklestad *et al.*, 1989). A pure strain of *Chaetoceros affinis* (CCAP 1010/27) was acquired from the Culture Collection of Algae and Protozoa (CCAP; Oban, Scotland). The strain was inoculated in 5 L flasks with sterilised artificial seawater (Table 7-1) containing trace elements and nutrients (f/2+Si medium) necessary for algae to grow rapidly and simulate an algal bloom. After 10-12 days exposure to artificial light (12:12h, light:dark regime) and continuous slow mixing condition at ambient temperature of 20±3∘C, the algal cells were separated from the medium containing algae-released biopolymer by allowing the cells to settle to the bottom of the flask within 24 hours. The supernatant solution was extracted and then further purified by 48-hour dialysis in 3.5 kDa MWCO membrane bags (Spectra/Por 3, SpectrumLabs) and ultrapure water as the dialysate. For a sample to dialysate ratio of 1:60 L, the dialysis process were able to remove >97% of dissolved inorganic ions in the medium (based on electrical conductivity measurement) and >75% of low molecular weight (LMW) organic material (based on LC-OCD analysis). The final dialysed solution dubbed from henceforth as "algal organic matter" or "AOM" comprised of 1.9 mg C/L of biopolymers. For the fouling experiments, different feedwater solutions were prepared from this solution by diluting part of it with ASW (with adjusted ion concentration depending on the dilution factor) to make up the biopolymer and ion concentration required for the experiments. To prevent variations in results due to storage, each set of experiments were performed on the same day.

7.3.3 Model biopolymers

Commercially-available polysaccharides, sodium alginate (Fluka 71240) and Xanthan gum (Sigma 1253), were purchased from Sigma-Aldrich. These biopolymers are often used as model polysaccharide for quantification of TEPs (Passow and Alldredge, 1995) and as model biopolymer foulant in various UF fouling studies (e.g., Katsoufidou et al., 2007; van de Ven et al., 2008; de la Torre et al., 2009). Sodium alginate is the sodium salt of alginic acid derived from the cell walls of brown macro-algae (e.g., seaweeds) while Xanthan gum ($C_{35}H_{49}O_{29}$) is derived from the cell wall of plant bacteria Xanthomonas campestris. Stock solutions were prepared by dissolving 20 mg of sodium alginate (SA) or xanthan gum (XG) in 1 L of ultra-pure water. Different solutions were prepared accordingly from the stock solutions following the same procedure with AOM.

7.3.4 Liquid chromatography – organic carbon detection (LC-OCD)

Biopolymer concentrations were measured using liquid chromatography – organic carbon detection (LC-OCD) analysis. The general protocol of this technique is described by Huber et al. (2011). LC-OCD was originally developed to characterise natural organic matter which can pass through 0.45µm pore size filter. For this study, the inline 0.45 µm filter was bypassed to also quantify a significant fraction of the particulate biopolymer materials present in the AOM and model biopolymer solutions. Without the inline filter, the approximate upper size limit of biopolymers it can analyse is 2µm based on the average pore size of the sinter filters of the chromatogram columns used in the analyses (Huber, 2012).

7.3.5 Transparent exopolymer particles (TEP)

TEP concentration was measured based on the spectrophotometric method modified from Passow and Alldredge (1995). The modified protocol is described in Chapter 5 of this thesis. With the current modification, TEP is measured using 0.1 µm pore size PC membranes (Whatman Nuclepore). $TEP_{0.1µm}$ were measured within 24 hours after dialysis treatment of AOM and after preparation of SA and XG stock solutions.

7.3.6 Dynamic light scattering measurements

The size distribution of TEP was estimated based on dynamic light scattering (DLS) measurements using a Malvern Zetasizer Nano ZS. This technique measures the diffusion of particles moving under Brownian motion and converted to size based on the Stokes-Einstein relationship. The obtained size is the diameter of a sphere with equivalent translational diffusion coefficient as the measured particle, called the hydrodynamic diameter. DLS measurements were performed for samples taken directly from the stock solutions of AOM, SA and XG without further pre-treatments. All measurements were performed at 25°C.

7.3.7 Atomic force microscopy

Interaction forces between biopolymer materials and clean or biopolymer-fouled UF membranes were measured using atomic force microscopy (AFM). In this study, the investigated biopolymer materials were allowed to adsorb to the surface of polystyrene microspheres attached to the tip of the AFM cantilever.

To adsorb biopolymers on AFM probe, about 10 ml of stock solutions of AOM and 5 mg C/L solutions of SA and XG in ASW were poured into separate petri-dishes (50 mm diameter). AFM cantilevers with attached 25 μm polystyrene microsphere tip (Novascan Technologies, Inc.) were submerged in the solutions by carefully positioning the cantilever (microsphere probes up) at the bottom of the petri-dish. The cantilevers were submerged for 5 days at 4°C to allow adsorption of biopolymers. The petri-dishes were then placed at room temperature (20°C) for >2 hours before AFM force measurements were performed.

Due to practical difficulties of performing AFM force measurement inside a capillary PES membrane, a disk polyethersulfone (PES) membrane (Omega 100 kDa MWCO, Pall Corp.) was used as the representative UF membrane surface. The membrane was first cleaned by soaking them in ultra-pure water for 24 hours and then filtering through >10 ml ultrapure water before the experiment. For biopolymer to membrane interaction measurements, a biopolymer-coated tip and a clean membrane were used. For biopolymer to biopolymer interaction measurements, the PES membrane was first coated with biopolymers by filtering through 5 ml of biopolymer solution in ASW. Filtration was performed at constant flux (60 L m^{-2}.h^{-1}) using a syringe pump (Harvard pump 33), 30 ml plastic syringe and 25 mm filter holder.

Force measurements were performed with the ForceRobot 300 (JPK Instruments) at room temperature (~20°C), using a small volume liquid cell, sealed with a rubber ring. In every test, the liquid cell is filled with ASW (TDS ~ 34 g/L; pH = 8±0.2). The cantilevers were calibrated with respect to their deflection sensitivity from the slope of the constant compliance region in the force curves obtained between the clean, hard surfaces and their spring constant using the thermal noise spectrum of the cantilever deflections (Ralston *et al.*, 2005; Spruijt *et al.*, 2012). Force-distance curves were recorded for an approach range of 10 μm, a tip velocity of 16-28 μm/s, 30 seconds surface delay and a sampling rate of 2 kHz. In each cycle of approach and retract, the probe is brought into contact with the surface with an average load force of 5 nN. The force-distance (F-D) curves were recorded in triplicates for each of the six points on the membrane surface within an area of 10 μm x 10 μm. The F-D curves were analyzed using the JPK data processing software to calculate the maximum retract force and total energy.

7.3.8 Filtration experiments

A lab-scale UF system was used for experiments to measure membrane fouling potential (i.e., MFI-UF) and backwashability of AOM, SA and XG. The UF system comprised of a small module of inside-out capillary UF membranes (150 MWCO, Pentair X-flow) and a continuous flow syringe pump (Harvard Pump 33) to provide constant permeate flux (Figure 7-3). To assess the rate of fouling, trans-membrane pressure (ΔP) was monitored using a pressure transmitter (Endress+Hauser PMD75 with ceramic sensor) and communication device connected to a computer with data acquisition software. Water temperature was also monitored to calculate water viscosity.

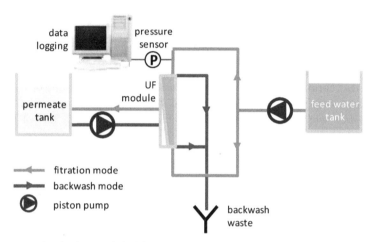

Figure 7-3: Graphical scheme of the lab-scale constant-flux UF system and modes of operation.

The lab-scale UF module was fabricated in an 8-mm diameter polyethylene tubes containing 4 capillaries (0.8 mm diameter; 150 kDa MWCO) of 15 cm in length. Both ends of the tubes (except the fibre ends) were sealed off with water-proofing resins. The feed solutions were fed from both ends of the fibres and the UF permeate flows through an outlet in the middle of the module. The system was operated at constant flux filtration of 80 L/m^2.h. For MFI-UF measurement, filtration was performed for 60 mins. For backwashability experiments, the system was operated for 6 hours with 1-minute backwashing (flux = 200 L/m^2/h) using UF filtrate performed for every 30 mins of filtration. New membrane modules were fabricated for each experiment. Before each experiment, the membranes were pre-conditioned by filling up the module and soaking the membranes for >10 mins in water (without biopolymers) with similar solution chemistry of the feedwater used in subsequent fouling experiments.

7.4 Results and discussion

7.4.1 Characteristics of biopolymers

Algal organic materials (AOM) extracted from marine diatom *Chaetoceros affinis* was used to investigate the fouling propensity of biopolymers produced during algal blooms in seawater. For comparison, polysaccharides which resemble the characteristics of micro-algal biopolymers such as sodium alginate (SA) and xanthan gum (XG) were also studied. Comparing the characteristics of AOM, SA and XG may provide indications on the fouling propensities of biopolymers produced by micro-algae, macro-algae and bacteria, respectively.

7.4.1.1 Biopolymers and TEP

Based on LC-OCD analysis, the dialysed AOM solution comprises 1.9 mg C/L of biopolymers and 0.6 mg C/L of residual low molecular weight organics from the culture medium (Figure 7-4). The organic nitrogen content of biopolymers was about 0.18 mg N/L, an indication that protein-like substances were also produced by *C. affinis*. An estimate of the organic carbon concentration of protein-like substances was 0.54 mg C/L which is about 28% of biopolymers. Hence, a substantial majority (~72%) of biopolymers in AOM solution are polysaccharides. SA and XG solutions comprise almost purely of biopolymers. They contain undetectable levels of organic nitrogen, which means they comprise mainly polysaccharide material.

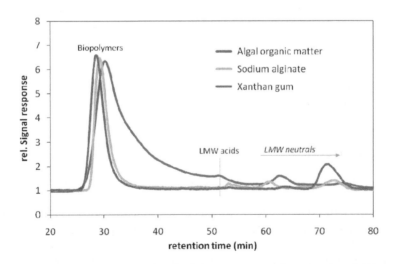

Figure 7-4: Representative organic carbon (OC) chromatograms of AOM and model biopolymers.

Alcian blue staining and microscopy analyses of the three polysaccharide samples showed indications of their diverse structures. As shown in Figure 7-5, Alcian blue stained aggregates of AOM showed striking similarities with the aggregates of XG. However, XG showed long thin fibrous units which have rather looser arrangement than AOM aggregates. On the other hand, large flocs of SA appear to be cloud-like which have denser structure than AOM and XG. At molecular level, SA ($C_6H_8O_6$) polymers comprise of simple and homogenous sequences of different types of uronic acids (e.g., mannuronic and guluronic acid; Haug *et al.*, 1974) while XG ($C_{35}H_{49}O_{29}$) are highly branched polymers comprised of repeated pentasaccharides containing multiple units of sugars (e.g., glucose, mannose, glucuronic acid; Garcia-Ochoa *et al.*, 2000). Although the molecular structures of biopolymers in AOM are still largely unknown, it has been reported to be very complex and highly branched (Passow, 2002), which possibly explain its similarities with XG.

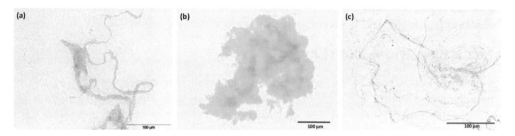

Figure 7-5: Microscope photographs of alcian blue stained aggregates of (a) AOM, (b) SA and (c) XG.

At identical biopolymer concentration of 0.6 mg C/L, $TEP_{0.1\mu m}$ concentrations were 0.5, 0.1 and 0.24 mg X_{eq}/L for AOM, SA and XG, respectively. The large differences in concentrations may be due to two factors:

1. TEP measurements were based on the anion density of the functional groups (e.g., carboxyl and sulphate groups) linked to these biopolymers, which means it does not account for their mass concentrations but rather the density of these functional groups; and

2. it does not cover all sizes of biopolymers as $TEP_{0.1\mu m}$ only covered those retained on 0.1μm PC filters. SA recorded the lowest $TEP_{0.1\mu m}$ concentration possibly because of its low molecular weight and poor retention in 0.1 μm filters during $TEP_{0.1\mu m}$ measurements.

7.4.1.2 Size distribution

The hydrodynamic size distributions of biopolymers were compared using dynamic light scattering (DLS). The size distribution of the three solutions appeared to be bimodal (Figure 7-6). AOM showed two peaks with mean hydrodynamic diameters of 540 nm and 2600 nm, respectively. SA also showed two peaks having mean diameters of 330 nm and 100 nm, respectively. XG showed two major peaks with mean diameters of 210 nm and 2350 nm, respectively. The latter was found to have the widest size range (25-3300 nm) as compared to AOM (200-3300 nm) and SA (53-7350 nm). Around 97% of AOM were between 200 and 1300 nm, ~82% of SA were in the range of 140-750 nm and ~90% of XG were in the range of 25-750nm.

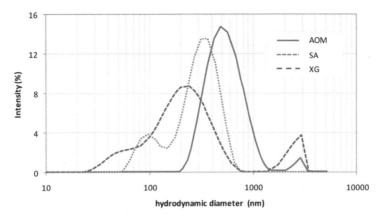

Figure 7-6: Size distribution of AOM and model biopolymers in low salinity solution (<1 g/L) based on DLS measurements.

The hydrodynamic diameter measured by DLS was derived from the diameter of an equivalent spherical particle with similar diffusion coefficient (Stokes-Einstein relationship) as the biopolymer. Hence, the results are just indications of their actual size. Biopolymers like TEPs are not spherical but generally fibrillar or branched in shapes which tend to change in configuration over time due to ageing or external forces (Passow, 2000). Because of this, the measured hydrodynamic size may have substantially underestimated the actual size distribution of biopolymers. Such measurement error is likely more pronounced for highly branched and extended polymers like AOM and XG. This might possibly explains why the largest fraction of XG showed a smaller mean diameter than SA.

7.4.1.3 UF membrane rejection

Filtration tests were performed to measure the apparent UF rejection of biopolymers (measured using LC-OCD) from the three prepared solutions at low ionic strength (0-0.3 g/L) conditions. The UF membrane rejected ~85% of AOM, ~70% of SA and ~90% of XG. The UF rejection of SA was higher than the range previously reported by Katsoufidou *et al.* (2007; 2010) which are between 0 and 60%. Considering that such studies used TOC measurement to measure the apparent size range of SA, it is possible that these were underestimated as possible contamination may have been introduced during filtration (e.g., organics leaching from the membrane). However using LC-OCD analysis to measure TOC, contamination can be detected and results can be corrected by excluding anomalies in the chromatogram peaks. $TEP_{0.1\mu m}$ concentrations of biopolymers in the permeate UF membrane were all below the detection limit of the method.

Although the DLS results showed XG having a lower average hydrodynamic diameter than AOM and SA, they were rejected better by the UF membrane. This could be due to the wider size range of XG (0.025-3.3µm) and the substantial presence of larger particles (>1µm). The large gels can immediately block the entrance of the pores at the start of filtration, possibly adsorbing the incoming smaller gels and preventing most of them from penetrating the pores. This might also explain why AOM has a higher UF rejection than SA, since the latter did not contain aggregates larger than 0.75 μm.

7.4.1.4 Cohesion and adhesion potential

To investigate the stickiness (cohesion/adhesion potential) of AOM compared to model biopolymers (SA and XG), interaction force measurements were performed using AFM. The force distance curves generated between clean/biopolymer-coated polystyrene microsphere probes and clean/biopolymer-fouled PES UF membrane are presented in Figure 7-7.

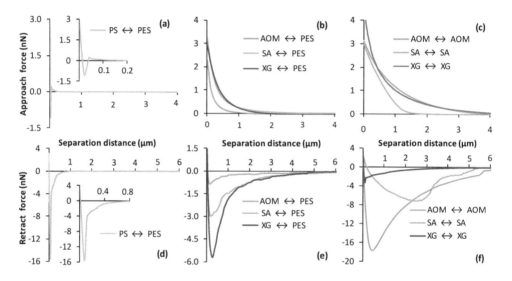

Figure 7-7: Typical force-distance curves for (a,d) clean PS probe and clean PES membrane, (b,e) biopolymer-coated probe and clean PES membrane, and (c,f) biopolymer-coated probe and biopolymer-fouled PES membrane. Graphs a-c are approach force curves while graphs d-f are retract force curves.

The generated approach force curves show that short-range attraction forces were observed between clean PS probes and clean membranes (Figure 7-7a). However after incubating the probes in AOM solution for 5 days, interaction shifted to rather long range repulsion (Figure 7-7b,c). This change indicates that indeed a layer of AOM was adsorbed to the surface of the polystyrene microspheres. The polystyrene particle (reported by supplier), the PES membrane (Ricq *et al.*, 1997) and biopolymers (Henderson *et al.*, 2008) are (slightly) negatively charged. However, electrostatic repulsion may not have a significant role in the surface interaction because the high ion concentration in the matrix screens the charges and limits the range of electrostatic interactions to short range distances of less than 1 nm (Mosley *et al.*, 2003). The long range (2-4 μm) gradual increase in approach force is the force needed to compress the soft and deformable AOM layer coating on the surface of the membrane and/or the probe.

Adhesion/cohesion forces of AOM and model biopolymers were investigated based on the retract force curves generated after 30 seconds of contact time (Figure 7-7d,e,f). With the exception of XG, the maximum retract forces were higher between biopolymer-coated probe and biopolymer-fouled PES membrane than biopolymer-coated probes and clean PES membrane. It was also observed that the retract force curves were generally broader when biopolymers were present on both the particle probe and the membrane surface (Figure 7-7f). This implies that not only a higher force but also a higher energy is needed to totally retract (pull-off) bonds between two biopolymer coated surfaces. In addition, it indicates that biopolymer layers can deform and stretch when pulled apart and only at distances of the order of several micrometers that they come apart. The softness and deformability of the biopolymer layers is corroborated by the approach curves in Figure 7-7b,c.

Hydrogen bonding may have played a major role in the adhesion/cohesion between investigated surfaces. Hydrogen bonding is a result of electron transfer between electronegative moieties and hydrogen atoms in biopolymers and on membrane surfaces. Moreover, cohesion between biopolymer-coated surfaces might also be influenced by polymer entanglement during contact between AOM-coated surfaces (Flemming et al., 1997). AOM polymers are generally flexible and elastic, so disentangling them may have occurred in a stepwise fashion and therefore requires higher energy to totally detach them as shown in Figure 7-7f.

The average maximum (adhesion) force and total energy needed to pull-off two investigated surfaces from contact are compared in Figure 7-8.

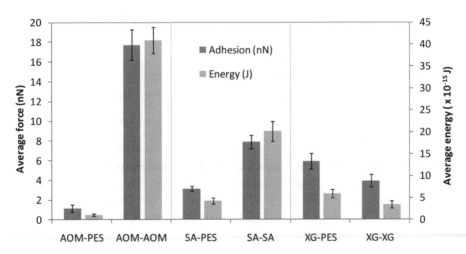

Figure 7-8: Average maximum retract forces and total energies needed to completely pull-off biopolymer-coated probe from clean/biopolymer-fouled membrane after 30 seconds of contact time.

Adhesion on clean PES membranes varied substantially with different biopolymers whereby the maximum retract force and total energy was highest for XG followed by SA and AOM, respectively. For AOM and SA, cohesion between the biopolymer-coated probe and the biopolymer-fouled PES membrane showed much higher magnitude (up 15 times) than adhesion between biopolymer-coated probe and clean PES membrane. On the contrary, adhesion of XG on clean PES membrane was higher than cohesion between XG-coated probe and XG-fouled membrane.

Although AOM showed a relatively lower adhesion potential on clean PES membranes compared to model biopolymers, it does not necessarily mean that it is easier to remove by hydraulic backwashing. The adhesion forces reported in this study is only based on measurements after 30 seconds of contact time between the biopolymer and the UF membrane. It was not feasible to perform a longer contact time with the current AFM set-up. In practice, the contact time can be much higher as filtration time between backwashing is typically 15-45 mins. Theoretically, a higher contact time can substantially increase hydrogen bonding between surfaces. Algal-derived biopolymers are usually made up of long highly branched fibres. Over time, these fibres may deform/re-assemble as they are pressed further towards the membrane surface due to cake compaction and/or water convection. Such scenario will increase the contact area between biopolymer fibres and the membrane surface, thereby increasing the bonding strength. Further studies are needed to verify this effect.

7.4.2 Membrane fouling potential and effect of solution chemistry

Organic fouling in UF membranes is largely influenced by the characteristics of the feedwater. This includes the concentration of biopolymers and the feed solution chemistry (e.g., pH, ionic strength and ionic composition). Moreover, anionic biopolymers such as TEPs may interact mainly with cations (e.g., Ca^{2+}, Na^+) in seawater. Investigations on the influence of these factors on the membrane fouling potential of AOM are discussed in the following sections.

7.4.2.1 Effect of biopolymer concentration

The feed solutions used in the filtration experiments were diluted from AOM and standard biopolymer (SA and XG) solutions. Dilution with artificial seawater was necessary as the initial biopolymer concentrations were extremely high compared to what is reported in natural seawater. Latest studies show that biopolymer concentration in the North Sea is typically very low during winter (<0.1mgC/L) but it can reach as high as 0.6 mgC/L during the spring period (Schurer et al., 2013; Salinas-Rodriguez, 2011). To cover this range, the membrane fouling potential (based on MFI-UF) of feedwater solutions containing biopolymer concentrations between 0.1 and 1.5 mgC/L were measured.

The fouling potential of the three biopolymers varies widely (Figure 7-9). For instance, at biopolymer concentration of 0.6 mg C/L, the fouling potential of AOM was ~50% higher than SA and ~5 times higher than XG. The membrane fouling potential of AOM and XG showed a direct linear correlation with biopolymer concentration. However, the fouling potential of SA best described as a power relationship with biopolymer concentration. Based on projections of these trend lines, the fouling potential of SA is expected to be higher than AOM for concentrations >1.6 mg C/L.

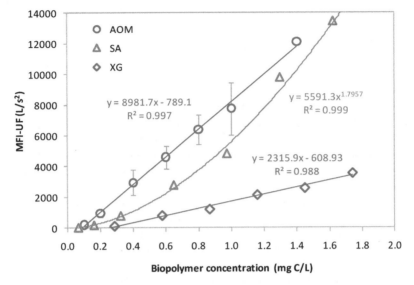

Figure 7-9: Fouling potential of different concentrations of AOM, XG and SA in seawater.

The linear relationship between biopolymer concentration in seawater and MFI (as in the case of AOM and XG) is similar to what was previously reported in surface waters (Salinas-Rodriguez, 2011; Boerlage et al., 2003a). This can be explained based on direct proportionality of the fouling index (I) with the foulant concentration (Eq. 3). However, this direct proportionality is based on the assumption that the specific cake resistance (α) is constant and that the membrane rejects 100% of the foulant materials, which might not be true in this study. The UF rejections of AOM (85 %) and XG (90 %) as discussed in Section 7.4.1.3 were higher than SA (70 %), which possibly explains the difference in trend. Another explanation for the exponential trend in SA is the possibly of higher degree of compression in SA gels than for AOM or XG.

At higher concentrations, biopolymers have higher collision frequencies than at lower concentrations, hence, higher coagulation possibilities. The coagulation process which may have occurred even before (during storage) and during the filtration experiment can increase the size of particles but may also allow more biopolymers to be rejected by the membrane. Hence, membrane rejection may increase with increasing biopolymer concentration and a non-linear increase in fouling potential may occur (as in the case of SA).

Theoretically, an increase in particle size translates to a decrease in specific cake resistance as explained in the Carman-Kozeny equation (Eq. 4). For SA, the effect of increased membrane rejection may have dominated the effect of decreasing α. Biopolymers are amorphous hydrocolloids which contradict the Carman-Kozeny model assumption of a cake layer comprising of solid spherical particles. Biopolymer materials can spontaneously deform and possibly rearrange during membrane cake formation. Hence, increased aggregation of polysaccharides might have insignificant effect on the α and the resulting increase in membrane rejection and/or cake compression may have a dominant effect on its fouling potential.

7.4.2.2 Effect of pH

In seawater UF-RO plants, lowering the pH of the feedwater may be performed in order to optimise performance of inline coagulation system or to minimise the possibility of calcium carbonate scaling in the downstream RO system. Considering that the feedwater pH in the plant may vary from pH 6 to 8, the investigation focused on measuring the fouling propensity of biopolymers covering this range. For direct comparison, the ionic strength of the feedwater was kept constant at 35 g/L. The feed water concentrations of the different solutions were 0.6 mg C/L, 0.65 mg C/L and 1.45 mg C/L for AOM, SA and XG, respectively. To minimise the effects of variability in the measurements, it was decided to keep all MFI results higher than 2000 s/L^2. As shown in Figure 7-9, a substantially higher concentration of XG should be present in the feedwater to reach an MFI higher than 2000 s/L^2. Consequently, a higher XG concentration was selected. To normalise the results, the specific membrane fouling potential are expressed in terms of MFI per mg C/L of biopolymers in the feedwater. The specific membrane fouling potential of biopolymers at different solution pH is presented in Figure 7-10.

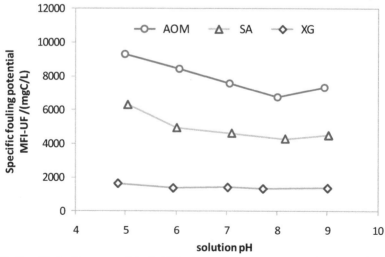

Figure 7-10: Specific fouling potential of AOM, SA and XG in seawater (TDS=35 g/L) at different solution pH.

In general, the fouling potentials of all three biopolymers were higher at lower solution pH until around pH 8. Increasing the pH until pH 9 did not show substantial change in fouling potential. A lower fouling potential at higher solution pH can be attributed to deprotonation of the polysaccharide functional groups (e.g., carboxyl groups), in which the macromolecules are more negatively charged and electrostatic repulsion is higher. These may have resulted in higher inter-particle distance and a looser cake layer with lower hydraulic resistance to permeate flow. Lee S. *et al.* (2006) reported that the carboxyl groups of SA are mainly deprotonated at pH>5 and found substantial SA fouling in RO membrane at lower pH than at higher pH. The same finding was also demonstrated in this study whereas the UF fouling potential of SA was higher at pH 5 and that only slight variations of fouling potential were observed between pH 6 and pH 9 (Figure 7-10). For XG, which also contains carboxyl groups, a similar trend was also observed. AOM on the other hand, which comprise mainly sulphated functional groups rather than carboxyl groups, showed a slightly different trend where a gradual decrease of fouling potential with increasing pH was observed between pH 5 and pH 8. Inter-particle repulsion between sulphated polysaccharides like carrageenan only becomes weaker when the pH is closer to 2, which is the pKa value of its anionic sulphate groups (Gu *et al.*, 2005). Therefore, the effect of pH on sulphated polysaccharides like AOM is expected to be similar to SA or XG. However, unlike SA or XG, AOM does not comprise only polysaccharide but it also proteins (~28%). These proteins were reported to comprise of arginine, asparagine, tyrosine and isoleucine amino acids (Myklestad *et al.*, 1989), which have pKa values ranging from 4 to 12.5. This means that a large fraction of these proteins are still positively charged between pH 5 and 8. The presence of proteins in AOM possibly explains the gradual decrease of MFI-UF at this pH range.

7.4.2.3 Effect of ionic strength

Filtration experiments of biopolymer solutions with different salinities were performed to measure the effect of ionic strength on the fouling potential of biopolymers. The observed UF membrane fouling by biopolymers at low and high salinities are presented in Figure 7-11.

For low ionic strength solutions, the mechanisms of UF fouling by biopolymers were generally in two phases. Membrane fouling by AOM showed a brief period (first ~3 mins) of pore blocking followed by cake/gel filtration (Figure 7-11a). On the other hand, fouling by SA and XG showed longer periods (first 10-15 mins) of pore blocking then slowly shifting to cake/gel filtration (Figure 7-11b,c). Katsoufidou *et al.* (2007) reported that more than one fouling mechanisms can occur simultaneously (standard and/or complete pore blocking) or consecutively (pore blocking followed by cake filtration) during filtration of biopolymers at low ionic strength. Pore blocking may have occurred whereas smaller biopolymers adsorb to or deposit on the pores of the membrane (standard blocking) while larger materials can block the entrance of the pores (complete blocking). As more pores are blocked over time, the transition to cake filtration follows.

The low ionic strength condition is expected to keep the size of biopolymers smaller, looser and flexible due to high electrostatic repulsion. De la Torre *et al.* (2009) reported that at low ionic strength solutions, SA can vary in size between 1kDa and 154 kDa while XG is known to be much larger, between 500-500,000 kDa. The size range of SA is basically smaller than the nominal molecular weight cut-off (150 kDa) of the UF membrane used in this study. This means they can accumulate in the pores of the membrane and cause severe pore blocking. The difference in size is likely the reason for the shorter period of pore blocking with XG than SA. However, the flexibility of biopolymers and their ability to squeeze through membrane pores smaller than their apparent size can be a major factor as well. On the other hand, the size of biopolymers in AOM is larger than SA and XG at low ionic strength solution based on Figure 7-6. This may have resulted in fast transition from blocking to cake filtration much faster than SA and XG. Furthermore, AOM and XG fibrils were observed to have longer physical structure than SA (Figure 7-5) and so it is likely that they accumulate more on the surface of the membrane than in the pores.

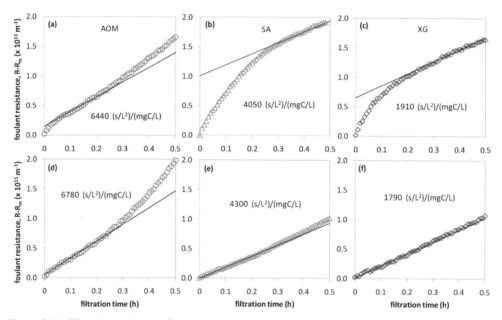

Figure 7-11: Filtration resistance due to accumulated biopolymers during filtration of (a,d) 0.6 mg C/L AOM, (b,e) 0.65 mg C/L SA and (c,f) 1.45 mg C/L XG solutions with (a,b,c) low and (d,e,f) high salinities. Solid lines represent the minimum slope of the curve and values represent calculated specific MFI.

At high ionic strength (e.g., TDS=35 mg/L), the mechanism of membrane fouling is mainly gel filtration (Figure 7-11d,e,f). As explained in Eq. 14, Debeye screening length (κ^{-1}) or the distance of influence of a certain particle is inversely proportional to the ionic strength of the solution. At high ionic strength, the electrical double layer surrounding each biopolymer material is compressed, minimising repulsion and enhancing aggregation between materials. The resulting large aggregates can rapidly form a gel layer on the surface of the membrane and minimise pore blocking. An apparent non-linear increased was observed during filtration of AOM which means the cake is compressible. However, cake compression was not apparent during filtration of SA and XG.

The fouling potential of biopolymers was further investigated for a wider range of salinity (TDS=0.2-65 g/L) to represent different types of water ranging from fresh surface water to seawater to RO concentrate. The different feedwater solutions were prepared while maintaining the fractional composition of inorganic ions in seawater (Table 7-1) and solution pH 8±0.2. The fouling potentials of the different solutions were relatively low at near zero salinity <<0.5 g/L but increasing the salinity up to about 1 g/L for AOM and SA and up to 5 g/L for XG resulted in significant increase of fouling potential (Figure 7-12). Increasing further the ionic strength however, resulted in decline of fouling potential. This declining trend was observed until TDS of 20 g/L, 5 g/L and 35 g/L for AOM, SA and XG solutions, respectively. At TDS higher than these values, the fouling potential appeared to increase with ionic strength. The highest specific MFI was recorded at TDS of 1, 65 and 5 g/L for AOM, SA and XG, respectively.

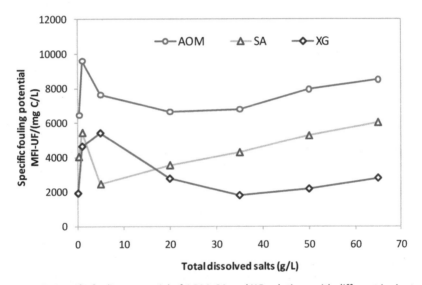

Figure 7-12: Specific fouling potential of AOM, SA and XG solutions with different ionic strength.

Salinas-Rodriguez (2011) reported a substantial increase in MFI-UF when increasing the TDS of seawater samples (20 to 70 g/L) which is similar to what was observed in this study. The lower fouling potential observed at very low TDS (<1 g/L) was possibly a result of lower membrane rejection and/or the formation of loosely arranged cake layers due to high inter-particle repulsion forces. A lower membrane rejection at low ionic strength is expected because the size of biopolymers is expected to be smaller and more flexible, which might allow them to squeeze through the pores of the UF membrane.

Although a higher membrane rejection is expected at higher ionic strength, it was not always the case (in this study) that the solution with the highest TDS concentration has the highest fouling potential. Peaks of MFI were recorded at TDS 1 g/L for AOM and SA, and 5 g/L for XG (Figure 7-10). A similar trend was also reported by Boerlage et al. (2003b) and Bacchin et al. (1996). Boerlage et al. (2003b) measured MFI-UF of tap water spiked with different concentrations of NaCl (0.5-12 g/L) and observed a peak value at TDS ~6 g/L. On the other hand, Bacchin et al. (1996) measured the specific cake resistance of bentonite in solutions containing 0.0007 to 22 g/L of KCl and recorded a maximum resistance at TDS ~ 0.6 g/L. Bacchin et al. (1996) explained that the initial increase in specific cake resistance was a result of decrease in cake porosity as the inter-particle distance in the cake is reduced. They further proposed that a "critical concentration of coagulation" (CCC) is reached when a further increase in ionic strength resulted in subsequent decrease in specific cake resistance. CCC is the critical ionic strength at which the zeta potential is reduced to about zero. Beyond this level would result in the formation of larger aggregates and thus, lower specific cake resistance. The possible CCC of AOM and SA were in the same range observed by Bacchin et al. (1996) for bentonite clay while XG was similar to what was reported by Boerlage et al. (2003b) for tap water. The variation was likely because of the different foulant composition which requires different electrolyte concentration to induce coagulation.

The range of ionic strength investigated in this study extended beyond the range (>20 g/L) explored by Bacchin et al. (1996) and Boerlage et al. (2003b). Interestingly, an apparent increase in MFI-UF were observed at TDS higher than 20 g/L, 5 g/L and 35 g/L for AOM, SA and XG, respectively. The increase of MFI-UF above the "critical ionic strength" is possibly the point where cake resistance is dominated by the effect of increased membrane rejection rather than the effect of increased in particle size in the cake. Increased coagulation above the "critical ionic strength" is expected to increase the rejection of biopolymers which also means a higher rate of increase in cake thickness.

In summary, the fouling potential of biopolymers is affected by the inter-particle charge repulsion at ionic strength below the so called "critical concentration of coagulation". Above such ionic strength, the electrical double layer is compressed and the cake resistance is governed by either the effect of increased particle size or the increased membrane retention due to increased coagulation of biopolymers.

7.4.2.4 Influence of mono- and di- valent cations

The influence of major mono- and di-valent cations in seawater on the membrane fouling potential of biopolymers was investigated. The different feedwater solutions were prepared based on the typical concentrations of individual cations in seawater while maintaining the solution at pH 8 (see Table 7-2). Figure 7-13 shows the comparison of MFI results for the different matrices.

The fouling potential of AOM in solutions dominated by a single type of cation (Na+, K+, Ca2+ or Mg2+) were not substantially different. On the contrary, the changes in cation composition had a rather drastic effect on the fouling potential of SA and XG. For example, when Ca2+ or Na+ was added , the fouling potential of SA was about 20-

25 times lower than the reference (only with $NaHCO_3$ buffer) while fouling potential was comparable and roughly 3 times higher than the reference when Mg^{2+} and K^+ were added, respectively. For XG, the lowest fouling potential (30% lower than the reference) was recorded when Mg^{2+} was added, while the fouling potential was highest (3 times higher than the reference) when only K^+ was present in the solution.

Table 7-2: The ion composition of dissolved salts in the feed solutions used in the experiments. To maintain comparable solution pH (pH ~6.0), all solutions contain 0.26 mg/L of $NaHCO_3$ buffer.

Feed solution	Na^+	K^+	Ca^{2+}	Mg^{2+}	Cl^-	SO_4^-	HCO_3^-	TDS
reference	0.11	-	-	-	-	-	0.15	0.26
K	0.11	0.38	-	-	-	-	0.15	0.64
Ca	0.11	-	0.30	-	-	-	0.15	0.56
Mg	0.11	-	-	1.17	-	-	0.15	1.43
Na	10.75	-	-	-	18.85	2.69	0.15	32.44
Na+K	10.75	0.38	-	-	18.85	2.69	0.15	32.83
Na+Ca	10.75	-	0.30	-	18.85	2.69	0.15	32.75
Na+Mg	10.75	-	-	1.17	18.85	2.69	0.15	33.62
Na+Ca+Mg+K	10.75	0.38	0.30	1.17	18.85	2.69	0.15	34.30

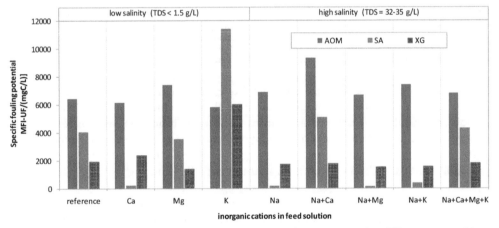

Figure 7-13: Specific fouling potential of AOM, SA and XG solutions containing different compositions of mono- and di- valent cations.

Various studies have demonstrated the influence of individual cations on the fouling potential of SA (e.g., van de Ven *et al.*, 2008; Katsoufidou *et al.*, 2007). The results of this study supported some of the earlier findings. For example, van der Ven *et al.* (2008) also found that Ca^{2+} has a negative effect while K^+ has a positive effect on the hydraulic resistance of SA solutions. The former can be explained based on the "egg-box" model, in which Ca^{2+} ions bond with oxygen atoms of carboxylate groups and form bridges between adjacent polymers, the result of which is a very structured and highly permeable gel network on the surface of the membrane. On the other hand, the presence of K^+ promotes self-entanglement of alginate polymers due charge screening effect leading to reduced electrostatic repulsion. As a result, a denser and less permeable gel layer is formed on the surface of the membrane. These effects however, were not clearly observed with AOM and XG solutions. Only K^+ had a significant effect on XG while no apparent effect by either Ca^{2+} or K^+ or Na^+ on AOM was observed.

In solutions containing more than two types of cations, only the combination of Na^+ and Ca^{2+} had a substantial effect (increased specific MFI-UF) on the fouling potential of AOM. No apparent variations in the fouling potential of XG at high salinity, regardless of the combination of cations investigated. SA on the other hand, was drastically affected by the different combinations of cations. When Na^+ was combined with Ca^{2+}, the fouling potential of SA was about 25% higher than the reference solution. However, the combination of Na^+ and Mg^{2+} or K^+ resulted in substantially lower fouling potential (10-30 times lower than the reference). The fouling potential of SA in feed solution containing all the cations was comparable with the reference solution and solution containing only Na^+ cation.

The combination of Na^+ and Ca^{2+} consistently demonstrated a strong influence on the fouling potential of biopolymers at high ionic strength conditions. Na^+ ions itself have a "charging up" effect which essentially promotes the deprotonation of carboxyl functional groups possibly resulting in dispersion of polysaccharide aggregates into smaller size (Jin *et al.*, 2009). This may have resulted in significant reduction of UF membrane rejection and thus, lower membrane fouling potential (as in the case of SA). In the presence of Ca^{2+}, the "charging up" effect of Na^+ can be greatly reduced, as Ca^{2+} ions can react with the deprotonated carboxyl groups (Jin *et al.*, 2009).

The combined effects of Na^+ and Ca^{2+} may have led to higher rejection of biopolymers due to calcium-carboxylate complexation as well as the formation of a dense TEP gel structures resulting from reduced intra-molecular repulsion as consequence of charge screening and electrical double layer compression of biopolymer molecules at high electrolyte concentration. This might explains the relatively high fouling potential of AOM and SA solutions containing both Na^+ and Ca^{2+}.

Combining all major cations of seawater in one solution resulted in a slightly lower fouling potential. This can be explained as a result of competition between Ca^{2+} and other cations (Mg^{2+} and K^+) for binding sites and possible stabilization effects by SO_4^{2-} and HCO_3^- (Jin *et al.*, 2009). Considering that carboxyl groups have stronger affinity to Ca^{2+} ions than to other cations might explain why its effect was dominant over other major cations in seawater.

Finally, it is apparent that the membrane fouling potential of AOM is significantly different from those of SA and XG. This indicates that using these model polysaccharides in fouling studies might result in outcomes being quite different from the fouling behaviour of seawater AOM and possibly of microbially-derived polysaccharides in general. It is therefore not sufficient to just rely on these model substances to interpret the fouling behaviour of biopolymers produced in natural aquatic systems (de la Torre *et al.*, 2009).

7.4.3 Reversibility of biopolymer fouling by hydraulic backwashing

During long term operation of a dead-end UF system, fouling reversibility or backwashability rather than filterability is a more important measure of fouling. In this case, the efficiency of hydraulic backwashing is dependent on the ability of foulant materials to firmly attach to membrane pores and/or surfaces. Considering that the reported stickiness of algal-derived biopolymers is about 2-4 orders of magnitude more than most other particles (Passow, 2002), their fouling propensity in terms of backwashability is expected to be an additional critical issue during UF operation.

7.4.3.1 Backwashability at different ionic conditions

The backwashability of AOM and SA solutions with similar biopolymer concentration was compared at constant filtration flux of 80 $L/m^2/h$ with periodic hydraulic backwashing for 1 min at 200 $L/m^2/h$ every 30 mins of filtration. Moreover, the two types of biopolymers were prepared in selected matrices to investigate the effect of solution composition. Filtration experiments for AOM were only conducted at high ionic strength conditions to focus more on their behaviour in seawater. For comparison, two different solutions of SA (with and without cations) were also tested. The results of the filtration experiments and the recorded minimum slope in every filtration cycle are presented in Figure 7-14.

The initial permeability of the UF membrane during fouling with SA was effectively restored by backwashing. Similar fouling behaviour was also observed during filtration of XG at similar biopolymer and ionic concentrations (results not shown). On the contrary, filtration of AOM feed solutions exhibited an increasing fouling rate over subsequent filtration cycles and incomplete restoration of initial membrane permeability by backwashing. It was also observed that the progression of fouling rates (minimum slope) in succeeding filtration cycles was influenced by the cation composition of the feedwater. The average slope was around 50% lower when Ca^{2+} was excluded from the synthetic seawater matrix while no substantial decrease was observed when K^+ and Mg^{2+} were also excluded from the matrix. Remarkably, non-backwashable fouling rates with SA were similar with and cations in the matrix.

The observed reversible fouling during filtration of SA and XG may be attributed to weak association of these biopolymers to the surface and the pore walls of the membrane. Such reversible fouling behaviour was also observed by van der Ven *et al.* (2008) for SA solutions with and without Ca^{2+} and K^+ ions. Without cations, SA aggregates are rather small due to electrostatic repulsion between molecules. The same repulsive force restricts SA molecules from attaching to the membrane which is also negatively-charged. Moreover, smaller biopolymers have higher back diffusion potential which means it has a strong tendency to form a concentration polarisation layer rather than a gel layer near the surface of the membrane. This layer can be easily removed when the convective flow is reversed during backwashing.

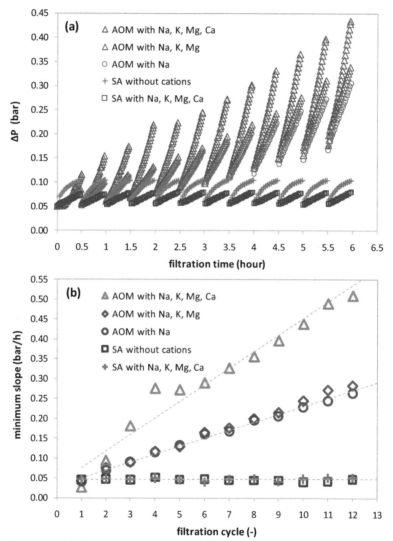

Figure 7-14: Filtration of AOM and SA solution with different cation composition: (a) trans-membrane pressure (ΔP) development over 12 filtration cycles and (b) minimum slope of ΔP increase per filtration cycle.

The partially irreversible fouling observed during filtration of AOM may be attributed to the stronger membrane affinity of TEP gels. Unlike SA or XG, the acidic functional groups which comprise AOM are mainly half-ester sulfates and not carboxyl (Myklestad *et al.*, 1972). Mopper and co-workers reported that biopolymers with covalently bounded half-ester sulfate groups are often linked to their high stickiness (Mopper *et al.*, 1995; Zhou *et al.*, 1998). This intrinsic stickiness may have led to strong attachment of AOM biopolymers on the surface and pores of the membrane by cation bridging (e.g., Ca-bridging). De la Torre *et al.* (2009) investigated the fouling propensity of various polymers and observed higher levels of non-backwashable fouling caused by sulphated polysaccharides (i.e., agarose and carrageenan) than other biopolymers. Furthermore, AOM biopolymers are highly branched (Haug and Myklestad, 1976) and likely having a more complex structure than SA or XG. This provides these biopolymers to have a higher surface area to interact with the membrane by hydrogen bonding. Although hydrogen bonding is rather weak, the huge area of binding sites between the membrane and the highly branched polymer network of AOM biopolymers may have provided significant combined adhesive force to resist the shear forces applied during backwashing.

The fact that there was no large difference in non-backwashable fouling by AOM in solutions with and without Ca^{2+} or Mg^{2+} may indicate that divalent cation bridging between the acidic functional groups of AOM and the negatively-charged membrane was not the main reason for their strong attachment on the membrane. In a recent study, Li *et al.* (2011) found that backwashing a fouled membrane with water containing no Ca^{2+}, Mg^{2+} and K^+ (ultrapure water) was not effective in improving fouling backwashability than when these ions were present in the backwash water. This further supports the theory that hydrogen bonding between AOM and the UF membrane, which is possibly enhanced by charged screening of Na^+ ions, has a major role in non-backwashable fouling in UF membranes by AOM.

7.4.3.2 Effect of biopolymer concentration on backwashability

The backwashability of AOM at different levels of biopolymer concentrations, were measured in seawater matrix. The pH (8.0±0.2), ionic strength (TDS~35 g/L) and ionic composition of the feedwater solutions were maintained for comparison. For the lowest biopolymer concentration (0.2 mgC/L), pore blocking was apparent at the first ~10 mins of filtration followed by cake filtration (Figure 7-15a). At higher biopolymer concentration, the effect of pore blocking occurred instantaneously in the first few minutes of the filtration cycle. In general, the slope of ΔP increase within each filtration cycle illustrates gel filtration with compression as the dominant fouling mechanism.

In terms of fouling reversibility, both backwashable and non-backwashable fouling rates increased with increasing biopolymer concentration. This relationship was not linear but showed an exponential trend (Figure 7-15b). For instance, the total fouling rate (f_T) for the lowest biopolymer concentration (0.2 mgC/L) was 0.015 bar/h. Tripling this concentration to 0.6 mgC/L increased the fouling rate by a factor of four (0.063 bar/h). Moreover, around half of the total fouling rate (42-55%) was due to non-backwashable fouling. Considering that there was a consistent increase in the minimum slope over the succeeding filtration cycles, this may indicate reduced surface porosity of the membrane due to complete plugging of some pores. Most likely, the biopolymers responsible for plugging these pores were not totally removed by backwashing, resulting in a lower membrane porosity and hence, lower initial membrane permeability for the succeeding filtration cycle. This effect appeared to be cumulative in time.

The reduction of membrane porosity due to non-backwashable fouling may cause localised increase in flux on the membrane. Considering that pressure drop through cake layer is directly proportional to the square of filtration flux (Eq. 5), the localise increase in flux has a major effect on the ΔP increase. Salinas-Rodriguez (2011) reported that the effect of cake compression and flux during filtration of coastal seawater are significant at flux >20 L/m²/h and >60 L/m²/h, respectively. Since filtration was performed at flux of 80 L/m²/h, both of these effects may have contributed to the fouling. This combined effect may have caused further adsorption of flexible biopolymers into the pores and surface of the membrane, making them more difficult to remove by subsequent backwashing.

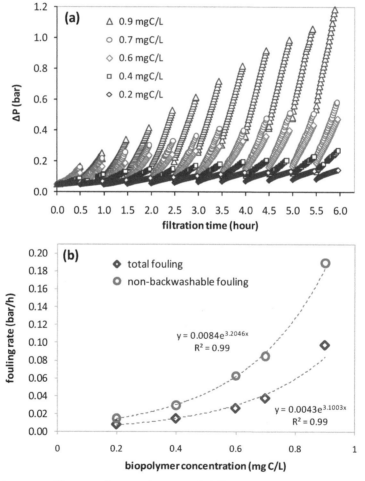

Figure 7-15: Long-term filtration of AOM solutions with different biopolymer concentrations: (a) trans-membrane pressure development over 12 filtration cycles and (b) relationship between feedwater biopolymer concentration and fouling rates (total and non-backwashable fouling).

7.4.4 Projected chemical cleaning interval

Non-backwashable fouling in dead-end UF systems is often mitigated by chemically enhanced backwashing (CEB) or in some cases by cleaning-in-place (CIP). The frequency of chemical cleaning is dependent on the cleaning criteria recommended by the membrane supplier or as set by the plant operator. Some UF membrane suppliers recommend a threshold of 1 bar of ΔP increase to perform CIP cleanings while they allow system operators to set their plant-specific CEB cleaning criterion (e.g., DOW, 2011). A number of UF plants perform CEB cleaning when the permeability goes down to about 100-200 $L/m^2.h.bar$ (Schippers et al., 2004; Schurer et al., 2013). In this study, a conservative threshold of 200 $L/m^2.h.bar$, which is equivalent to a ΔP increase of 0.4 bar at 80 $L/m^2.h$ of permeate flux, was adopted.

To calculate the period of membrane operation before chemical cleaning is needed, the model described in Section 7.2.2 was applied based on the fouling rates measured from various backwashability tests (6-8 hour duration) with feedwaters containing different concentrations of AOM in seawater matrix. For comparison, the model was also applied for similar tests performed with SA and XG (Table 7-3).

Table 7-3: Comparison of the fouling rates and projected chemical cleaning interval during filtration of AOM, SA and XG in seawater.

Feedwater	Biopolymer (mg C/L)	$TEP_{0.1\mu m}$ (mg X_{eq}/L)	MFI-UF [a] s/L^2	Backwashability (bar/h) BW	nBW	Projected chemical cleaning interval [b] (h)
AOM	0.2	0.17	771	0.007	0.008	27
	0.4	0.35	2352	0.015	0.015	14
	0.6	0.52	5116	0.036	0.026	6.9
	0.7	0.61	7053	0.047	0.038	5.2
	0.9	0.78	8942	0.092	0.097	2.6
SA	0.6	0.10	2266	<0.001	<0.001	>483
XG	0.6	0.24	595	0.001	0.002	135

(a) Calculated based on the minimum slope of the first filtration cycle.
(b) Based on chemical cleaning criterion where cleaning is performed at ≥0.4 bar ΔP increase (200 $L/m^2.h.b$ permeability)

The projected chemical cleaning (CEB or CIP) interval for AOM ranged from every 2.6 (0.9 mg C/L) to 27 hours (0.2 mgC/L). These cleaning intervals were high (20-70 times) compared to the projected values for SA and XG. TEP concentrations of the different AOM solutions showed a negative power correlation ($R^2 \sim 0.86$) with the projected chemical cleaning interval (Figure 7-16). This further illustrates the major role of TEP in non-backwashable fouling in dead-end UF membranes.

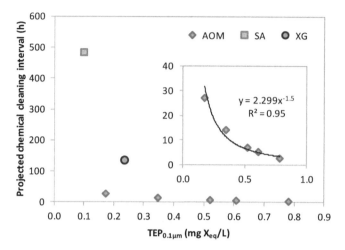

Figure 7-16: Relationship between $TEP_{0.1\mu m}$ concentration and projected chemical cleaning interval during filtration of AOM and model biopolymers in seawater.

7.5 Conclusions

1. The fouling propensity (MFI-UF and non-backwashable fouling) of AOM was substantially higher than model biopolymers (SA and XG) at concentrations typical in seawater algal blooms (<1 mg C/L).

2. Lowering the pH of feedwater containing AOM and model biopolymers in seawater matrix resulted to an increase in the specific MFI-UF per mg C/L.

3. Substantial variations in MFI-UF observed when varying the ionic strength in the feedwater. The membrane fouling potential of AOM is characterised by rapid increase between 0 and 1 g/L of salinity followed by a decreasing trend until 20 g/L of salinity and then gradual increase was observed between 20 and 65 g/L salinity. A similar trend was observed with SA and XG.

4. Among the major mono- and di- valent cations in seawater, only Ca^{2+} demonstrated a significant influence on the fouling propensity (MFI-UF and backwashability) of AOM. However, other cations (e.g., Na^+, K^+ and Mg^{2+}) also showed a pronounced effect on the fouling propensity of model biopolymers (especially SA).

5. Filtration of AOM in seawater caused severe non-backwashable fouling in UF membranes and the rate of fouling increase exponentially with increasing feedwater concentration. On the contrary, negligible non-backwashable fouling was observed with model biopolymers (SA and XG). Consequently, these model biopolymers cannot be used as model foulants in seawater.

6. A simple model was formulated to incorporate both membrane fouling potential (MFI-UF) and backwashability in calculating the projected operation time of UF system before chemical cleaning is required. The projected values were inversely

proportional to the non-backwashable fouling rates and further demonstrated the higher UF fouling propensity of AOM than model biopolymers.

7. Overall, the findings of this study emphasise the importance of understanding the specific fouling potential of algal-derived biopolymers in seawater because their fouling behaviour might not be reliably simulated by just using model biopolymers (e.g., SA and XG).

References

Alldredge AL (1999) The potential role of particulate diatom exudates in forming nuisance mucilaginous scums. Ann Ist Super Sanita 35:397-400.

Alldredge, A. L., Passow, U., and Logan, B. E. (1993) The abundance and significance of a class of large, transparent organic particles in the ocean. *Deep Sea Research(Part I, Oceanographic Research Papers)*, 40(6), 1131-1140.

Amy, G. (2008) Fundamental understanding of organic matter fouling of membranes, *Desalination* 231, 44-51.

Azetsu-Scott, K., and Passow, U. (2004) Ascending marine particles: Significance of transparent exopolymer particles (TEP) in the upper ocean. Limnol. Oceanogr, 49(3), 741-748.

Bacchin P., Aimar P., Sanchez V. (1996) Influence of surface interaction on transfer during colloid ultrafiltration, J. Membr. Sci. 115, 49–63.

Bar-Zeev E, Berman-Frank I, Girshevitz O, and Berman T (2012) Revised paradigm of aquatic biofilm formation facilitated by microgel transparent exopolymer particles. PNAS 109(23):9119–9124.

Berman T, Holenberg M (2005) Don't fall foul of biofilm through high TEP levels. Filtration & Separation 42(4):30-32.

Boerlage S.F.E., Kennedy M.D., Tarawneh Z., De Faber R., Schippers J.C. (2004) Development of the MFI-UF in constant flux filtration. Desalination 161 (2):103–113.

Boerlage, S. F. E., Kennedy, M. D., Aniye, M. P., Abogrean, E., Tarawneh, Z. S., & Schippers, J. C. (2003a). The MFI-UF as a water quality test and monitor. Journal of Membrane Science, 211(2), 271-289.

Boerlage, S. F. E., Kennedy, M., Aniye, M. P., & Schippers, J. C. (2003b). Applications of the MFI-UF to measure and predict particulate fouling in RO systems. Journal of Membrane Science, 220(1-2), 97-116.

Braghetta A., DiGiano F.A. and Ball W., (1997) Nanofiltration of natural organic matter: pH and ionic strength effects, J. Environ. Eng. 6, 628–641.

Dam, H. G., and Drapeau, D. T. (1995). Coagulation efficiency, organic-matter glues, and the dynamics of particles during a phytoplankton bloom in a mesocosm study. Deep-Sea Research II, 42, 111–123.

de La Torre, T., Harff, M., Lesjean, B., Drews, A., & Kraume, M. (2009). Characterisation of polysaccharide fouling of an ultrafiltration membrane using model solutions. Desalination and Water Treatment, 8(1-3), 17-23.

de la Torre, T., Lesjean, B., Drews, A., and Kraume, M. (2008) Monitoring of transparent exopolymer particles (TEP) in a membrane bioreactor (MBR) and correlation with other fouling indicators. Water Science and Technology 58 (10), 1903–1909.

DOW (2011) Ultrafiltration product manual, version 3. Online access (16 August 2013): http://msdssearch.dow.com/PublishedLiteratureDOWCOM/dh_08d9/0901b803808d961b.pdf?filepath=liquidseps/pdfs/noreg/795-00022.pdf&fromPage=GetDoc

Engel, A. (2000). The role of transparent exopolymer particles (TEP) in the increase in apparent particle stickiness (alpha) during the decline of a diatom bloom. Journal of Plankton Research, 22, 485–497.

Faibish, R. S., Elimelech, M., & Cohen, Y. (1998). Effect of interparticle electrostatic double layer interactions on permeate flux decline in crossflow membrane filtration of colloidal suspensions: An experimental investigation. Journal of Colloid and Interface Science, 204(1), 77-86.

Flemming, H.C., Schaule, G., Griebe, T., Schmitt, J., & Tamachkiarowa, A. (1997) Biofouling - the achilles heel of membrane processes. Desalination, 113(2-3), 215-225.

Frank, B. P., & Belfort, G. (2003). Polysaccharides and sticky membrane surfaces: Critical ionic effects. Journal of Membrane Science, 212(1-2), 205-212.

García-Ochoa, F., Santos, V. E., Casas, J. A., & Gómez, E. (2000). Xanthan gum: Production, recovery, and properties. Biotechnology Advances, 18(7), 549-579.

Gasia-Bruch E., Sehn P., García-Molina V., Busch M., Raize O., Negrin M. (2011) Field experience with a 20,000 m3/d integrated membrane seawater desalination plant in Cyprus. Desalination & Water Treatment 31:178–189.

Gu, Y. S., Decker, E. A., & McClements, D. J. (2005). Influence of pH and carrageenan type on properties of β-lactoglobulin stabilized oil-in-water emulsions. Food Hydrocolloids, 19(1), 83-91.

Haug, A., & Myklestad, S. (1976). Polysaccharides of marine diatoms with special reference to chaetoceros species. Marine Biology, 34(3), 217-222.

Haug, A., Larsen, B., & Smidsrød, O. (1974). Uronic acid sequence in alginate from different sources. Carbohydrate Research, 32(2), 217-225.

Henderson, R.K., Baker, A., Parsons, S.A. and Jefferson, B. (2008) Characterisation of algogenic organic matter extracted from cyanobacteria, green algae and diatoms. Water Research 42, 3435-3445.

Huber S. (2012) personal communication.

Huber S. A., Balz A., Abert M. and Pronk W. (2011) Characterisation of aquatic humic and non-humic matter with size-exclusion chromatography – organic carbon detection – organic nitrogen detection (LC-OCD-OND), Water Research 45 (2), 879-885.

Hung, C.C., Guo, L., Santschi, P. H., Alvarado-Quiroz, N., and Haye, J. M. (2003). Distributions of carbohydrate species in the gulf of mexico. Marine Chemistry, 81(3-4), 119-135.

Jin, X., Huang, X., & Hoek, E. M. V. (2009). Role of specific ion interactions in seawater RO membrane fouling by alginic acid. Environmental Science and Technology, 43(10), 3580-3587.

Katsoufidou K, Yiantsios S.G., Karabelas A.J.: Experimental study of ultrafiltration membrane fouling by sodium alginate and flux recovery by backwashing Journal of Membrane Science 300 (2007) 137–146

Katsoufidou, K. S., Sioutopoulos, D. C., Yiantsios, S. G., & Karabelas, A. J. (2010). UF membrane fouling by mixtures of humic acids and sodium alginate: Fouling mechanisms and reversibility. Desalination, 264(3), 220-227.

Kennedy, M. D., Muñoz Tobar, F. P., Amy, G., & Schippers, J. C. (2009). Transparent exopolymer particle (TEP) fouling of ultrafiltration membrane systems. Desalination and Water Treatment, 6(1-3), 169-176.

Lee, N., Amy, G., & Croué, J. P. (2006). Low-pressure membrane (MF/UF) fouling associated with allochthonous versus autochthonous natural organic matter. Water Research, 40(12), 2357-2368.

Lee, S., Ang, W. S., & Elimelech, M. (2006). Fouling of reverse osmosis membranes by hydrophilic organic matter: Implications for water reuse. Desalination, 187(1-3), 313-321.

Leppard G.G. (1997) Colloidal organic fibrils of acid polysaccharides in surface waters: electron-optical characteristics, activities and chemical estimates of abundance. Colloids and Surfaces A: Physicochemical and Engineering Aspects 120:1-15.

LeRoux, M. A., Guilak, F., & Setton, L. A. (1999). Compressive and shear properties of alginate gel: Effects of sodium ions and alginate concentration. Journal of Biomedical Materials Research, 47(1), 46-53.

Li, S., Heijman, S. G. J., Verberk, J. Q. J. C., Le Clech, P., Lu, J., Kemperman, A. J. B., Amy, G.L. and van Dijk, J. C. (2011). Fouling control mechanisms of demineralized water backwash: Reduction of charge screening and calcium bridging effects. Water Research, 45(19), 6289-6300.

Logan, B. E., Passow, U., Alldredge, A. L., Grossart, H.-P., & Simon, M. (1995). Rapid formation and sedimentation of large aggregates is predictable from coagulation rates (half-lives) of transparent exopolymer particles (TEP). Deep-Sea Research II, 42, 203–214.

Mopper, K., Zhou, J., Ramana, K. S., Passow, U., Dam, H. G., & Drapeau, D. T. (1995). The role of surface-active carbohydrates in the flocculation of a diatom bloom in a mesocosm. Deep-Sea Research II, 42, 47–73.

Mosley, L. M., Hunter, K. A., & Ducker, W. A. (2003). Forces between colloid particles in natural waters. Environmental Science and Technology, 37(15), 3303-3308.

Myklestad, S. M. (1995). Release of extracellular products by phytoplankton with special emphasis on polysaccharides. Science of the Total Environment, 165, 155-164.

Myklestad, S., Haug, A., & Larsen, B. (1972). Production of carbohydrates by the marine diatom Chaetoceros affinis var. willei (Gran) Husted II. Preliminary investigation of the extracellular polysaccharides. Journal of Experimental Marine Biology and Ecology, 9, 137–144.

Myklestad, S., Holm-Hansen, O. Varum, K.M. and Volcani, B.E. (1989) Rate of release of extracellular amino acids and carbohydrates from marine diatom Chaetoceros Affinis. *Journal of Plankton Research* 11 (4), 763-773.

Passow, U. (2000) Formation of transparent exopolymer particles (TEP) from dissolved precursor material. Marine Ecology Progress Series 192, 1-11.

Passow, U. (2002) Transparent exopolymer particles (TEP) in aquatic environments, Progress In Oceanography 55(3-4), 287-333.

Passow, U., and Alldredge, A. L. (1995) A Dye-Binding Assay for the Spectrophotometric Measurement of Transparent Exopolymer Particles (TEP). Limnology and Oceanography 40(7), pp. 1326-1335.

Ralston J., Larson I., Rutland M., Feiler A., Kleijn M. (2005) Atomic force microscopy and direct surface force measurements (IUPAC Technical Report), Pure and Applied Chemistry, 77, 2149-2170.

Ricq, L., Pierre, A., Bayle, S., & Reggiani, J.C. (1997). Electrokinetic characterization of polyethersulfone UF membranes. Desalination 109(3), 253-261.

Salinas-Rodriguez S.G. (2011) Particulate and organic matter fouling of SWRO systems: Characterization, modelling and applications. Ph.D. thesis, UNESCO-IHE/TUDelft, Delft.

Schippers, J. C., Hanemaayer, J. H., Smolders, C. A., & Kostense, A. (1981). Predicting flux decline of reverse osmosis membranes. Desalination, 38(C), 339-348.

Schippers, J. C., Kruithof, J. C., Nederlof, M. M., Hofman, J. A. M. H., and Taylor, J. S. (2004) Integrated Membrane Systems, AWWA Research Foundation and American Water Works Association.

Schippers, J.C. and Verdouw, J. (1980) The modified fouling index, a method of determining the fouling characteristics of of water. Desalination 32, 137-148.

Schurer R., Tabatabai A., Villacorte L., Schippers J.C., Kennedy M.D. (2013) Three years operational experience with ultrafiltration as SWRO pre-treatment during algal bloom. Desalination & Water Treatmen 51 (4-6), 1034-1042.

Schurer, R., Janssen, A., Villacorte, L., Kennedy, M.D. (2012) Performance of ultrafiltration & coagulation in an UF-RO seawater desalination demonstration plant. Desalination & Water Treatment 42(1-3), 57-64.

Sioutopoulos D.C, Yiantsios S.G., Karabelas A.J. (2010) Relation between fouling characteristics of RO and UF membranes in experiments with colloidal organic and inorganic species - Journal of Membrane Science 350, 62–82

Spruijt E., van den Berg S.A., Cohen-Stuart M.A., and van der Gucht J. (2012) Direct Measurement of the Strength of Single Ionic Bonds between Hydrated Charges. ACS Nano 6 (6), 5297–5303.

van de Ven, W. J. C., Sant, K. v., Pünt, I. G. M., Zwijnenburg, A., Kemperman, A. J. B., van der Meer, W. G. J., & Wessling, M. (2008). Hollow fiber dead-end ultrafiltration: Influence of ionic environment on filtration of alginates.Journal of Membrane Science, 308(1-2), 218-229.

van Nevel S., Hennebel T., De Beuf K., Du Laing G., Verstraete W., Boon N. (2012) Transparent exopolymer particle removal in different drinking water production centers. Water Research 46:3603-2611.

Verdugo, P., Alldredge, A. L., Azam, F., Kirchman, D. L., Passow, U. and Santschi, P. H. (2004) The oceanic gel phase: a bridge in the DOM–POM continuum. Marine Chemistry 92, 67-85.

Villacorte, L. O., Kennedy, M. D., Amy, G. L., and Schippers, J. C. (2009a) The fate of Transparent Exopolymer Particles (TEP) in integrated membrane systems: Removal through pretreatment processes and deposition on reverse osmosis membranes. Water Research 43 (20), 5039-5052.

Villacorte, L. O., Kennedy, M. D., Amy, G. L., and Schippers, J. C. (2009b) Measuring Transparent Exopolymer Particles (TEP) as indicator of the (bio)fouling potential of RO feed water. Desalination & Water Treatment 5, 207-212.

Villacorte, L. O., Schurer, R., Kennedy, M., Amy, G., Schippers, J.C. (2010a) The fate of transparent exopolymer particles in integrated membrane systems: a pilot plant study in Zeeland, The Netherlands. Desalination & Water Treatment 13:109-119.

Villacorte, L.O, Schurer, R., Kennedy, M., Amy, G. and Schippers, J.C. (2010b) Removal and deposition of Transparent Exopolymer Particles (TEP) in seawater UF-RO system. IDA Journal 2 (1), 45-55.

Villacorte, L.O., Ekowati, Y., Winters, H., Amy, G.L., Schippers, J.C. and Kennedy, M.D. (2013) Characterisation of transparent exopolymer particles (TEP) produced during algal bloom: a membrane treatment perspective. Desalination & Water Treatment 51 (4-6), 1021-1033.

Wilkinson, K.J., Joz-Roland, A., Buffle, J., 1997. Different roles of pedogenic fulvic acids and aquagenic biopolymers on colloid aggregation and stability in freshwaters. Limnol. Oceanogr. 42, 1714–1724.

Ye, Y., Le Clech, P., Chen, V., Fane, A. G., & Jefferson, B. (2005). Fouling mechanisms of alginate solutions as model extracellular polymeric substances. Desalination, 175, 7-20.

Zhou, J., Mopper, K., & Passow, U. (1998). The role of surface-active carbohydrates in the formation of transparent exopolymer particles by bubble adsorption of seawater. *Limnology and Oceanography, 43*(8), 1860-1871.

Fouling potential of algae in inside-out capillary UF membranes

Contents

Abstract..214

8.1 Introduction...215

    8.1.1 Fouling in capillary MF/UF...216
    8.1.2 Goal and objectives 218

8.2 Theoretical considerations 219

    8.2.1 Hydrodynamic conditions in capillary UF..........................219
    8.2.2 Particulate fouling in capillary UF membranes.................220
    8.2.3 Particle transport through capillary membrane 224
    8.2.4 Model reference conditions...227

8.3 Experimental methods...228

    8.3.1 Algal culture 228
    8.3.2 Cell counting and TEP measurement................................228
    8.3.3 Lab-scale UF unit..228

8.4 Results and discussion 229

    8.4.1 Membrane fouling potential of uniformly deposited algae 229
    8.4.2 Plugging potential of non-uniformly deposited algae.....................235
    8.4.3 Non-backwashable fouling in capillary membrane........................242
    8.4.4 Fouling mechanisms..243

8.5 Conclusions...244

References 245

This chapter is based on:

Villacorte L. O., Gharaibeh M., Schippers J.C. and Kennedy M.D. (2013) Fouling potential of algae in inside-out driven capillary UF membranes. *In Preparation for Journal of Membrane Science.*

Abstract

During severe algal blooms, membrane fouling and plugging may occur in capillary ultrafiltration (UF) membranes due to accumulation of algal cells and algal organic matter (AOM). In this chapter, theoretical and experimental investigations were performed to determine the membrane fouling and plugging potential of microscopic algae on the operation of dead-end inside-out capillary UF membranes during filtration of seawater in severe algal bloom conditions. The membrane fouling potential is measured based on the modified fouling index model (MFI-UF) and the plugging potential was calculated based on the tendency of algal cells (mainly depending on their size) to deposit in the dead-end section of the capillary membrane. Moreover, the potential impact of plugging on the hydraulic performance of capillary membranes and how these can be mitigated was also investigated.

The theoretical membrane fouling potential (in terms of MFI-UF) of 16 species of bloom-forming algae were 100 to 10,000 times lower than what was observed in filtration experiments with two laboratory grown species (Chaetoceros affinis and Microcystis sp.). This observation can be attributed to the presence of AOM produced by algae (e.g., transparent exopolymer particles, TEP). Furthermore, the experimental data showed lower correlation (R²<0.5) between algae concentration and MFI-UF than what was observed between TEP concentration and MFI-UF (R²>0.8). Experiments with a UF test unit demonstrated severe non-backwashable fouling when free AOM and/or AOM attached to algal cells are present in the feed solution.

Further analyses were performed to investigate the transport and deposition of algal cells through capillary membrane by combining the effect of different particle back-transport mechanisms due to shear. The deposition of algae is mainly influenced by shear inertial lift and induced diffusion along the length of the capillary. The larger the algal cell, the higher is the possibility that it will accumulate on the end or in the middle of the capillary (depending on the UF module design). Capillary plugging due to accumulation of algal cells in the dead-end section of the capillary can cause increase in membrane and cake resistance due to localised increase in flux. Such potential problems can be addressed by shortening the filtration cycle, lowering the flux and introducing a small cross-flow (bleed) at the dead-end of the capillary.

8.1 Introduction

Algal blooms can have drastic consequences on the operation of conventional and advanced water treatment facilities. A remarkable example is the severe "red tide" occurrence caused by the dinoflagellate *Cochlodinium polykrikoides* in the Middle East Gulf region in 2008-2009 (Richlen *et al.*, 2010). At least five seawater reverse osmosis (RO) plants were forced to reduce or shutdown operation mainly due to clogging problems in the pre-treatment systems (i.e., granular media filters, GMF) and due to concerns that the high silt density index (SDI) of the GMF effluent would irreversibly foul RO membranes (Berktay, 2011; Pankratz, 2008). This incident exposes the vulnerability of granular media filters (GMF) to such operational problems during severe algal bloom situations.

Several options have been proposed to solve the problem of rapid clogging of GMF during algal blooms, namely:

- installing a dissolved air flotation (DAF) system (preceded by a coagulation step) before the GMF;
- installing an advanced coagulation/sedimentation system (e.g., micro-sand dosing); and
- reducing the rate of filtration in the GMF.

To solve the problem of poor quality of the pre-treated water (GMF effluent), a few options have been proposed. Incorporating and/or increasing the dose of coagulant in front of the GMF may improve the effluent water quality. However, an increase in coagulant dose will further increase the rate of clogging in GMF. Installing a DAF or advanced coagulation system in front of GMF may enable increase in coagulant dosage to improve the effluent quality while reducing clogging problems in the GMF. Another option is to install ultrafiltration (UF) or microfiltration (MF) membrane systems to replace the GMF. UF/MF pre-treatment can guarantee an RO feed water with low SDI even during severe algal bloom. However, some concerns have been expressed regarding the rate of fouling in the UF/MF membrane systems (e.g., backwashable and non-backwashable fouling) even when the option of in-line coagulation is in place (Schurer *et al.*, 2012; 2013). To overcome this concern, a DAF system preceding a UF/MF system has been recommended.

Over the last decade, the application of MF/UF is gaining ground as a more reliable alternative to granular media filtration. UF/MF membranes are generally effective in removing particulate and colloidal material from RO feed water than other widely used RO pre-treatment technologies. Thus, it is expected that UF/MF will be more reliable in maintaining RO feed water with a low particulate fouling potential even during an algal bloom period.

During algal bloom in surface water, suspended particulate materials largely comprise living and dead microorganisms (e.g., algae, bacteria) and algal-derived biopolymers. Hard cell debris such as silica frustules from diatomaceous algae and cellulose thecal plates from armoured dinoflagellates may also be abundant. A high concentration of algae in surface water is often followed with a substantial concentration of algal organic materials (AOM) such as transparent exopolymer particles (TEP), which they actively release, shed off from their cell wall or the remains of deteriorating dead cells.

Bloom-forming algae are particulate material, generally larger than 2 μm. TEPs, as the name suggests, are often considered as particulate material but since they can retain water and only allow some water to pass through, they can also be classified as hydrogels (Verdugo et al., 2004). Like other hydrogels, TEPs are extremely hydrated and likely containing more than 99% water (Azetsu-Scott and Passow, 2004). This means that they can bulk-up to more than 100 times their solid volume. So far, no theoretical and limited experimental investigations have been reported on the combined effect of these materials on MF/UF performance during filtration of algal bloom impacted waters.

8.1.1 Fouling in capillary MF/UF

MF/UF membranes in capillary configuration are dominating the market in surface water treatment. Membrane elements are mainly manufactured as capillaries 0.5–5 mm diameter and 1-2.3 long. A membrane element usually contains hundreds and up to thousands of capillaries bundled together with one or both ends of the capillaries potted with sealing agents such as epoxy resins, polyurethanes and silicone rubber (Lerch, 2008). Capillary membrane modules are either operated "inside-out" or "outside-in" filtration mode. During filtration of surface of water, dead-end MF/UF membranes are frequently backwashed to remove the accumulated suspended and colloidal matter. This study focused on fouling – under algal bloom conditions - of capillary MF/UF membrane operated in dead-end, inside-out filtration mode.

During filtration of water suffering from severe algal bloom through inside-out capillary membranes, different fouling mechanisms might occur simultaneously such as:

- Uniform accumulation of small particles (e.g., TEP-like materials) on the membrane surface;
- Non-uniform deposition of larger particles (e.g., algal cells) along the length of the capillary channel. Large particles tend to deposit at one end of the capillaries (if water is fed from one end) or in the middle of the capillaries (if water is fed from both ends).
- Deposition of large particles at the entrance of the capillaries, resulting in partial/complete blockage of the feed water channel. In practice, however, micro-strainers (50 – 150 μm) are usually installed in front of the membrane system to avoid this type of fouling.

Fouling in MF/UF membranes due to uniform deposition of particles/colloids (e.g., algae, AOM) is illustrated in Figure 8-1 and further discussed in the succeeding sections.

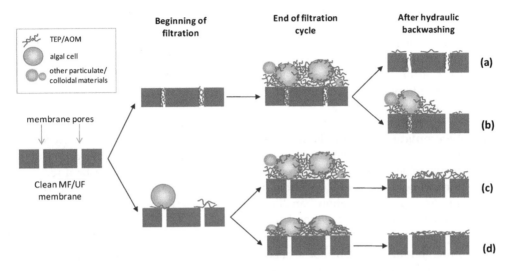

Figure 8-1: Possible fouling mechanisms involved during filtration of algal bloom impacted waters through MF/UF membrane.

(a) **Constriction of pores**. Colloidal AOMs can enter into the narrow pores of MF/UF membranes, some of which will adsorb to the pore wall and eventually cause partial blockage of permeate flow. This can cause rapid increase in trans-membrane pressure during the initial stage of filtration. Meanwhile, algal cells and large AOM can form a cake/gel layer on the surface of the membrane. Colloidal AOMs and other colloids will then fill up the large interstitial voids of the cake, narrowing the voids during the process. This may result in substantial increase in cake resistance due to the gradual reduction in cake porosity. During hydraulic backwashing, the adhesive AOM materials which accumulated inside the membrane pores may not be completely removed, resulting in only partial recovery of the initial membrane permeability (Figure 8-1a).

(b) **Substantial loss in effective filtration area**. Colloidal AOM can accumulate inside the membrane pores while algal cells and large AOM can accumulate at the entrance of the pores. In both cases, some pores may be completely blocked by the materials and the active filtration area (membrane surface porosity) is substantially reduced, resulting in higher localised flux for the remaining active pores. A similar effect may also occur if a substantial section of the capillary channel is plugged by algal cells and AOM (see Section 8.4.3). An increase in flux can cause proportional increase in membrane resistance while a more substantial increase in cake resistance can be expected due to cake compression. Additionally, non-backwashable fouling can occur if the material blocking the pores are not effectively removed by subsequent hydraulic backwashing (Figure 8-1b).

(c)   **Incomplete cake/gel removal during backwashing**. Since algal cells and a substantial fraction of AOM are much larger than the pores of the membrane, cake/gel formation may be mainly responsible for the increase in trans-membrane pressure. Accumulated AOM material (like TEPs) can be very sticky and tend to adhere strongly to the membrane surface. During backwashing, a layer of the cake may remain on the surface of the membrane, which will then cause additional filtration resistance in the succeeding filtration cycle (Figure 8-1c).

(d)   **Accumulated cake/gel is compressible.** Filter cake/gel comprising AOM and algal cells (soft-bodied) can be compressed at high cake pressure drop due to increased localised flux (Figure 8-1d). Such increase in flux can be a result of narrowing of pores and/or substantial loss in effective filtration area. Hence, this fouling phenomenon can likely occur in combination with other fouling mechanisms (e.g., mechanisms a and b).

Membrane fouling due to uniform deposition of particles/colloids is widely studied while fouling problem due to uneven deposition of particles (e.g., plugging) capillary membranes has not been adequately studied. More investigations are therefore needed to measure the effect of non-uniform deposition of particles on capillary membranes, especially during algal bloom situations.

8.1.2 Goal and objectives

The goal of this chapter is to illustrate the role of algal cells (with and without AOM) in the fouling of capillary MF/UF membranes during algal bloom based on experimental and theoretical studies. The following are the specific objectives:

2. Calculate the fouling potential of bloom-forming algae based on the cake filtration model (i.e., MFI-UF);

3. Measure the effect of bloom-forming algae with and without free TEP-like materials on the operational performance of capillary UF;

4. Discuss the various mechanisms controlling the transport of algal cells through capillary UF membrane and illustrate the potential consequences; and

5. Investigate the effect of unequal deposition of algal cells – under algal bloom conditions – on the performance of capillary UF systems.

8.2 Theoretical considerations

8.2.1 Hydrodynamic conditions in capillary UF

Feedwater flow over the length of the capillary membrane is not uniform. It is generally characterised by the higher radial and axial velocity near the entrance of the capillary and lower radial and zero axial velocity at the dead-end side. Therefore, the condition is not strictly dead-end but rather cross-flow filtration especially near the entrance of the capillary. Panglisch (2001) demonstrated that the pressure and velocity profiles over the axial (z) direction of the capillary membrane can be calculated as a function of capillary dimensions, average flux and membrane resistance. In this study, the equations developed by Panglisch (2001) were adopted to calculate the hydrodynamic profiles in capillary UF membrane.

The pressure profile along the longitudinal axis of a capillary membrane considering both trans-membrane pressure and feed channel pressure losses is described by the following equations:

$$P(z) = P_p + A \cosh(az) + B \sinh(az) \hspace{3cm} \text{Eq. 1}$$

$$A = \frac{8\eta Q_{in}}{\pi r_c^4} \frac{1}{a \tanh(aL_c)}$$

$$B = -\frac{8\eta Q_{in}}{\pi r_c^4} \frac{1}{a}$$

$$a = \sqrt{\frac{16}{r_c^3 R_m}}$$

where P(z) is the pressure at a specific section along the capillary length, P_p is the permeate pressure, z is the distance along the axial direction from the inlet of the capillary, r_c is the internal radius of the capillary, L_c is the length of the capillary, η is the dynamic water viscosity, Q_{in} is the inlet water flow and R_m is the resistance of the membrane.

The feed pressure (P_f) is the maximum pressure at z=0. Assuming P_p is negligible, P_f is calculated as follows:

$$P_f = A = \frac{8\eta Q_{in}}{\pi r_c^4} \frac{1}{a \tanh(aL_c)} \hspace{3cm} \text{Eq. 2}$$

The average axial fluid velocity ($\bar{v}_z$) at any point along the longitudinal axis (z) of capillary membrane is calculated as follows:

$$\bar{v}_z(z) = -\frac{r_c^2}{8\eta}[Aa \sinh(az) + Ba \cosh(az)] \hspace{3cm} \text{Eq. 3}$$

The average radial water velocity (J_p) at any point along the longitudinal axis (z) of a capillary membrane is calculated as follows:

$$J_p(z) = \frac{1}{\eta\, R_m} [A \cosh(az) + B \sinh(az)] \qquad\qquad \text{Eq. 4}$$

8.2.2 Particulate fouling in capillary UF membranes

The following sections explain fouling in inside-out capillary UF membranes for two different scenarios: uniform and non-uniform deposition of particles.

8.2.2.1 Uniform deposition of particles

Clean water flux through an MF/UF membrane is inversely proportional to membrane resistance. During dead-end filtration of natural waters, suspended particulate and colloidal materials accumulate on the pores and/or the surface of the membrane, eventually generating additional hydraulic resistance to filtration.

$$J_p = \frac{\Delta P}{\eta\,(R_m + R_b + R_c)} \qquad\qquad \text{Eq. 5}$$

where J_P is the membrane permeate flux, ΔP is the total trans-membrane pressure drop, η is the dynamic water viscosity, R_m is the membrane resistance, R_b is blocking resistance and R_c is the cake or gel resistance. In this model, the deposition of particles is expected to be uniform on the membrane surface.

During an algal bloom, it is expected that gel/cake filtration is a dominant fouling mechanism in UF/MF membranes because of the high concentration of particles larger than the membrane pores (e.g., algae cells, TEPs). However during subsequent filtration cycles the restoration of the initial permeability is not complete as a result of different fouling mechanisms including pore blocking. Due to complexity of these mechanisms, the focus of this chapter is on cake/gel filtration in order to simplify the model. Consequently, the resistance due to pore blocking (R_b) was neglected from Eq. 5.

Typically, UF/MF membrane systems operate in constant flux filtration mode. Under such conditions, the increase in hydraulic resistance through the cake/gel layer is directly proportional to the concentration of particles in the bulk solution.

$$R_c = a\, C_b J_p\, t \qquad\qquad \text{Eq. 6}$$

where a is the specific cake resistance and C_b is the bulk particle concentration.

Schippers and Verdouw (1980) proposed measuring the membrane fouling potential of the feed water by defining the fouling index (I) as follows:

$$I = a\, C_b \qquad\qquad \text{Eq. 7}$$

Substituting Eq. 6 and Eq. 7 to Eq. 5 and assuming a constant permeate flux results in,

$$\Delta P = J_p \, \eta \, R_m + J_p^2 \, \eta \, l \, t \qquad \text{Eq. 8}$$

Simplifying Eq. 8 in terms of pressure drop through the clean membrane (ΔP_m) and pressure drop through the filter cake/gel (ΔP_c),

$$\Delta P = \Delta P_m + \Delta P_c \qquad \text{Eq. 9}$$

The pressure gradient through the filter cake is defined based on the Kozeny-Carman equation:

$$\frac{dP}{dL} = \frac{180 \, \eta \, V_o}{\psi^2 d_p^2} \frac{(1-\varepsilon)^2}{\varepsilon^3} \qquad \text{Eq. 10}$$

where V_o is the superficial velocity, ε is the porosity of the cake, ψ sphericity of particles and d_p diameter of particles comprising the cake. The particle sphericity is calculated based on the volume of the particle (v_p) and its surface area (A_p):

$$\psi = \frac{\pi^{\frac{1}{3}} (6 v_p)^{\frac{2}{3}}}{A_p} \qquad \text{Eq. 11}$$

From Eq. 10, the pressure drop (ΔP_c) through a cake of known thickness (L) can be calculated as follows:

$$\Delta P_c = \frac{180 \, \eta \, J_P}{\psi^2 d_p^2} \cdot \frac{(1-\varepsilon)^2}{\varepsilon^3} L \qquad \text{Eq. 12}$$

During constant flux dead-end filtration, the change in cake thickness over time is directly proportional to the concentration of particles (C_p) and their specific volume (v_p) in the bulk solution. Note: C_p is the particle concentration in terms of particle count per ml while C_b is in terms of mg/L.

$$\frac{dL}{dt} = \frac{J_P \, C_p v_p}{1-\varepsilon} \qquad \text{Eq. 13}$$

Assuming that foulant materials are spherical particles,

$$\frac{dL}{dt} = \frac{\pi \, J_P \, C_p d_p^3}{6 \, (1-\varepsilon)} \qquad \text{Eq. 14}$$

Integrating the cake thickness over time,

$$L = \frac{\pi \, J_P \, C_p d_p^3}{6 \, (1-\varepsilon)} t \qquad \text{Eq. 15}$$

Substituting Eq. 15 to Eq. 12,

$$\Delta P_c = \frac{30\pi\, d_p \eta\, J_p^2\, C_p t}{\psi^2} \cdot \frac{(1-\varepsilon)}{\varepsilon^3} = \eta\, J_p^2\, I\, t \qquad \text{Eq. 16}$$

Simplifying and rearranging Eq. 16 to calculate the fouling index (I),

$$I = \frac{30\pi\, d_p C_p}{\psi^2} \cdot \frac{(1-\varepsilon)}{\varepsilon^3} \qquad \text{Eq. 17}$$

In practice, the fouling index (I) is further normalised based on the standard reference conditions set by Schippers and Verdouw (1980), officially known as the modified fouling index (MFI).

$$\text{MFI} = \frac{\eta_{20^\circ C}\, I}{2\, \Delta P_0 A_0^2} \qquad \text{Eq. 18}$$

where $\eta_{20^\circ C}$ is the water viscosity at 20°C (0.001003 Pa.s), ΔP_o is the standard feed pressure (2 bar) and A_o is the standard membrane area (13.8 x10-4 m²).

Substituting Eq. 17 to Eq. 18,

$$\text{MFI} = \frac{15\pi\, d_p C_p}{\psi^2} \cdot \frac{(1-\varepsilon)}{\varepsilon^3} \cdot \frac{\eta_{20^\circ C}}{\Delta P_0 A_0^2} \qquad \text{Eq. 19}$$

Initially, MFI was measured using membranes with 0.45 or 0.05 µm pore sizes and at constant pressure. However, it was later found that particles smaller than the pore size these membranes most likely play a dominant role in particulate fouling. In addition, it became clear that the predictive value of MFI measured at constant pressure is limited. For these reasons, MFI test measured at constant flux with ultra-filtration membranes was eventually developed (Boerlage et al, 2004; Salinas-Rodriguez et al., 2012). In this chapter, MFI-UF constant flux is applied.

To determine the experimental membrane fouling potential of water samples, constant flux MFI-UF measurement was performed using 15 cm long and 0.8 mm diameter capillary membranes (150 kDa MWCO, Pentair X-Flow). Sample filtration was set at 80 L/m².h flux and the increase in trans-membrane pressure (ΔP) was monitored over time for up to 30 mins. The fouling index (I) was then calculated based on the minimum slope of the ΔP versus time plot.

In dead-end MF/UF systems, filtration time between hydraulic cleanings (backwashing) is rather short (15-60 mins). Assuming that cake/gel filtration is the dominant fouling mechanism, the calculated MFI of the feed water can be used to predict the increase of cake pressure drop at the end of the first filtration cycle. This model was used in this study to measure the fouling potential of algal cells and investigate their role in the fouling of MF/UF system during algal bloom.

8.2.2.2 Non-uniform deposition of particles

Non-uniform deposition of particles might also occur in capillary membranes. In extreme situation, particles will not deposit evenly on the membrane surface but will accumulate at the end (if fed from one end) or in the middle (if fed from both ends) of the capillary. This can cause (partial) plugging of the feed channel. Consequently, the active membrane area will be reduced. When the permeate flow is kept constant, the flux of the remaining active membrane area will increase and a higher feed pressure is needed (Figure 8-2).

Considering a scenario where all algal cells from the feed water accumulate on the dead-end side of a capillary;

$$V_{plugged} = \frac{V_{algae}}{1-\varepsilon} \qquad\qquad\qquad Eq.\ 20$$

$$V_{plugged} = \frac{L_{plugged}}{L_c} \cdot V_{capillary} \qquad\qquad\qquad Eq.\ 21$$

$$V_{algae} = \phi_b Q_{in} t \qquad\qquad\qquad Eq.\ 22$$

$$\frac{L_{plugged}}{L_c} \approx \frac{A_{loss}}{A_m} \qquad\qquad\qquad Eq.\ 23$$

where $V_{plugged}$ is the plugged volume of capillary, V_{algae} is total volume of accumulated algae, $V_{capillary}$ is total volume of capillary channel, $L_{plugged}$ is the length of the section of capillary which is plugged with algal cells, L_c is the capillary length, ε is cake porosity, ϕ_b is the volume fraction of algae in the feed water, Q_{in} is the feed flow, t is the filtration time, A_{loss} is the effective membrane area reduction and A_m is the total membrane area.

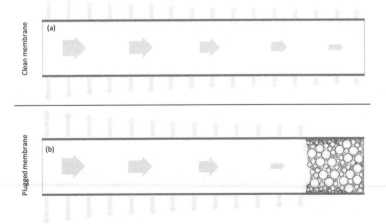

Figure 8-2: Graphical illustration of plugging in dead-end inside-out capillary UF membranes by algal cells.

By combining Eq. 20 to Eq. 22, the filtration time (t) can be calculated as follows;

$$t = \frac{A_{loss}}{A_m} \cdot \frac{V_{capillary}}{\phi_p} \cdot \frac{(1-\varepsilon)}{Q_{in}}$$

$$t = \frac{A_{loss}}{A_m} \cdot \frac{3\, d_c^2\, L_c}{2\, d_p^3\, C_p} \cdot \frac{(1-\varepsilon)}{Q_{in}} \qquad\qquad\qquad Eq.\ 24$$

where d_c is the capillary diameter, d_p is algal cell diameter and C_p is algal cell concentration.

The effect of plugging on the permeate flux (J_p) is explained as follows;

$$J_p = \frac{Q_{in}}{A_m} \qquad\qquad \text{(for clean membrane)} \qquad\qquad Eq.\ 25$$

$$J_p = \frac{Q_{in}}{\left(1-\frac{A_{loss}}{A_m}\right)A_m} \qquad\qquad \text{(for plugged membrane)} \qquad\qquad Eq.\ 26$$

The effect of plugging on the pressure gradient along the length of the capillary are: (1) increase in trans-membrane pressure due to decrease in effective membrane area and (2) reduction in feed channel pressure drop due to shortening of the effective channel length. The feed pressure (P_f) considering these two effects with respect to the reduction of effective membrane area is calculated using Eq. 27, which is derived from the equations discussed in Section 8.2.1.

$$P_f = \frac{8\eta Q_{in}}{\pi r_c^4} \cdot \frac{1}{a \tanh\left[\left(1-\frac{A_{loss}}{A_m}\right)aL_c\right]} \qquad\qquad Eq.\ 27$$

8.2.3 Particle transport through capillary membrane

Particle transport through the feed channel of an inside-out capillary membrane is governed by;

- the drag force due to permeation;
- the gravity force (sedimentation or buoyancy); and
- the shear forces imposed by the tangential flow facilitating back-transport of particles away from the membrane wall (Figure 8-3).

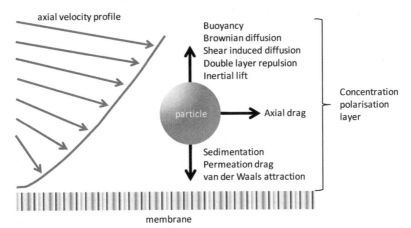

Figure 8-3: Schematic representation of forces affecting the deposition of particles in capillary membranes at laminar flow conditions (adopted from Panglisch, 2003).

Davis (1992) identified the 3 dominant mechanisms affecting back-transport of particles near the surface of the membrane, namely:

1. brownian diffusion,
2. shear-induced diffusion
3. inertial lift

Other less dominant particle transport mechanisms such as particle to particle and particle to membrane interactions (e.g., electrostatic and van der Waals interactions) are not considered in this study (Figure 8-3; Panglisch, 2003, Lerch, 2008). Particle-particle and particle-membrane interactions were excluded in the investigation because their effects are only significant for concentrated colloidal (0.001-1um) suspensions (Davis, 1992; Bowen and Sharif, 1998). Most bloom-forming algae are much larger than 2 μm.

The effect of **sedimentation and buoyancy** of particles are directly related to its size and the deficit between particle and water densities. The settling velocity of spherical particles of uniform density can be calculated based on the Stoke's equation:

$$J_{SB} = -\frac{1}{18}\frac{d_p^2}{\eta}\left(\rho_p - \rho_w\right)g \qquad\qquad\qquad \text{Eq. 28}$$

where d_p is the particle diameter, ρ_p is the particle density, ρ_w is the water density and g is acceleration due to gravity. A positive and negative J_{SB} means the particle is ascending (buoyant) and settling, respectively.

Brownian diffusion can cause displacement of sub-micron particles away from the membrane as a result from random movement of particles through their interaction with fast moving water molecules. The particle back-transport for dilute solutions due to Brownian diffusion is calculated according to the equation reported by

Davies (1992) which is based from the numerical and asymptotic results of Trettin and Doshi (1980).

$$J_B = 1.31 \sqrt[3]{\frac{\gamma_w D_B^2}{L_c} \frac{\phi_w}{\phi_b}}$$

Eq. 29

Where γ_w is the nominal shear rate at the surface of the membrane, D_B is the Brownian diffusivity, L_c is the capillary length, $_w$ and ϕ_b are the volume fractions of particles at the surface of the membrane and the bulk solution, respectively.

For parabolic laminar flows (Re<2300), the nominal shear rate at the wall of a narrow tube (e.g., capillary) is estimated as,

$$\gamma_w = \frac{4Q_z}{\pi r_c^3} = \frac{4 \bar{v}_z}{r_c}$$

Eq. 30

where Q_z and $\bar{v}_z$ are the volumetric flow rate and average axial velocity at a specific point along the longitudinal axis of the capillary, respectively.

The Brownian diffusivity of particle in a fluid of viscosity η is calculated using the Stokes-Einstein relationship,

$$D_B = \frac{kT}{3\pi\eta \, d_p}$$

Eq. 31

where k is Boltzmann constant (1.38×10^{-23} kg.m^2/s^2.K) and T is the absolute temperature (293.15 K).

Shear-induced diffusion is a back-transport mechanism characterised by the random displacements of individual particles from the streamlines in a shear flow as they interact with and tumble over other particles (Davis, 1992). The shear-induced hydrodynamic diffusivity model was first proposed by Zydney and Colton (1986) and further developed by Davis and Sherwood (1990). For dilute suspensions (ϕ_b < 0.1) of mono-disperse rigid spherical particles with a maximum packing density in the boundary layer of $\phi_w \approx 0.6$, the back-transport velocity due to this phenomenon can be estimated as,

$$J_S = 0.0286 \, \gamma_w \sqrt[3]{\frac{d_p^4}{L_c} \frac{\phi_w}{\phi_b}}$$

Eq. 32

The third major back-transport mechanism is due to **inertial lift** (tubular-pinch-effect). First introduced by Green and Belfort (1980), this mechanism is expected to mainly affect large particles at high flow rate. For dilute suspension of spherical particles under fast laminar flow conditions, the approximate back-transport velocity due to inertial lift is,

$$J_I = 0.0045 \, \frac{\rho_w d_p^3 \gamma_w^2}{\eta}$$

Eq. 33

Since the aforementioned dominant particle back-transport mechanisms can occur simultaneously, Jiang *et al.* (2007) proposed summing up the effects of these mechanisms to investigate deposition of particles in cross-flow tubular membranes. The same assumption was adopted in this study but incorporating vertical transport due to either sedimentation or buoyancy (J_{SB}) in the calculation (Eq. 28).

$$J_T = J_{SB} + J_B + J_s + J_I \qquad\qquad\qquad \text{Eq. 34}$$

The described model above assumes that all algae cells are mono-dispersed, non-adhesive, solid and spherical particles and transported under laminar hydrodynamic condition (Re<2300). It also assumes that all particles reaching the surface of the membrane will deposit on the membrane and will not re-suspend back into the bulk solution.

8.2.4 Model reference conditions

A capillary UF membrane (Pentair X-flow, the Netherlands), widely used for seawater RO pre-treatment, was used as the reference membrane for this study. The membrane has an internal diameter of 0.8 mm and molecular weight cut-off of 150 kDa (~0.03um pore size). The membrane modules can be installed in either Seaguard or Aquaflex system configuration.

In the Seaguard system, 4 elements, 1.5 m in length and an internal diameter of 0.8 mm are placed in horizontal pressure vessels. During filtration, water is feed from both ends of the capillary in each element. The hydrodynamic conditions in the first and second half of the capillary length are expected to be symmetrical. Hence, for simplicity, theoretical calculation for this configuration only covered the first half of the capillary length.

In the Aquaflex system, each element (2.2 m long) is placed inside a vertically positioned pressure vessel. Water is feed only from the top end of the capillaries in each element. This configuration provides an option to operate at cross flow mode (bleed) by allowing a small fraction of the feed water to flow out of the lower end of the fibre and then re-circulated back to the feedwater.

For simplicity, the investigation focuses on the Seaguard configuration. The list of reference parameters and variables used in the theoretical calculations are summarized in Table 8-1. The reference values were used in all calculations except where otherwise stated.

Table 8-1: List of reference values and range of different parameters used in theoretical calculations.

Parameters	Units	Reference values	Range of simulation
capillary diameter, d_c	mm	0.8	-
capillary length, l_c	m	0.75	-
average flux, J_{ave}	L/m^2.h	80	20-300
permeate pressure, P_p	bar	0	-
membrane resistance, R_m	m^{-1}	2.24×10^{11}	-
water density, ρ_w	kg/m^3	1030	-
particle density, ρ_p	kg/m^3	1030	-
particle diameter, d_p	μm	15	2-1000
particle sphericity, ψ	-	1	0.1-1
cake porosity, ε	-	0.6	0.1-0.6
volume fraction (bulk), ϕ_b	-	0.0005	0.00001-0.1
volume fraction (membrane), ϕ_m	-	0.6	-

8.3 Experimental methods

8.3.1 Algal culture

A severe algal bloom situation was simulated in the laboratory by growing batch cultures of two species of algae: *Chaetoceros affinis* (CA) and *Microcystis sp.* (MSp). Within a 20-day incubation period, water samples were collected from the batch cultures to measure algal cell concentration (direct counting), TEP concentration and MFI-UF.

8.3.2 Cell counting and TEP measurement

Algal cell counting was performed in triplicates using a Thoma counting chamber and a light microscope. TEP measurement was performed following the modified protocol of Passow and Alldredge (1995) but using 0.1μm polycarbonate membranes (Whatman Nuclepore) to collect TEP from the water (see Chapter 5 for details).

8.3.3 Lab-scale UF unit

A lab-scale UF set-up was used for experiments to measure trans-membrane pressure during filtration of TEP materials from algal cultures. The UF system is a constant flux backwashable unit with a capillary UF module containing 0.8 mm capillary UF (150 KDa MWCO, Pentair X-Flow). To assess the performance of the membrane, trans-membrane pressure (ΔP) was monitored using a sensitive pressure transmitter (Endress+Hauser PMD75) with communication device connected to a computer. The UF module was fabricated in the laboratory containing 6 capillaries of 15 cm in length. Both ends of the module were sealed with water-proofing resins. The feed solutions were fed from both ends of the capillaries (inside-out filtration) and the UF permeate were collected through an outlet in the middle of the module. The system was operated at constant flux of 80

L/m²/h. Every 30 minutes, the system was backwashed with UF permeate at 200 L/m²/h for 1 minute. New membrane modules were fabricated for each experiment.

Some filtration experiments were performed using flat sheet regenerated cellulose membranes (30 kDa MWCO, Millipore) at 80 L/m².h flux. The ΔP was measured over time (up to 50 minutes) without backwashing.

8.4 Results and discussion

8.4.1 Membrane fouling potential of uniformly deposited algae

The fouling potential of algal suspension (without AOM) was calculated based on the model discussed in Section 8.2.2.1. The variables taken into account were size, algal cell concentration (C_p), total cell volume, cake/gel porosity (ε) and cell sphericity (Ψ). The results are shown in Figure 8-4.

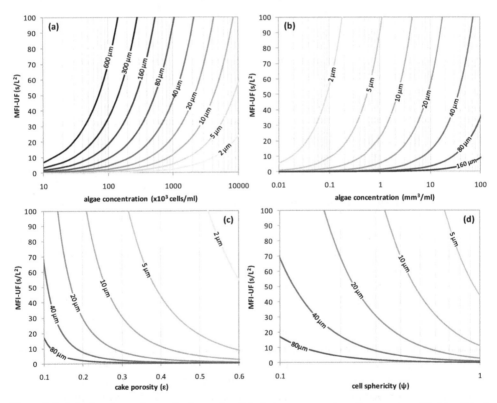

Figure 8-4: Calculated membrane fouling potential of algae suspensions for a range of (a-b) cell and volume concentrations (ε=0.4; ψ=1), (c) cake porosity (C_p= 0.5 mm³/ml; ψ=1) and (d) cell sphericity (C_p= 0.5 mm³/ml; ε=0.4).

Increasing algal concentration in the bulk solution was predicted to result in a linear increase of MFI. In terms of cell count (#/ml), MFI was higher for larger diameter algae than for smaller algae (Figure 8-4a). During bloom situations, peak cell concentrations may vary widely for different groups of algae. Cyanobacterial (blue-green algae) blooms may reach concentrations up to several millions of cells per ml, diatom blooms are from 10s to 100s of thousands of cells per ml and dinoflagellate blooms can reach peak concentrations between few hundreds to thousands of cells per ml. The wide range of bloom concentrations is largely due to differences in cell size and growth characteristics. To consider differences in cell size, MFI was also calculated based on volume concentrations of algae (Figure 8-4b). In this approach, the MFI increase depends on the differences in algae cell size, whereas smaller sized algae have higher fouling potential than larger sized algae.

A natural algal bloom rarely exceeds a volume concentration of 2 mm^3/ml. The highest cell concentration (63.7 mm^3/ml) was reported during the *Noctiluca scintillans* bloom in New Zealand in 2000 (Chang, 2000 as cited by Fonda-Umani *et al.*, 2004). For comparison, the major *Cochlodinium polykrikoides* bloom in the Oman Gulf in 2008/2009 reported only a maximum algal concentration of ~0.5 mm^3/ml (UAE-Interact, 2009; Richlen *et al.*, 2010).

Various species of bloom forming algae such as Phaeocystis globosa, athecated dinoflagellates (e.g., *Cochlodinium polykrikoides*, *Karenia brevis*, *Noctiluca scintillans*) and those belonging to the algal group cyanobacteria and raphidophytes have soft cells which may possibly deform during membrane filtration. In such case, the sphericity of algal cells and the porosity of the accumulated cake may decrease during filtration, resulting in higher cake/gel resistance. Furthermore, flux-driven gradual rearrangement of algal cells (without deformation) resulting in higher cake packing density may also occur (Salinas-Rodriguez *et al.*, 2012; Salinas-Rodriguez, 2011). As shown in Figure 8-4c, a decrease in cake porosity is predicted to result in exponential increase of MFI. A similar effect was observed when decreasing the cell sphericity (Figure 8-4d).

The Intergovernmental Oceanographic Commission of UNESCO identified about 300 species of microscopic algae that were reported to cause blooms (IOC-UNESCO, 2013). Sixteen common species representing different major groups of algae were investigated in this study. These algal species cover a wide range of shapes and sizes (Table 8-2). Typical cells are spherical or ellipsoidal but some are cylindrical (e.g., most diatoms) or elongated (e.g., Pseudo-nitzschia spp.). To normalise for such shape heterogeneities, the equivalent spherical diameter was calculated based the estimated cell volume of the cells. The equivalent cell diameters of the 16 species range from 4μm (Microcystis spp.) to 400μm (Noctiluca scintillans) and a median size of 12 μm. The cell sphericities range from 0.391 to 1.0, where about 90% of which was between 0.8 and 1.0. The maximum recorded concentrations of each species are also shown in Table 8-2. The species with the highest and lowest recorded concentration are *Microcystis spp.* (14,800,000 cells/ml) and *Noctiluca scintillans* (1,900 cells/ml), respectively. The calculated fouling potential (MFI-UF) of algae cells considering the maximum recorded cell concentrations ranged from 0.4 to 70 s/L^2 (Table 8-2).

Table 8-2: Cell characteristics, recorded severe bloom concentrations and calculated membrane fouling potentials of 16 species of common bloom-forming algae.

Bloom-forming algae	Cell shape $(\mu m)^{(a)}$	Eq. diam. (μm) [b]	Sphericity (-)	Severe bloom situation	
				cells/ml [c]	MFI-UF (s/L^2)
Dinoflagellates					
Alexandrium tamarense	RE	32	0.995	10,000	0.38
Cochlodinium polykrikoides	RE	33	0.985	27,000	1.07
Karenia brevis	RE	36	0.984	37,000	1.60
Noctiluca scintillans	Sp	400	1.000	1,900	0.88
Prorocentrum micans	FE	44	0.914	50,000	3.06
Diatoms					
Chaetoceros affinis	OC	15	0.968	900,000	16.76
Pseudo-nitzschia spp.	0.8*PP	7	0.391	19,000	1.01
Skeletonema costatum	Cy	5	0.808	88,000	0.78
Thalassiosira spp.	Cy	12	0.867	100,000	1.86
Cyanobacteria					
Nodularia spp.	Cy	21	0.543	605,200	50.77
Anabaena spp.*	Sp	6	1.000	10,000,000	69.80
Microcystis spp.*	Sp	4	1.000	14,800,000	68.87
Haptophytes					
Emiliania huxleyi	Sp	5	1.000	115,000	0.67
Phaeocystis globosa	0.9*Sp	6	0.933	52,000	0.42
Raphidophytes					
Chattonella spp.	Co+0.5*Sp	15	0.665	10,000	0.39
Heterosigma akashiwo	Sp	20	1.000	32,000	1.68

* Non-marine species of algae; (a) Equivalent geometric dimensions of algal cells based on Olenina et al. (2006); (b) Equivalent diameter of a sphere with similar volume as the cell; (c) Maximum recorded concentrations reported in various literatures. RE=rotational ellipsoid; Sp=sphere; FE=flattened ellipsoid; OC=oval cylinder; PP=parallelepiped; Cy=cylinder; Co=cone.

In general, the fouling potential of algae themselves is low (<100 s/L^2) even during severe bloom situations compared to what was reported in surface water, which are typically higher than 1000 s/L^2 (Salinas-Rodriguez, 2011). The remarkable difference may be attributed to:

- presence of particles much smaller than algae (e.g., TEP–like materials); and/or
- algae are surrounded by TEP–like materials which are blocking the gaps (interstitial voids) between deposited algal cells.

8.4.1.1 Comparison of calculated and measured fouling potential

The calculated fouling potential of algal suspensions (without AOM) showed substantially low MFI in comparison to what is generally observed in surface water. A comparison was conducted between the results of the laboratory simulated algal bloom experiments and the theoretical calculations. Results of the comparison are presented in Table 8-3.

MFI-UF values from lab-scale experiments (algae and AOM) were 100 to 10,000 times higher than the theoretical data (algae only). The calculated pressure drops after 30 minutes of constant flux filtration through a filter cake of accumulated algae cells (first cycle) were very low (<0.000014 bar) while the experimental data showed pressure drops of up to 3.8 bars when AOM were also present in the solution (Table 8-3). These observations indicate that the contribution of algal cell suspensions, without being surrounded with TEP-like materials, to the fouling potential of the water is low and might be negligible.

Table 8-3: Results of the simulated bloom experiments in comparison with theoretical calculations.

Age (days)	Algae conc. (#/mL)	TEP$_{0.1\mu m}$ (mg X$_{eq}$/L)	Experimental results (Algae + AOM)		Theoretical calculations (Algae)	
			MFI-UF [a]	ΔP_c (mBar)*	MFI-UF [b]	ΔP_c (mBar)*
Chaetoceros affinis						
1	31300	1.43	5929	40	0.5	0.004
6	32800	3.21	28088	192	0.5	0.004
8	56300	3.78	43483	297	0.9	0.007
10	55000	5.06	160228	1093	0.9	0.007
13	82500	6.63	81509	556	1.3	0.010
15	92500	11.61	489787	3341	1.4	0.012
17	77500	12.76	557085	3800	1.2	0.010
20	57500	11.94	531279	3624	0.9	0.007
Microcystis sp.						
3	30000	0.22	203	1	0.1	0.001
6	113000	0.07	292	2	0.4	0.004
10	310000	0.32	26828	183	1.2	0.010
13	355000	0.36	15985	109	1.4	0.011
17	285000	1.90	76491	522	1.1	0.009
20	435000	2.12	184022	1255	1.7	0.014

(a) Calculated from experimental data (batch algal cultures); (b) Calculated from theoretical data: ϕ_s=1.0 (MSp) & 0.968 (CA); ε=0.4; d_p= 4µm (MSp) & 15µm (CA); J = 80 L/m^2/h.; * Cake pressure drop at t=30 mins.

It is important to note that the TEP$_{0.1\mu m}$ concentration in algal suspensions covers both free TEPs and those bound (attached) to algal cells. For both cultures, the TEP$_{0.1\mu m}$ concentration and fouling potential have a linear correlation with MFI (R^2>0.82) based on the data presented in Table 8-3. On the other hand, a not very pronounced correlation (R^2<0.5) was observed between cell concentration and MFI. This observation indicates that the algal cell itself are likely not responsible for the high fouling potential of the water during algal bloom situations but rather due to AOM (including TEPs) itself or in combination with algal cells.

8.4.1.2 Filtration resistance of algal cells and AOM

To investigate the role of TEP–like materials (AOM) on UF fouling, filtration experiments were performed with solutions collected from 2 batch cultures of bloom-forming algae (CA: *Chaetoceros affinis* and MSp: *Microcystis sp.*) with and without algal cell suspensions. The different feed solutions were prepared according to the protocol illustrated in Figure 8-5.

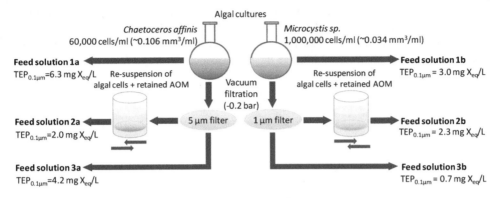

Figure 8-5: Overview of the protocol used to prepare the feed solutions for filtration experiments and their corresponding algal cell and $TEP_{0.1\mu m}$ concentrations.

Based on $TEP_{0.1\mu m}$ measurements, the feed solutions all contain AOM. It was not possible to totally separate AOM from algal cells because some AOM were bound to the cell walls and some formed large aggregates bigger than the pores of the filters (5 µm or 1µm PC filters) used to separate the two materials. Feed solution 1 comprises both algal cells and AOM. Feed solution 2 comprises algal cells, AOM bound/attached to the cells and free AOM retained on 5 µm (for CA) or 1µm (for MSp) polycarbonate (PC) filters (Whatman). Finally, feed solution 3 comprises only free AOM that were not retained on the PC filter. About 30% of $TEP_{0.1\mu m}$ measured in raw CA solution was retained by 5µm filter while 75% of $TEP_{0.1\mu m}$ in raw MSp solution was retained by 1µm filter.

Constant flux filtration tests were performed using flat sheet UF membranes (30 kDa MWCO) to compare fouling rates with the different feed solutions (Figure 8-6). Filtration with raw CA solution (Feed 1a) showed more than 2 bar ΔP increase after 25 minutes of filtration while a similar ΔP increase was recorded with untreated MSp solution (Feed 1b) only after about 50 minutes of filtration. A higher fouling rate with CA solution (Feed 1a) was expected considering that it contains twice as much $TEP_{0.1\mu m}$ and 3 times more algae volume concentration than MSp solution (Feed 1b).

In tests with feed solutions from CA culture, the ΔP increase (after 25 mins. filtration) reduced by ~15% when all algal cells and AOMs larger than 5µm were removed, an indication that CA cells and large AOMs have much lower filtration resistance than free AOMs (Figure 8-6a). The rate of ΔP increase during filtration with free AOMs (<5 µm) illustrate possible pore blocking within the first 2 mins of filtration followed by cake filtration as indicated by the constant slope of the ΔP curve and then followed by cake filtration with enhanced compression and/or increasing rejection as indicated by the increasing slope (Figure 8-6c). The membrane fouling potential (MFI) of the feed solutions based on the minimum fouling rates are shown Table 8-4. Despite containing no algal cells and ~30% lower $TEP_{0.1\mu m}$ concentration, the fouling potential of Feed 3a (free AOM>5µm) is comparable to Feed 1a (algae and total AOM). This can be attributed to the composition of the cake accumulated during filtration. A filter cake comprising heterogenous layers of CA cells and AOM might have higher porosity than a cake comprising only AOM. CA cells, like most diatomaceous algae, are encased in rigid silica shells called "frustules" (Figure 8-7b).

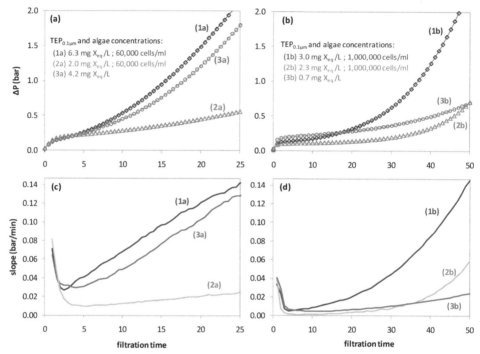

Figure 8-6: Trans-membrane pressure (a,b) and fouling rate developments (c,d) during filtration through 50 kDa membranes with solutions comprise of algal cells and/or AOM from CA (a,c) and MSp (b,d) cultures.

Table 8-4: Summary of MFI-UF, cell and $TEP_{0.1\mu m}$ concentrations of the 3 feed solutions.

Feed water	Cell concentration (cells/ml)		$TEP_{0.1\mu m}$ (mg X_{eq}/L)		MFI-UF (s/L^2)	
	CA	MSp	CA	MSp	CA	MSp
Feed 1	60,000	1,000,000	6.3	3.0	119,200	20,900
Feed 2	60,000	1,000,000	2.0	2.3	43,000	4,600
Feed 3	0	0	4.2	0.7	129,400	19,400

In the tests with feed solutions from MSp culture, the fouling rate with Feed 1b (algae and total AOM) was initially comparable to Feed 3b (free AOM<1μm) but the former rapidly increased after 20 minutes of filtration while the latter only slightly increased (Figure 8-6d). Filtration with Feed 2b (algae and AOM>1μm) showed the lowest fouling rate in the first 25 minutes of filtration but then increased rapidly and eventually exceeding the fouling rate of Feed 3a. The variations in fouling rate can be mainly attributed to the differences in TEP or AOM concentrations.

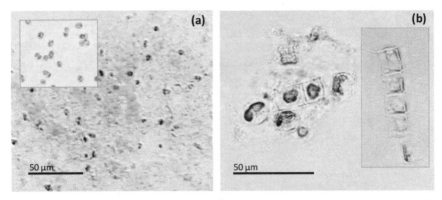

Figure 8-7: Light microscope photographs of flocs of Microcystis sp. (a) and Chaetoceros affinis (b) before (inset photos) and after staining with Alcian blue. Algal cells are dark brown to brownish-green color while TEP-like materials are stained blue.

Based on $TEP_{0.1\mu m}$ measurement, a substantial fraction of AOM produced by MSp was either larger than $1\mu m$ and/or bound to algal cells. Cell bound AOM in MSp has been reported in various studies (e.g., Li *et al.*, 2012; Qu *et al.*, 2012). As illustrated in Figure 8-7a, bound and free AOM > $1\mu m$ were detected in terms of TEP. These large AOM materials can directly contribute to the high fouling rates of feed solutions containing algal suspensions by filling the gaps between accumulated algal cells. Furthermore, MSp cells have softer cells than CA. It might deform/compress at higher cake pressure drop resulting in a decrease in cake porosity and an increase in hydraulic resistance (Figure 8-6d). It is also possible that breakage of MSp cells and the release of intracellular material might have contributed significantly to membrane fouling.

In summary, it was demonstrated in the filtration tests that AOM (bound and/or free) are likely to be a major cause of fouling in UF membranes during filtration of algal bloom impacted waters. However, algae with soft cells (like MSp) may contribute substantially to the fouling due to cake compression and/or due to release of intracellular AOMs.

8.4.2 Plugging potential of non-uniformly deposited algae

The potential effect of capillary plugging due to non-uniform deposition of algal cells was investigated taking into account the variations in hydrodynamic conditions in the capillary membrane.

8.4.2.1 Particle transport in dead-end capillary membrane

Particle transport and deposition in capillary membrane is primarily controlled by the shear forces imposed by the axial flow and the drag force due to permeation (radial flow). The calculated velocity profiles along the length of a clean 0.8 mm diameter capillary UF (Seaguard) at different average fluxes are presented in Figure 8-8. The hydrodynamic calculations were based on the equations discussed in Section 8.2.2.

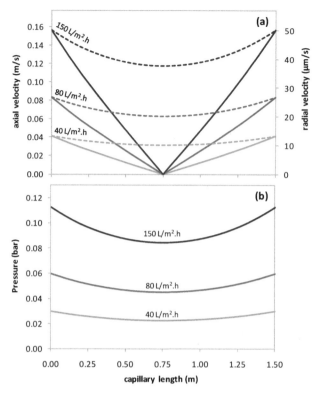

Figure 8-8: Calculated radial and axial fluid velocities (a) and channel pressure over the length of a clean inside-out capillary membrane (d_c=0.8mm) at various fluxes. In (a) solid lines represent average axial velocities and dashed lines represent average radial velocities.

Increasing the average flux (set flux) will increase the feed flow and will result in higher variations in axial velocities along the capillary channel (Figure 8-8a). Higher axial velocities will cause increase in pressure drop due to friction losses. As a consequence, the pressure profile along the feed channel will also increase (Figure 8-8b). Finally, the pressure profile defines the variations of average radial velocity (flux) along the capillary length. Such variations in pressure and radial velocity along the length of capillary are expected to reduce when the capillary membrane started to foul. The Reynolds number at the maximum axial velocity is about 150, which means the hydrodynamic condition is laminar. This indicates that the model explained in Section 8.2.3 is a valid approach for investigating the transport of algal cells in capillary UF membranes.

Depending on particle size and the fluid shear rate, Brownian diffusion, shear-induced diffusion and inertial lift (lateral migration) is considered to be the dominant mechanism for the back-transport of particles near the surface (concentration polarization layer) of the membrane (Davies, 1992). The back-transport velocity of particles of various sizes was calculated based on these three mechanisms. At an average axial velocity of 0.08 m/s, the back-transport velocity due to Brownian diffusion primarily affects colloidal particles (<0.5 µm) while the effect of shear induced diffusion dominates for intermediate sized particles (0.5-50 µm) and inertial lift affects those larger than 50 µm (Figure 8-9). The lowest total back-transport velocities were recorded for particles between 0.1 and 1 µm in size. This is consistent with what was predicted by Davies (1992) for neutrally buoyant particles and Jiang *et al.* (2007) for activated sludge suspensions. Considering that most algal cells are much larger than 1 µm, the influence of Brownian diffusion on algae deposition is negligible.

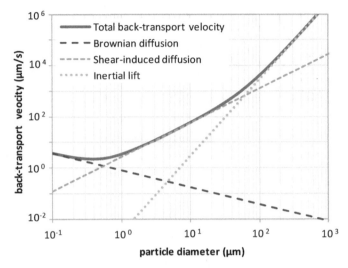

Figure 8-9: Typical variations in back-transport velocities of different particles sizes due to Brownian diffusion, shear-induced diffusion and inertial lift in an inside-out capillary UF membrane (d_c=0.8mm; L_c = 0.75m; J_p=80 L/m^2.h; ρ_p = 1030 kg/m^3; v_z= 0.08m/s).

Assuming algal cells are spherical particles and neutrally buoyant in seawater (ρ_p = 1030 kg/m^3), the total back-transport velocities were calculated as the sum of the effects of the three mechanisms (brownian diffusion, shear-induced diffusion and inertial lift). The total back-transport velocities of algal cells at various axial velocities are presented in Figure 8-10.

In general, the effect of inertial lift on the back-transport velocity of algae is more substantial than shear-induced diffusion. Inertial lift affects both small and large algae at high axial velocities as well large algae at low axial velocities.

As illustrated in Figure 8-8, the axial and radial fluid flow in capillary membrane varies along the longitudinal axis. Theoretically, if the back-transport velocity (J_T) of an algal cell approaching the surface of the membrane (concentration polarisation layer) is lower than the radial velocity (J_p), the cell might deposit on the membrane surface. However if $J_T > J_p$, the cell will not deposit on the membrane surface but will remain in suspension while being transported further towards the axial direction. To illustrate this, the total back-transport velocities of algae (different sizes) were plotted together with the radial velocity (J_p=80 L/m^2.h) along the length of a capillary UF (Figure 8-11).

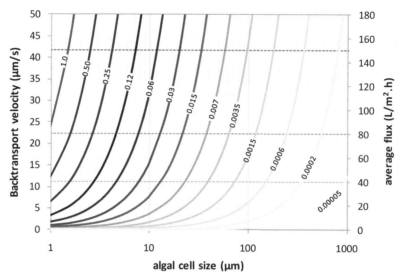

Figure 8-10: Back-transport velocities (solid lines) at various axial velocities for different sizes of algae (ρ_p = 1030 kg/m³). Dash lines in indicate the average radial velocities for 3 different membrane fluxes (40, 80 and 150 L/m².h)

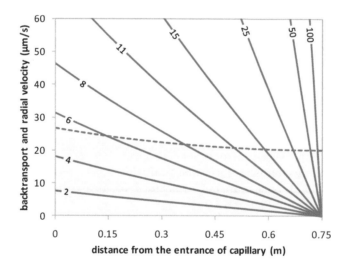

Figure 8-11: Back-transport velocities (solid lines) of algae (ρ_p = 1030 kg/m³) with respect to the distance from the water inlet of capillary membranes. Dash lines indicate the radial velocity profile for an average membrane flux of 80 L/m².h.

Algal cells are expected not to deposit before the point where the back-transport velocity intersects with the radial velocity. At an average flux of 80 L/m².h and algal cell density of 1030 kg/m³, algae smaller than 5 µm will deposit anywhere between the water inlet and the dead-end side of the capillary membrane while larger algae do not deposit until they reach a certain distance from the water inlet (Figure 8-11). In addition, algae larger than 25 µm are likely to deposit on the dead-end side (middle of Seaguard membrane) of the capillary as their higher back-transport velocity can induce lateral migration of cells.

In summary, theoretical calculations illustrate that algae larger than 5 μm are likely to deposit unevenly along the length of capillary membrane. Moreover, it is likely that only a part of the algal cell arriving at the membrane surface will remain attached to the membrane during filtration, due to shear forces near the membrane surface. An uneven deposition of large algal cells inside the capillary may cause full or partial accumulation on the dead-end (or middle if fed from both sides) of capillary channels (plugging). Considering that majority of bloom-forming species of algae are larger than 5 μm (Table 8-2), it is not unlikely that most of the algal cells will accumulate at the end of the capillaries and cause plugging on that part of the membrane.

8.4.2.2 Calculating the effect of capillary plugging

The potential effect of capillary plugging on the operation of UF membrane was calculated based on the model described in Section 8.2.2.2. The calculation assumed that 100% of the algal cells arrive at the end (Aquaflex) or at the middle (Seaguard) of the capillary. The expected loss of active surface area and increase of required feed pressure for a clean membrane as a function of number of algal cells per ml, membrane flux and filtration time is presented in Figure 8-12.

Typical operation of a dead-end MF/UF system is not continuous since periodic hydraulic cleanings (backwashing) are performed every 15-60 minutes. Considering a filtration time of 60 minutes and feed concentration of 250,000 cells/ml, the expected loss in effective membrane area due to plugging is 30% after the filtration cycle (Figure 8-12a). At this scenario, the actual membrane flux is about 115 L/m².h, 44% higher than the initial flux (Figure 8-12b). Additionally, the feed pressure will increase from 60 mbar to about 80 mbar (Figure 8-12c).

The increase in flux due to capillary plugging can have a drastic effect on membrane fouling when AOM are also present in the feedwater. AOM, particularly free AOM, are generally smaller than algal cells and they are expected to deposit more evenly along the length of the capillary membranes. When the remaining active membrane area is fouled by AOM, an increase in flux would mean substantial increase in cake resistance. Based on the cake filtration model (Eq. 8), a 44% increase in flux will result in a 107% increase in cake resistance. In addition, the filter cake is more compressible at a higher flux, so cake compression can further increase the fouling rate.

To further illustrate the potential effect of plugging, the expected loss of active membrane area and the localised flux increase was calculated for different species of algae at severe algal bloom situations. Results are presented in Table 8-5. Based on the calculations, some bloom-forming species can cause severe plugging problems in capillary UF. Severe blooms caused by three species of algae (*Noctiluca scintillans, Prorocentrum micans, Nodularia spp.*) can cause complete plugging of capillary within 30 minutes of filtration (J_p=80 L/m².h). Two other species (*Chaetoceros affinis, Anabaena spp.*) caused more than 50% loss in active membrane area and 2-5 times increase in average flux of the remaining active membrane area. These findings indicate that large and high biomass producing algae can cause blooms which may have severe implications to the operation of capillary UF due to plugging.

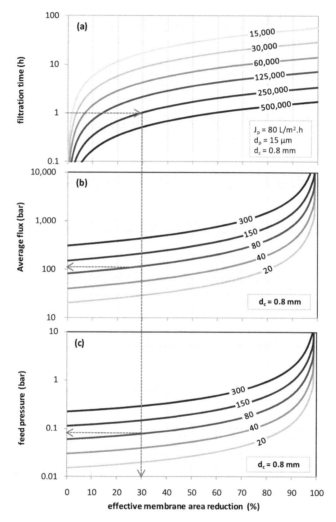

Figure 8-12: Filtration time (a), average local flux (b) and feed pressure (c) in relation to membrane area reduction in capillary membrane (Seaguard configuration) due to plugging by 15 µm algae without AOM. Isolines in (a) are results for different cell concentrations of algae in cells/ml at average flux of 80 L/m².h while isolines in (b) and (c) represent 5 different initial (clean membrane) fluxes in L/m².h.

Table 8-5: Calculated loss in effective membrane area and localised increase in average flux due to capillary plugging by 16 species of algae during severe bloom situations.

Bloom-forming algae	Diameter (µm)	Cell conc. (cells/ml) [a]	Active membrane area lost (%) [b]	Ave. flux after plugging (L/m^2.h)[c]
Alexandrium tamarense	32	10,000	9	87.5
Cochlodinium polykrikoides	33	27,000	25	107.2
Karenia brevis	36	37,000	45	146.0
Noctiluca scintillans	400	1,900	----- completely plugged -----	
Prorocentrum micans	44	50,000	----- completely plugged -----	
Chaetoceros affinis	15	900,000	80	390.7
Pseudo-nitzschia spp.	7	19,000	0.2	80.1
Skeletonema costatum	5	88,000	0.3	80.2
Thalassiosira spp.	12	100,000	5	83.8
Nodularia spp.	21	605,200	----- completely plugged -----	
Anabaena spp.	6	10,000,000	57	184.1
Microcystis spp.	4	14,800,000	25	106.4
Emiliania huxleyi	5	115,000	0.4	80.3
Phaeocystis globosa	6	52,000	0.3	80.2
Chattonella spp.	15	10,000	1	80.7
Heterosigma akashiwo	20	32,000	7	85.8

(a) Maximum recorded concentrations reported in various literatures; (b) Membrane area lost as a percentage of the initial clean membrane area after 30 seconds of filtration at J_p = 80 L/m^2.h; and (c) is the average flux of the remaining active membrane area after 30 minutes of filtration and keeping the permeate flow constant.

In summary, theoretical calculations show that plugging of capillaries by algal cells during severe blooms may substantially increase membrane fouling with the presence of AOM. Plugging can also cause/enhance non-backwashable fouling if the accumulated algae are not effectively removed from the feed channel by hydraulic cleanings.

Backwashing and even chemical cleaning may not be effective in removing foulants in plugged capillaries (Heijman et al., 2005; 2007). In such case, the plugged section of the capillaries will continue to increase in the succeeding filtration cycles and cause severe non-backwashable fouling. To minimise plugging problems in UF membranes, the following are recommended:

- shortening the filtration cycle (e.g., from 30 mins. to 15 mins.);
- reducing the average flux (Note: various UF plants are operating at a more conservative flux - 50 L/m².h - than what was used in this study);
- introducing a small cross-flow at the dead-end side of the capillary (This is feasible for capillaries in the Aquaflex system).

According to Panglisch (2003), introducing a cross-flow velocity of 0.05 m/s at the end of the capillary can prevent plugging. This means the UF system will be operating at 83% recovery (excluding backwash water) at a flux of 80 L/m².h. The concentrate waste (or a part of it) may also be recovered by recycling it back to the UF feed water tank. Moreover, introducing a cross-flow for few minutes at the end of a filtration cycle (before hydraulic backwashing) might be sufficient to clean the plugged portion of the capillary.

Other concepts such as partial backwashing (van de Ven *et al.*, 2008) and forward flushing (Kennedy *et al.*, 1998) to remove plugging foulants from the capillary channel has been introduced over the years. However, further studies are needed to investigate the effectiveness of these concepts in cleaning capillary channels plugged with high concentration of algal cells and AOM.

8.4.3 Non-backwashable fouling in capillary membrane

The role of algae and AOM in non-backwashable fouling in capillary UF membranes was investigated experimentally. Filtration experiments with periodic backwashing were performed with 0.8 mm diameter capillary membranes (150 kDa MWCO). The average filtration flux was set at 80 L/m².h and a 1-min backwashing at 200 L/m².h was performed every 30 minutes of filtration. The filtration experiments were performed with three different feed solutions, namely:

1) Solution 1: Untreated sample from a batch culture of *Chaetoceros affinis* diluted with artificial seawater down to cell concentration of about 15,000 cells/ml;
2) Solution 2: Part of Solution 1 filtered through 5µm PC filter to remove algal cells and;
3) Solution 2: Part of Solution 2 filtered further through 0.4µm PC filter.

The filtration pre-treatments for preparing Solutions 1 and 2 removed substantial concentrations of AOM from Solution 1 based on reduction of $TEP_{0.1µm}$ (Figure 8-13a). $TEP_{0.1µm}$ concentrations of the 3 feed solutions were 0.72, 0.43 and 0.30 mg X_{eq}/L, respectively. This result indicates that an important part of $TEP_{0.1µm}$ is larger than 5µm.

As shown in Figure 8-13b, hydraulic cleaning was not effective in restoring the initial permeability of UF, which means not all accumulated foulants were removed. To compare the rate of non-backwashable fouling, the feed pressure at the beginning of each filtration cycle right after hydraulic cleaning was plotted against filtration time in Figure 8-13c. The rate of feed pressure increase in all three experiments increased with each succeeding filtration cycle, which is attributed to remaining fouling after each backwashing. The non-backwashable fouling caused by untreated water (Solution 1) is about 3 and 5 times higher than with the two pre-treated feedwaters (Solutions 1 and 2), which is quite significant.

To investigate the possible role of algal cells on UF fouling, theoretical calculations were performed based on the effect of plugging discussed in Section 8.2.2.2. The theoretical feed pressure during filtration of 15,000 cells/ml algal suspensions (without AOM) was calculated based on Eq. 27 with the assumptions that they all deposit on the dead-end side of the capillary and the deposits were not removed by the subsequent backwashing (Figure 8-13c). The increase in predicted feed required pressure is negligible compared to the observed feed pressure increase during filtration with Solution 1 (algae + AOM). The predicted loss in effective membrane area due to algae plugging after 4 hours of filtration is about 7%, resulting in an increase in average flux from 80 to 86 L/m².h (calculated using Eq. 26). Based on the cake filtration model (Eq. 8), such increase in flux is expected to increase the cake resistance of deposits (e.g., AOM) in the remaining active membrane area by ~15%.

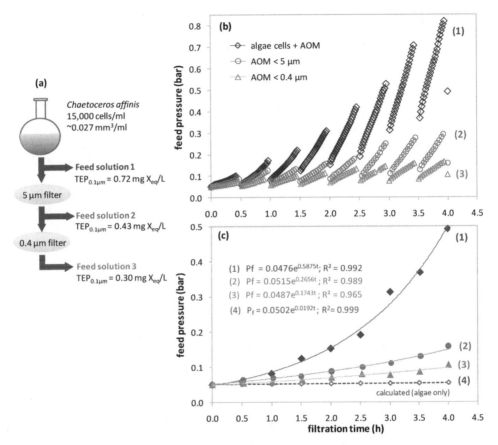

Figure 8-13: Filtration performance of capillary membrane during filtration of samples with and without algal cells: (a) algae and $TEP_{0.1\mu m}$ concentrations of the feed solutions (b) trans-membrane pressure over time (c) trans-membrane pressure after hydraulic cleanings. Dash line in (b) represents theoretical feed pressure after backwashing due to plugging with 15,000 cells/ml algal suspensions without AOM.

In summary, capillary plugging by algal cells may theoretically enhance backwashable and non-backwashable fouling in capillary UF membranes. However, theoretical calculations show that in this case the effect of plugging is marginal.

8.4.4 Fouling mechanisms

The dominant membrane fouling mechanisms possibly involved during filtration of algal bloom impacted waters were identified based on the conducted theoretical and experimental investigations. During filtration of solutions containing AOM with and without algal cells, the observed change in feed pressure after backwashing consistently showed an exponential trend (Figure 8-13c). This trend might be the result of the following phenomena/mechanisms:

1) narrowing of pores of the membrane and/or the cake due to deposition of AOM and other colloids;
2) substantial loss in effective filtration area resulting in much higher local flux;
3) partial removal of the cake during backwashing (e.g., lifting and partial dispersion of breaking cake/gels);
4) accumulated cake/gel is compressible;

In Figure 8-6, the feed pressure increase within a filtration cycle illustrates that pore blocking and cake filtration with compression occurred sequentially. Similar mechanisms also occurred in the absence of algal cells. This is an indication that fouling due to pore blocking and the compressible character of the cake/gel can be mainly due to the presence of AOM. AOMs which accumulated in the membrane pores can be very sticky, flexible and elastic hydrogels (e.g., TEPs). During backwashing, these hydrogels which are partially or fully blocking the interior of the membrane pores may not be removed effectively from the membrane (Figure 8-13b). As a result, the effective filtration area for the succeeding filtration cycle is reduced and higher feed pressure is required to maintain the permeate flow. The effective filtration area may further decrease in the succeeding filtration cycles as more AOM remain in the membrane pores and/or membrane surface and backwashing were becoming less effective in removing them. This scenario might explain the exponential trend of non-backwashable fouling observed in the filtration experiments (Figure 8-13c).

8.5 Conclusions

- The theoretical membrane fouling potentials (measured as MFI-UF) of 16 species of bloom-forming algae are between 100 to 10,000 times lower than what was recorded in natural seawater. The substantial difference can be attributed mainly to the presence of algal-derived organic material in real seawater (e.g., TEPs).

- In experimental studies with 2 batch cultures of algae (*Chaetoceros affinis* and *Microcystis sp.*), a reasonable correlation ($R^2 > 0.82$) was observed between membrane fouling potential (MFI-UF) and TEP concentration while a much weaker correlation ($R^2 < 0.5$) was observed between MFI-UF and algae concentration.

- Pore constriction, substantial loss in filtration area, cake compression and partial cake/gel removal during backwashing are the possible UF membrane fouling mechanisms involved during filtration of algal bloom impacted waters.

- Algal cell transport through the feed channel of capillary membrane is mainly influenced by inertial lift and shear-induced diffusion. Large algae (>25 μm) tend to arrive at the middle or end of the capillary depending on the design and operation mode of the UF module. Smaller algal cells tend to attach partly on the membrane surface and partly at the end or at the middle of the capillary.

- Theoretical calculations indicate that plugging by algal cells at the dead-end section of capillary UF may cause severe problems due to increase in membrane and cake resistance resulting from the flux increase in the remaining active membrane area. However, shortening the length of filtration cycles, reducing the average flux and introducing a cross-flow (or bleed) at the end of the capillary can minimise or eliminate its potential effect.

References

Azetsu-Scott, K., and Passow, U. (2004) Ascending marine particles: Significance of transparent exopolymer particles (TEP) in the upper ocean. Limnol. Oceanogr, 49(3), 741-748.

Berktay, A. (2011). Environmental approach and influence of red tide to desalination process in the middle-east region. International Journal of Chemical and Environmental Engineering 2 (3), 183-188.

Boerlage S.F.E., Kennedy M.D., Tarawneh Z., De Faber R., Schippers J.C. (2004) Development of the MFI-UF in constant flux filtration. Desalination 161 (2):103–113.

Bowen, W. R., & Sharif, A. O. (1998). Long-range electrostatic attraction between like-charge spheres in a charged pore. *Nature, 393*(6686), 663-665.

Chang, F. H. (2000) Pink blooms in the springs in Wellington Harbour. Acquacult. Update, 24, 10–12.

Davis, R. H. (1992). Modeling of fouling of crossflow microfiltration membranes. Separation and Purification Methods, 21(2), 75-126.

Davis, R. H., & Sherwood, J. D. (1990). A similarity solution for steady-state crossflow microfiltration. Chemical Engineering Science, 45(11), 3203-3209.

Fonda Umani S., Beran A., Parlato S., Virgilio D., Zollet T., De Olazabal A., Lazzarini B., and Cabrini M. (2004) Noctiluca scintillans Macartney in the Northern Adriatic Sea: long-term dynamics, relationships with temperature and eutrophication, and role in the food web, J. Plankton Res. 26(5), 545-561.

Green G. and Belfort G (1980) Fouling of ultrafiltration membranes—Lateral migration and the particle trajectory model, Desalination 35, 129-147.

Heijman, S. G. J., Kennedy, M. D., & van Hek, G. J. (2005). Heterogeneous fouling in dead-end ultrafiltration. Desalination, 178(1-3), 295-301.

Heijman, S. G. J., Vantieghem, M., Raktoe, S., Verberk, J. Q. J. C., & van Dijk, J. C. (2007). Blocking of capillaries as fouling mechanism for dead-end ultrafiltration. Journal of Membrane Science, 287(1), 119-125.

IOC-UNESCO (2013) What are Harmful Algae? Accessed 28 August 2013, Available from: http://hab.ioc-unesco.org/index.php?option=com_content&view=article&id=5&Itemid=16

Jiang, T., Kennedy, M. D., Yoo, C., Nopens, I., van der Meer, W., Futselaar, H., Schippers, J.C., Vanrolleghem, P. A. (2007). Controlling submicron particle deposition in a side-stream membrane bioreactor: A theoretical hydrodynamic modelling approach incorporating energy consumption. Journal of Membrane Science, 297(1-2), 141-151.

Kennedy, M., Kim, S., Mutenyo, I., Broens, L., & Schippers, J. (1998). Intermittent crossflushing of hollow fiber ultrafiltration systems. Desalination, 118(1-3), 175-188.

Lerch, A. (2008) Coating layer formation by flocs in inside-out driven capillary ultrafiltration membranes. In: Dissertation an der Universita t Duisburg-Essen, Dresden: Eigenverlag—ISBN/ISSNurn:nbn:de:hbz:464-20080505-111100-1.

Lerch, A. (2008). Introducing a 'slow' cross-flow at the capillary outlet in comparison to conventional dead-end mode - A trajectory analysis of the effects. Water Science and Technology: Water Supply 7 (4), 37-47.

Lerch, A., Uhl, W., & Gimbel, R. (2007). CFD modelling of floc transport and coating layer build-up in single UF/MF membrane capillaries driven in inside-out mode. Water Science & Technology: Water Supply 7(4), 37–47.

Li, L., Gao, N., Deng, Y., Yao, J., & Zhang, K. (2012). Characterization of intracellular & extracellular algae organic matters (AOM) of microcystic aeruginosa and formation of AOM-associated disinfection byproducts and odor & taste compounds. Water Research, 46(4), 1233-1240.

Olenina, I., Hajdu, S., Edler, L., Andersson, A., Wasmund, N., Busch, S., Göbel, J., Gromisz, S., Huseby, S., Huttunen, M., Jaanus, A., Kokkonen, P., Ledaine, I. and Niemkiewicz, E. (2006) Biovolumes and size-classes of phytoplankton in the Baltic Sea, HELCOM Balt. Sea Environ. Proc. 106, ISSN 0357-2994.

Panglisch, S. (2001). Zur Bildung und Vermeidung schwer entfernbarer Partikelablagerungen in Kapillarmembranen bei der Dead-End Filtration, Dissertation, IWW-Mülheim.

Panglisch, S. (2003) Formation and prevention of hardly removable particle layers in inside-out capillary membranes operating in dead-end mode. Water Science and Technology: Water Supply 3, (5-6), 117-124.

Pankratz, T. (2008), "Red Tides Close Desal Plants", Water Desalination Report, 44 (44).

Passow, U., and Alldredge, A. L. (1995) "A Dye-Binding Assay for the Spectrophotometric Measurement of Transparent Exopolymer Particles (TEP)". Limnology and Oceanography 40(7), pp. 1326-1335.

Qu, F., Liang, H., He, J., Ma, J., Wang, Z., Yu, H., & Li, G. (2012a). Characterization of dissolved extracellular organic matter (dEOM) and bound extracellular organic matter (bEOM) of microcystis aeruginosa and their impacts on UF membrane fouling. Water Research, 46(9), 2881-2890.

Richlen M. L., Morton S. L., Jamali E. A., Rajan A., Anderson D. M. (2010), "The Catastrophic 2008–2009 Red Tide in the Arabian Gulf Region, with Observations on the Identification and Phylogeny of the Fish-killing Dinoflagellate Cochlodinium Polykrikoides", Harmful Algae, 9(2), pp. 163-172.

Salinas-Rodriguez S.G. (2011) Particulate and organic matter fouling of SWRO systems: Characterization, modelling and applications. Ph.D. thesis, UNESCO-IHE/TUDelft, Delft.

Salinas-Rodríguez S.G., Kennedy, M.D., Amy, G.L., & Schippers, J.C. (2012) Flux dependency of particulate/colloidal fouling in seawater reverse osmosis systems. Desalination and Water Treatment, 42(1-3), 155-162.

Schippers, J.C. and Verdouw, J. (1980) The modified fouling index, a method of determining the fouling characteristics of of water. Desalination 32, 137-148.

Schurer R., Tabatabai A., Villacorte L., Schippers J.C., Kennedy M.D. (2013) Three years operational experience with ultrafiltration as SWRO pre-treatment during algal bloom. Desalination & Water Treatment 51 (4-6), 1034-1042.

Schurer, R., Janssen, A., Villacorte, L., Kennedy, M.D. (2012) Performance of ultrafiltration & coagulation in an UF-RO seawater desalination demonstration plant. Desalination & Water Treatment 42(1-3), 57-64.

Trettin D. R. and Doshi M. R., Chem. Eng. Commun. 4, 507-522 (1980).

UAE-Interact (2009) Red tide turns into regional issue, UAE-Interact, 13 January 2009. http://www.uaeinteract.com/docs/Red_tide_turns_into_regional_issue/33710.htm.

van de Ven, W. J. C., Pünt, I. G. M., Zwijnenburg, A., Kemperman, A. J. B., van der Meer, W. G. J., & Wessling, M. (2008). Hollow fiber ultrafiltration: The concept of partial backwashing. Journal of Membrane Science, 320(1-2), 319-324.

Verdugo, P., Alldredge, A. L., Azam, F., Kirchman, D. L., Passow, U. and Santschi, P. H. (2004) The oceanic gel phase: a bridge in the DOM–POM continuum. Marine Chemistry 92, 67-85.

Villacorte, L.O., Ekowati, Y., Winters, H., Amy, G.L., Schippers, J.C. and Kennedy, M.D. (2013) Characterisation of transparent exopolymer particles (TEP) produced during algal bloom: a membrane treatment perspective. Desalination & Water Treatment 51 (4-6), 1021-1033.

Zydney A.L., Colton C.K. (1986) A concentration polarization model for the filtrate flux in cross-flow microfiltration of particulate suspensions, Chem. Eng. Commun. 47 1-21.

Biofouling in cross-flow membranes facilitated by algal organic matter

Contents

Abstract 248
9.1 Introduction 249
9.2 Materials and methods 250
    9.2.1 Algal culture and AOM extraction 251
    9.2.2 Liquid chromatography – organic carbon detection 251
    9.2.3 Model seawater bacteria 251
    9.2.4 Bacterial growth potential test 252
    9.2.5 Force measurement using atomic force microscopy 253
    9.2.6 Cross-flow capillary membrane set-up 254
    9.2.7 Membrane fouling simulator 255
    9.2.8 Biofouling experiments 256
    9.2.9 Hydraulic calculations 257
9.3 Results and discussion 258
    9.3.1 Cohesion and adhesion potential 258
    9.3.2 Bacterial growth potential 260
    9.3.3 Effect of AOM in the feedwater 264
    9.3.4 Effect of AOM deposits on the membrane surface 265
    9.3.5 Effect of AOM deposits in spiral wound membrane 268
9.4 Conclusions 272
    References 272
    Annexes: Supporting information and results 276

This chapter is an extended version of:

Villacorte L.O., Nyambi Z., Keseilius V., Calix-Ponce H.N., Ekowati Y., Prest E., Kleijn M., Vrouwenvelder J.S., Winters H., Amy G., Schippers J.C., and Kennedy M.D. (2013) Biofouling in cross-flow membranes facilitated by algal organic matter. *In Preparation for Water Research.*

Abstract

Algal-derived organic matter (AOM) such as transparent exopolymer particles (TEP) has been suspected to initiate or enhance biofilm development in aquatic systems. This chapter aims to investigate the possible role of AOM on biofouling of reverse osmosis (RO) membranes treating seawater suffering from algal bloom. Lab-scale experiments were performed with AOM extracted from bloom-forming marine alga (Chaetoceros affinis) to assess its adhesion/cohesion potential, bacterial growth potential and effect on biofouling development in cross-flow membranes. Adhesion force measurements using atomic force microscopy showed better adhesion of AOM on clean polyamide RO membrane than on clean polyethersulfone UF membrane. Moreover, it was observed that AOM can attach stronger on AOM-fouled membranes than on clean membranes. The latter may indicate that bacteria with TEP-like substances on their cell wall can attach and colonise more effectively on RO membranes fouled with AOM.

Batch growth potential tests of natural seawater spiked with different concentrations of AOM showed net growth of seawater bacteria corresponding to the concentration of spiked AOM, supporting the initial hypothesis that AOM enhances bacterial activity in seawater. Long-term (8-20 days) biofouling experiments (simulating similar conditions as in seawater RO systems) were performed using cross-flow capillary membranes and membrane fouling simulators (MFS) to demonstrate the effect of AOM in fouling development. Results show that under non-nutrient limited conditions, AOM can substantially accelerate biofouling when present in high concentrations, either suspended in the feed water or attached to the membrane surface/spacer. A substantially lower fouling rate was observed when only nutrients or only AOM was present in the feed water. This illustrates that the role of AOM may be more as an initiator/enhancer of biofilm development rather than as a nutrient source for bacteria.

9.1 Introduction

Accumulation of organic and biological material can cause serious operational problems in reverse osmosis (RO) systems (Baker and Dudley, 1998; Nguyen et al., 2012). Organic fouling in reverse osmosis (RO) membranes usually occurs due to adsorption and accumulation of organic material from the feed water. Biological fouling or biofouling is due to growth and accumulation of microorganisms (mainly bacteria) and the extracellular polymeric substances (EPS) they produced in the RO system.

Organic and biofouling in RO membranes are closely related in many ways. Both types of fouling can cause increase in hydraulic resistance in the system resulting in an increase in feed channel pressure drop and/or decrease in normalised flux of the membrane. The hydraulic resistance of biofilms on RO membranes are mainly due to accumulation of organic material (i.e., EPS) released by biofilm bacteria rather than the bacteria themselves (Drezer et al., 2013). Most organic foulants in seawater are of biological origin such as algal organic material (AOM) produced by phytoplankton. Furthermore, some fraction of these organic substances can be degraded or utilised as nutrients by bacteria and may therefore accelerate biological growth in the RO system.

The various processes involved in the occurrence of biofouling in RO membrane systems can be divided into 4 phases (Flemming and Schaule, 1998):

1. **Surface conditioning**. Initial adsorption of macromolecules (e.g., humic substances, lipopolysaccharides and other microbial-derived substances) on the surface of clean membrane and spacers eventually leading to the formation of a "conditioning film". Winters et al., (1983) reported that these fast adsorbing substances are mainly anionic biopolymeric materials which are capable of increasing the capacity of the surface to absorb and concentrate nutrients from the feedwater.

2. **Bacterial adhesion and colonization**. Fast adhering bacteria from the raw water first colonise the "conditioned surface" and eventually form micro-colonies. Bacterial cells which produce more exopolysaccharides are more likely to be the primary colonisers (Allison and Sutherland, 1987; Winters and Isquith, 1979).

3. **Biofilm formation**. Further growth of primary colonizers and subsequent adhesion of more species of bacteria led to rapid increase in bacterial colonies and accumulation of excreted EPS, eventually forming a layer of slime known as biofilm (Characklis, 1981).

4. **Biofouling**. When biofilm thickness or their hydraulic resistance increase over time until an operationally defined "level of interference" on the system operation is surpassed, this is considered as biofouling (Flemming, 2002). In RO membrane systems, the level of interference is often defined as >15% increase in feed channel pressure drop and/or >15% decrease in normalized flux.

Winters *et al.* (1983) suspected that Alcian blue stainable substances in seawater are mainly involved in the surface conditioning of RO membranes. Alcian blue is a cationic dye known for its specific affinity to acidic polysaccharides and glycoproteins (Ramus, 1977). Passow and Alldredge (1995) developed a spectrophotometric assay to measure Alcian blue stainable substances in seawater which they called transparent exopolymer particles (TEP). This technique has been extensively applied in the field of oceanography and limnology for many years but not for desalination and water treatment studies. In 2005, Berman and Holenberg implicated TEPs as "major initiators" of biofilm in RO membranes (Berman and Holenberg, 2005). This paved the way for various investigations about TEPs in desalination and water treatment (dela Torre *et al.*, 2008; Bar-Zeev *et al.*, 2009; Kennedy *et al.*, 2009; Villacorte *et al.*, 2009a,b; Berman *et al.*, 2011; van Nevel *et al.*, 2012; Discart *et al.*, 2013).

Recently, Bar-Zeev *et al.*, (2012) proposed a "revised paradigm of aquatic biofilm formation facilitated by TEPs" where they emphasise the important role of TEPs in the conditioning and bacterial colonisation of surfaces exposed to seawater. TEPs are likely involved in the surface conditioning phase because of their high adhesion potential. Their reported stickiness is about 2-4 orders of magnitude higher than suspended particles (Passow, 2002). Furthermore, TEPs are gel-like substances which are often associated with or tend to absorb proteins, lipids, trace elements and heavy metals from surface water (Passow, 2002a), making them a nutritious platform and hotspot for bacterial activity. In fact, various studies found significant percentages (up to 68%) of bacterial populations in seawater that were attached to or embedded in TEP gels (e.g., Alldredge *et al.*, 1993, Passow and Alldredge, 1994).

The abundance of TEPs in seawater is generally associated with the occurrence of algal blooms as they are a major component of algal organic matter (AOM). Algal blooms such as "red tide" recently became a pressing issue in the desalination industry (Pankratz, 2008; Caron *et al.*, 2010). In RO desalination plants, granular media filters (GMF) are usually installed to pre-treat seawater prior to the RO system. During algal blooms, GMF can minimise breakthrough of algal cells but a substantial fraction of the AOM can still pass through the pre-treatment systems and can potentially initiate/enhance biofouling in the downstream RO membrane.

The adverse impacts of AOM generated during algal blooms have already been demonstrated on the operation of dead-end (Villacorte *et al.*, 2010a,b, Schurer *et al.*, 2012; 2013) and cross-flow (Berman *et al.*, 2011) capillary membrane systems. However, these studies only focused on the short-term direct effect of AOM (organic fouling) rather than their long term potential role on biofouling. The objective of this study is to perform controlled experiments to demonstrate and explain the possible role of AOM on biofouling in seawater RO system.

9.2 Materials and methods

Various investigations were performed to assess the biological fouling potential of AOMs extracted from common species of bloom-forming marine algae based on their adhesion/cohesion potential (stickiness), bacterial growth potential and impact on biofouling development in a cross-flow operated membrane system.

9.2.1 Algal culture and AOM extraction

A common species of bloom-forming marine algae, *Chaetoceros affinis*, was selected as the source of AOM for this study. A pure strain (CCAP 1010/27) of the alga was inoculated in 5 L flasks with sterilised synthetic seawater (TDS ~ 34 g/L; pH = 8±0.2) containing trace elements and nutrients (f/2+Si medium) necessary for them to grow rapidly and simulate an algal bloom. After 10-12 days under controlled light (12:12h, light:dark) and continuous slow mixing at ambient temperature (20±3°C), the algal cells were separated from the medium containing AOM by allowing the cells to settle for 24 hours. The supernatant solution was then extracted and filtered through 5 μm polycarbonate filters (Whatman Nuclepore) to remove the remaining algal cells in suspension. The algal culture is not totally axenic which means bacteria may be present in the AOM solution. Nevertheless, bacteria are always present during algal blooms in natural waters.

9.2.2 Liquid chromatography – organic carbon detection

To measure the concentrations of the different fractions of AOM, water samples were sent to DOC-Labor (Karlsruhe, Germany) for analyses using liquid chromatography - organic carbon detection (LC-OCD) method. The LC-OCD analysis fractionates the constituents of organic materials in terms of biopolymers, humic substances, building blocks, low molecular weight (LMW) organic acids and organic neutrals. The specification of the equipments used and how the different fractions were quantified are described elsewhere (Huber *et al.*, 2011). In this study, LC-OCD analyses were performed without pre-filtration and water samples were directly fed to two chromatogram columns in series (HW65S and HW50S). The theoretical maximum chromatographable size without sample pre-filtration is 2 μm, which is based on the pore size of the sinter filters of the columns used (S. Huber, *per. com.*).

TEP substances are a major component of biopolymeric AOM (Villacorte *et al.*, 2013). Hence, biopolymer concentration was used in this study as an indicator of TEP concentration in AOM solutions. The biopolymer concentration of AOM solution used in this study was ~8 mg C/L. Feed AOM concentrations in all experiments were adjusted by diluting the TEP solutions with synthetic seawater (similar ionic strength as the medium).

9.2.3 Model seawater bacteria

A common species of marine bacteria, *Vibrio harveyi* (NCCB 79042) was used as model seawater bacteria. *Vibrio harveyi* is a gram-negative luminous bacteria which can be found free-swimming or in biofilms in tropical marine systems (Austin and Zhang, 2006). Recently, the bacteria were used as the standard bacteria for measuring assimilable organic carbon (AOC) in seawater (Weinrich *et al.*, 2011). In the current study, a pure strain of the bacteria was inoculated in sterilised saline (34 g/L) medium for lucibacterium and incubated at 30°C until the concentration reach ~100 million cells/ml. The solution was then diluted 100 times with artificial seawater for biofouling experiments. Cell concentration was measured based on the number of colonies grown and incubated at 37°C on peptone agar media for 24-48 hours.

9.2.4 Bacterial growth potential test

The AOM solution used in the growth potential tests was pre-filtered through 0.2 μm polycarbonate membranes (Whatman Nuclepore) to remove most bacteria in the AOM solution. To further remove low molecular weight constituents from the solution (e.g., dissolved nutrients in the culture medium), the filtered solution (0.5 L) was then dialysed for 3 days using 3.5 kDa MWCO RC membrane sacks (Spectra/Por 3, SpectrumLabs). The 20 L draw solution (ultra-pure water) was continuously stirred (using a magnetic stirrer) and replenished twice a day. LC-OCD analysis was performed on the dialysed AOM solution to measure the final biopolymer concentration.

A batch of seawater samples (20 L) were collected from the raw water of the Jacobahaven desalination plant (Zeeland, Netherlands) in November 2012. Part of the collected samples was filtered through a 0.2μm polycarbonate filter by vacuum filtration to remove bacteria. Eight unused dark glass bottles were soaked (>24 hours) and rinsed 3 times with ultra-pure water to desorb potential organic contaminants. Each of the 8 cleaned glass bottles was filled with 730 ml of filtered seawater. Different volumes of pre-treated AOM solutions and artificial seawater were added to the bottles to make up 800 ml solutions with different concentrations of AOM (0, 0.08, 0.4 and 0.8 mgC/L) in duplicates.

Total bacteria concentration was measured using flow cytometry at the Kluver Laboratory (TUDelft) based on the standard protocol proposed in the Swiss guideline for drinking water analysis (SLMB, 2012; Prest *et al.*, 2013). The staining solution was prepared with SYBR Green I (1:100 dilution in DMSO; Molecular Probes). Water sample (1 ml) were first incubated in the dark at 35°C for 5 mins, then stained with 10 μl of staining solution, vortexed and again incubated in the dark at 35°C for 10 mins. The samples were then vortexed just before flow cytometric analysis. The flow cytometer used was an Accuri™ C6 flow cytometer (BD Biosciences, San Jose, CA).

The measured total bacterial concentration in untreated seawater was about 1 million cells/ml. The seawater bacteria were inoculated by adding 8 ml of untreated seawater to each of the 8 prepared bottles. The bottles were then gently mixed and sampled to measure the initial total bacteria concentration. They were then tightly covered and stored at room temperature (20±3 °C). The concentration of bacteria was monitored in all bottles by collecting and analysing samples every 24 hours for 4 days and another set of samples were collected on the 7th day. To monitor changes in concentration and organic matter composition, LC-OCD analyses were performed for samples collected from the bottle containing 0.8 mg C/L at the beginning (day 0) and the 4th day of the test.

9.2.5 Force measurement using atomic force microscopy

To assess the stickiness of AOM, interaction forces between biopolymer materials and clean or AOM-fouled UF membranes were measured using atomic force microscopy (AFM). In this study, the investigated biopolymer materials were allowed to adsorb to the surface of polystyrene microspheres attached to the tip of the AFM cantilever. As the cantilever approached towards or retracted from the membrane surface, the AFM cantilever tip deflects depending on the direction and magnitude of the forces between the two surfaces (Figure 9-1). Interaction force curves as a function of the separation distance between the tip and the membrane surface are then generated based on Hooke's law:

$$F = -k\Delta z \qquad\qquad\qquad\qquad \text{Eq. 1}$$

where F is the approach/retract force, k is the cantilever spring constant and Δz is the deflection of the cantilever tip.

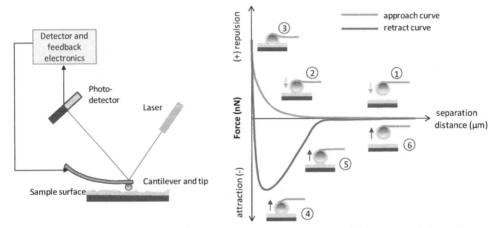

Figure 9-1: Illustration of interaction force measurement using AFM and the generated force-distance curves. The arrows indicate the movement of the particle attached to the cantilever relative to the surface.

Each force measurement cycle generates two force-distance (F-D) curves: the approach force curve and the retract force curve (Figure 9-1). The approach force curve shows the force interactions between the microsphere probe and the membrane surface at different separation distances. The retract force curve features the adhesion/cohesion force keeping the probe surface and the membrane surface together after contact. The adhesion (AOM to membrane) and cohesion (AOM to AOM-fouled membrane) forces are identified as the maximum negative forces recorded in the retract force curve. Furthermore, the total energy needed to completely separate two surfaces from contact was calculated by integrating the measured negative forces with respect to the separation distance in the retract force curve.

To adsorb AOM on AFM probe, ~10 ml of stock solutions of AOM were poured into a petri-dish (50 mm diameter). AFM cantilevers with attached 25 µm polystyrene microsphere tip (Novascan Technologies, Inc.) were submerged in the solutions by carefully positioning the cantilevers (microsphere probe up) at the bottom of the petri-dish. The cantilevers were submerged for 5 days at 4°C to allow adsorption of AOM. The petri-dishes were then placed at room temperature (20°C) for ~2 hours before AFM force measurements were performed.

AFM force measurements were performed on flat sheet polyethersulfone UF (Omega 100 kDa MWCO, Pall Corp.) and polyamide thin-film composite RO (Filmtec BW30, Dow) membranes. The UF membrane was first cleaned by soaking in ultra-pure water for 24 hours and then filtering through >10 ml ultrapure water before the experiment. Coupons (~1cm x 10cm) of polyamide RO membranes were stored in 1% sodium bisulphite solution for at least 24 hours and then rinsed with ultra-pure water before the experiment. For AOM to AOM interaction measurements, a PES membrane was coated with AOM by filtering 5 ml of AOM solution. Filtration was performed at constant flux (60 L m^{-2}.h^{-1}) using a syringe pump (Harvard pump 33), 30 ml plastic syringe and 25 mm filter holder.

Force measurements were performed with the ForceRobot 300 (JPK Instruments) at room temperature (~20°C), using a small volume liquid cell, sealed with a rubber ring. In every test, the liquid cell is filled with ASW (TDS ~ 34 g/L; pH = 8±0.2). The cantilevers were calibrated with respect to their deflection sensitivity from the slope of the constant compliance region in the force curves obtained between the clean, hard surfaces and their spring constant using the thermal noise spectrum of the cantilever deflections (Ralston et al., 2005; Spruijt et al., 2012). The measured cantilever spring constants range from 0.18-0.31 N/m although the nominal cantilever spring constant supplied by the manufacturer is higher (0.35 N/m). Force-distance curves were recorded for an approach range of 10 µm, a tip velocity of 16-28 µm/s, 30 seconds surface delay and a sampling rate of 2 kHz. In each cycle of approach and retract, the probe is brought into contact with the surface at an average load force of 5 nN. The force-distance (F-D) curves were recorded in triplicates for each of the six points on the membrane surface within an area of 10 µm x 10 µm. The F-D curves were analyzed using the JPK data processing software to calculate the maximum adhesion force and total energy.

9.2.6 Cross-flow capillary membrane set-up

Biofouling experiments were performed using a lab-scale capillary membrane system operated in cross-flow mode (Figure 9-2). The capillary membrane has an internal diameter of 0.8 mm which is similar to the typical channel height of a spiral wound RO membrane. The membrane used has a MWCO of a tight UF membrane (7 kDa) to allow retention of most biopolymers on the membrane.

The capillary membranes were provided by Pentair X-flow (Enschede, Netherlands). The cross-flow setup comprises a miniature UF module fabricated by inserting two 1 m long capillary membranes in a 8 mm diameter polyethylene tube. Both ends of the tubing (except the capillary ends) were sealed with water-proof resin. A permeate outlet was positioned in the middle of the module to allow inside-out filtration. The filtration system was equipped with a peristaltic pump (Masterflex L/S 77201-60) to provide constant feed flow, 2 pressure sensors (Endress+Hauser PMD70 and PMC51) to record pressure drop and concentrate pressure and a digital balance (Sartorius TE3102S) to measure permeate flow. The cross-flow velocity (CFV) and permeate flux were set to the required values by adjusting the speed of the feed pump and the valve on the concentrate side. Feed and concentrate pressures and permeate flow were recorded automatically by the computer via data acquisition software. All experiments were conducted at room temperature (20°C).

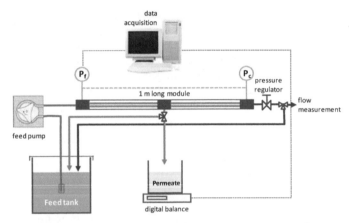

Figure 9-2: Graphical scheme of the cross-flow capillary membrane filtration set-up.

9.2.7 Membrane fouling simulator

Biofouling experiments were also performed using membrane fouling simulators (MFS) to simulate biofouling in spacer-filled RO membrane channels. The design specification of the MFS flow cells is described elsewhere (Vrouwenvelder *et al.*, 2006). The MFS set-up used in this study is presented in Figure 9-3.

A multi-channel peristaltic pump (Masterflex L/S 77201-60) was used to provide constant feed water flow to the MFS. To measure the pressure drop through the MFS cell, a differential pressure sensor (Endress+Hauser PMD75) was installed. To minimise air bubble entrainment in the system, the feed tank was elevated 0.6 m above the pump (Figure 9-3).

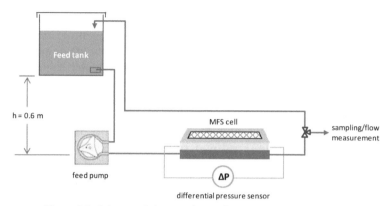

Figure 9-3: Scheme of the membrane fouling simulator set-up.

The spiral wound membrane and spacer (Trisep TS80) coupons were carefully cut into sheets (4 cm x 20 cm) to fit into the flow cell. Extra care was taken to ensure that the orientation of the spacers was the same for all experiments and to make sure that there were no visible scratches or irregular edges on both the membrane and spacer. The cut membranes/spacers were stored in 1% sodium bisulphite solution for at least 24 hours and then rinsed with ultra-pure water before the experiment. In all experiments, water coming out of the system was re-circulated back to the feed tank. Plastic mesh with 20-50 µm opening size was installed in the suction side of the feed tubing to prevent clogging in the MFS cell by large particulate material.

Pressure drop readings were recorded once or twice per day. Before each recording, the linear cross-flow velocity across the MFS cell was verified and adjusted (if necessary) by controlling the volume of water entering the cell over time. This was done by slowly adjusting the speed of the peristaltic pump, measuring the water volumetric flow coming out of the cell and then calculating the apparent linear flow velocity (Eq. 2) until a correct linear velocity is attained.

$$v = \frac{Q}{A_{eff}} = \frac{Q}{\varepsilon w h} \qquad\qquad\qquad\qquad \text{Eq. 2}$$

where Q is the feed flow (m^3/s), A_{eff} is the effective cross-sectional area of the channel (m^2), ε is spacer porosity (0.85, Vrouwenvelder *et al.*, 2009a), w is width of spacer (0.04m) and h is height of spacer (0.000787 m).

9.2.8 Biofouling experiments

Biofouling experiments were performed in both the cross-flow UF set-up (Section 9.2.6) and membrane fouling simulators (Section 9.2.7). Three sets of experiments were conducted to investigate the effect of (1) AOM in the feedwater with mixed bacteria from seawater, (2) AOM on the membrane surface with mixed bacteria from seawater and (3) AOM on the membrane surface with marine bacteria *Vibrio harveyi*. Sets 1 and 2 were performed with the cross-flow UF set-up while Set 3 was performed with both the cross-flow UF and MFS set-ups. Each set of experiments were performed with 4 modules or MFS cells running in parallel with varying conditions (e.g., AOM, nutrients) in the feed tank or the membrane surface. In all experiments, both the concentrate and permeate were recycled back to the feed tank. In all cases, except otherwise stated, nutrients (0.2 mgC/L, 0.04 mg N/L, 0.01 mg P/L) were added regularly to the feedwater to maintain biological growth in the system. To better simulate bacterial growth in seawater, raw seawater sample from a seawater RO plant or the standard marine bacteria *V. harveyi* (~1 million cells/ml) was introduced to the feed solution at the beginning of the biofouling experiment.

9.2.8.1 Membrane autopsy

Autopsies were performed for selected experiments using the capillary UF set-up and MFS. For the capillary UF membranes, the module containing capillaries were first frozen, cut into small pieces (3-5 cm) and soaked in plastic cups containing 100 ml of ultra-pure water, tightly covered, vigorously mixed and placed in a sonicator bath (Branson 2510E-MT) for 15 mins. For spiral wound membranes, both the fouled membrane and spacers were soaked in ultrapure water and sonicated for 15 mins. Sample solutions from the sonicated membrane samples were used to measure TOC, TEP and *V. harveyi* plate counting analysis. TOC was measured using the Shimadzu V_{CPN} TOC analyser. TEP measurement was based on the method by Passow and Alldredge (1995) with the latest modifications by Villacorte *et al.* (2013). Plate counting of *V. harveyi* from extracted biofilm samples was performed as described in Section 9.2.3.

9.2.9 Hydraulic calculations

The water flux through a semi-permeable membrane is inversely proportional to the hydraulic resistance of the membrane. When biofilm accumulates on the surface of the membrane, additional resistance is generated. This can be explained based on the resistance in series model derived from Darcy's Law.

$$J = \frac{\Delta P}{\eta \, (R_m + R_c)} \qquad\qquad \text{Eq. 3}$$

where J is the membrane permeate flux, ΔP is the differential pressure between the feed and permeate side of the membrane, η is the dynamic water viscosity, R_m is the membrane resistance and R_c is the gel or biofilm resistance.

Rearranging Eq. 3 to calculate the cake resistance;

$$R_c = \frac{\Delta P}{\eta J} - R_m \qquad\qquad \text{Eq. 4}$$

Biofilm accumulation in cross-flow membrane systems can also cause increase in the feed channel pressure drop due to partial clogging or constriction of some portions of the feed channel. The feed channel pressure drop (ΔP_{fc}) is measured as the difference between the recorded pressure in the feed (P_f) inlet and concentrate (P_c) outlet of the membrane module.

$$\Delta P_{fc} = P_f - P_c$$ Eq. 5

9.3 Results and discussion

The biological fouling potential of AOM was investigated based on adhesion force measurements, bacterial growth potential tests and (bio)fouling experiments using capillary and spiral wound membranes. The results of the investigations are discussed in the following sections.

9.3.1 Cohesion and adhesion potential

The first hypothesis linking AOM to biofilm formation is their ability to adhere to membrane surfaces and form a "conditioning layer" where bacteria can adhere to and grow more effectively. With its rather sticky characteristics (adhesive and cohesive strength), it allows bacteria to be strongly attached to the surface and possibly resist detachment due to shear forces imposed by the tangential flow. To evaluate these characteristics (cohesive and adhesive strength), AOM to membrane and AOM to AOM interactions were measured using atomic force microscopy (AFM). Force measurements using AFM had been applied in various studies to measure the affinity of known foulants to membrane surfaces by fixing a surface coated or modified microsphere on the AFM cantilever tip (Frank and Belfort, 2003; Li and Elimelech, 2004; Yamamura et al., 2008). In this study, the cohesive and adhesive strength of AOM was investigated by measuring interactions between 25µm polystyrene microspheres with adsorbed AOM on clean and AOM-fouled membranes.

The generated approach force curves show that short-range attraction forces were observed between clean polystyrene probes and clean membranes (Figure 9-4a). The attraction between clean polystyrene and clean membranes may be due to van der Waals interaction, hydrogen bonding and/or hydrophobic interaction. The PS particle is quite hydrophobic and probably prefers contact with the membrane surface over contact with water. After incubating the probes in AOM solution for 5 days, interaction between the probe and the membranes shifted to rather long range repulsion (Figure 9-4b,c). This change indicates that indeed a layer of AOM was adsorbed to the surface of the polystyrene microspheres. Although both the membranes (Hurwitz et al., 2010; Childress and Elimelech, 1996; Ricq et al., 1997) and AOM (Henderson et al., 2008) are (slightly) negatively charged, electrostatic repulsion does not play a significant role in the surface interaction because the high ion concentration in the matrix screens the charges and limits the range of electrostatic interactions to distances of less than 1 nm (Mosley et al., 2003). Therefore, the long range (1 µm) repulsion on probes with adsorbed AOM is not due to electrostatic repulsion but rather the resistance exerted by the soft AOM layer as it is being gradually squeezed in between the microsphere probe and the membrane surface.

Figures 9-4d to 9-4f illustrates the retraction force curves after clean and AOM-coated polystyrene tips were brought into contact with clean and AOM-fouled membranes for 30 secs. Clean polystyrene particles adhere more strongly on PES membranes than on polyamide membranes. However when AOM is adsorbed to the polystyrene probe, it adheres stronger on polyamide membranes than on PES membrane. Cohesion between AOM-coated polystyrene and AOM-fouled PES membranes was observed to be substantially stronger than adhesion between AOM and clean PES membranes.

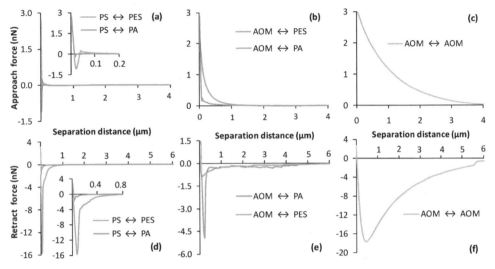

Figure 9-4: Typical force-distance curves showing interactions between different surfaces: (a-c) approach force curves; (d-f) retract force curves. PS=clean polystyrene tip; PES=polyethersulfone membrane; PA=polyamide membrane; AOM=algal organic matter fouled PES membrane.

It was also observed that the retraction force curves were generally broader when biopolymers were present on both the particle probe and the membrane surface (Figure 9-4f). These may imply that not only a higher force but also a higher energy is needed to totally retract (pull-off) bonds between two AOM surfaces. Moreover, the retract curve illustrate that AOM layers can deform and stretch when pulled apart and only at distances of the order of several micrometers that they come apart. The softness and deformability of the biopolymer layers is corroborated by the approach curves in Figures 9-4b and 9-4c, showing a gradual increase in force (needed for compression of the AOM layer) over an approach distances of several micrometers (1-4 µm).

The average maximum (adhesion/cohesion) force and average energy to totally pull-off two investigated surfaces from contact are presented in Figure 9-5. Based on maximum retract force, AOMs adhere on polyamide membrane almost 5 times more than on PES membrane. However, the energies needed to detach TEP from both membranes were comparable. The bond between AOM-coated probe and AOM-fouled membrane was at least ~3 times higher force and ~20 times more energy to separate completely. Overall, the measured adhesion/cohesion forces of AOM range from 1.2 to 18.8 nN. These values are 2 to 3 orders of magnitude higher than the typical lift forces occurring in cross-flow membrane reported by Kang et al. (2004). This may indicate that the adhesion force between AOM and the membrane surface can overcome the typical shear forces in an RO feed channel, making them more likely to remain on the surface of the membrane after being deposited.

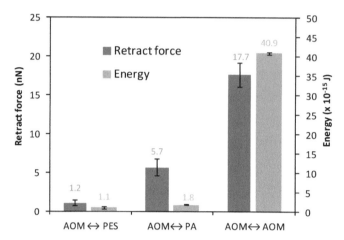

Figure 9-5: Comparison of average adhesion forces and energies between various surfaces. Values and standard deviations were based on at least 18 force-distance curves obtained from 6 different membrane surface locations.

Hydrogen bonding may have played a major role in the adhesion of AOM on membrane surfaces. Since hydrogen bonding is a result of electron transfer between electronegative moieties and hydrogen atoms on AOM and membrane surfaces, the bond strength might be influenced by the membrane surface roughness. The reported surface roughness of PA membranes is generally higher than PES membrane (Pieracci et al., 1999; Hurwitz et al., 2010). This might explain the higher adhesion force of AOM on PA membranes than on PES membrane as the actual surface interaction area is higher in rougher membranes. This is crucial considering that AOMs, particularly TEPs, are deformable and may likely follow the topography of the membrane surface. On the other hand, cohesion between AOMs may be influenced not only by hydrogen bonding but also by polymer entanglement (Flemming et al, 1997). AOM polymers are generally flexible and elastic, so disentangling them may have occurred in a stepwise fashion and therefore requires higher energy to totally detach them.

In general, the results of adhesion force measurements demonstrated the high potential of AOM to deposit on polyamide RO membrane. Once a substantial deposition of AOM is already in place, it may enhance deposition of bacteria and further deposition of more AOM. Moreover, most bacterial cells are coated with sticky TEP-like materials. Hence, a stronger bacterial adhesion is expected on the surface of the membrane with a deposited layer of AOM.

9.3.2 Bacterial growth potential

The growth potential of bacteria in seawater containing AOM was investigated based on bacterial growth potential tests. In this study, the bacterial communities and available nutrients in the intake water of a desalination plant was used as the base solution and flow cytometric analyses were employed to monitor their growth. Bacterial growth in duplicate batch solutions spiked with different concentrations (0, 0.08, 0.4 and 0.8 mg C/L) of purified AOM was monitored and compared. The aim of this experiment was to test if the presence of AOMs can enhance growth of bacteria and if they can be easily degraded/utilised by bacteria within a short period of time (4 days). The bacterial growth curves of solutions spiked with different concentrations of AOM is presented in Figure 9-6.

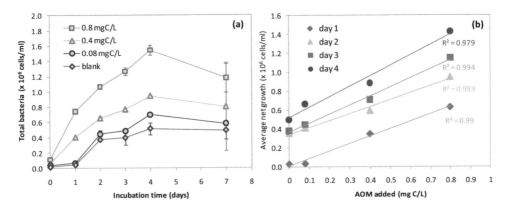

Figure 9-6: (a) Growth curves of live bacteria in batch seawater solutions spiked with different concentrations of AOM. (b) Linear regression between AOM concentration and average net growth for day 1-4.

Bacteria growing in filtered seawater without additional AOM (blank) showed a lag phase for approximately 24 hours and an exponential growth phase between day 1 and day 2 (Figure 9-6a). A similar trend was observed for solutions spiked with 0.08 mg C/L of AOM. For solutions spiked with higher concentrations of AOM (0.4 and 0.8 mg C/L), no apparent lag phase was recorded possibly due to the low frequency (every 24 hours) of sampling.

Despite having been inoculated with the same concentration of bacteria from seawater, the initial bacterial concentrations were higher with higher concentration of AOMs. The additional bacteria likely originated from the AOM solution itself. As described in Section 9.2.4, the AOM solution added to pre-filtered seawater was pre-treated by filtration through a 0.2µm PC membrane to remove bacteria and then dialysed (3.5 kDa membrane) to remove low molecular weight constituents in the solution, including excess nutrients and dissolved salts from the culture medium. This means that some contamination by freshwater bacteria might have occurred during the handling and preparation of the desalted AOM solution. These bacteria are expected be inactive in saline conditions when added to seawater. Thus, it can be assumed that the net growth in solutions spiked with AOM is due to multiplication of the inoculated seawater bacteria only.

In general, the growth of bacteria within the incubation period was directly proportional to the concentration of AOM added to the seawater. The net growth was determined as the difference between the bacterial concentration (in days 1, 2, 3 and 4) and the initial concentration at day 0. As shown in Figure 9-6b, the net growth of bacteria showed significant linear correlation ($R^2 > 0.97$) with the concentration of AOM in the solution. This may suggest that TEP indeed enhance growth of bacteria in seawater.

To investigate if AOM was possibly degraded and/or consumed by bacteria during the batch tests, LC-OCD analysis was performed on samples collected from the solution with the highest net growth (i.e., spiked with 0.8 mg C/L AOM). The analyses results are presented in Figure 9-7.

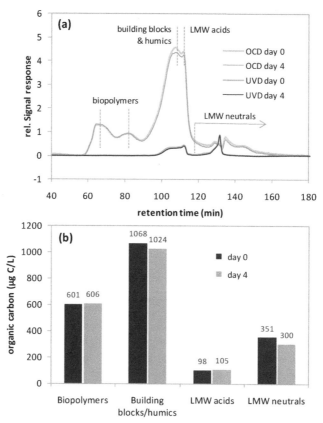

Figure 9-7: (a) OCD and UVD chromatograms of water samples collected at day 0 and day 4 of the growth potential test of seawater spiked with 0.6 mg C/L AOM. (b) Comparison of carbon concentrations of various fractions of organic matter in samples between day 0 and 4 as determined by LC-OCD.

The initial LC-OCD chromatogram of filtered seawater spiked with 0.8 mg C/L of AOM showed a characteristic double biopolymer peaks (63 and 81 mins), a humics/building block peak (106 mins) with a shoulder peak (110 mins) assigned to low molecular weight (LMW) organic acids and a series of peaks (115-180 mins) of LMW organic neutrals (Figure 9-7a). The spiked AOM was pre-filtered through 0.2 µm membranes and dialysed using 3.5 kDa MWCO membranes. Based on these pre-treament procedures, the size range of AOM used in the growth potential tests is between 3.5 kDa and 0.2µm. The purified AOM comprise 67% biopolymers, 25% building blocks, 5% LMW neutrals and 3% LMW acids (see Annex Figure 9-14). The filtered seawater base solution contains 1.15 mg C/L of organic matter, comprising 5% biopolymers, 47% humics, 16% building blocks, 22% LMW neutrals and 10% LMW acids (Annex Figure 9-14).

Comparison of the LC-OCD chromatograms between day 0 and day 4 showed no apparent shifting in retention time of the identified peaks, which means that there was no obvious change in molecular weight distribution of organic matter in the solution (Figure 9-7a). However, a slight decrease in OCD signal response was observed for the low molecular weight fractions (e.g., building blocks, organic neutrals). The carbon concentrations of the different fractions derived from the OCD chromatograms are presented in Figure 9-7b for comparison. The most remarkable change in concentration was observed for the humics/building blocks and organic neutral fractions, where the concentrations decreased by 44 and 51 µg C/L, respectively. A negligible increase (<7 µg C/L) in concentrations were observed for biopolymers and organic acid fractions. In total, the organic carbon concentration in the solution decreased by 83 µg C/L after 4 days of incubation.

The organic carbon lost after 4 days of incubation may represent the biodegradable fraction of organic matter which is often measured as biodegradable organic carbon (BDOC) or assimilable organic carbon (AOC). For comparison, the value found in the experiment is within the range of seawater AOC concentrations (30-360 µg Ac-C/L) reported by Weinrich et al. (2011) but lower than the range of seawater BDOC concentrations (190-290 µg/L) reported by de Vittor et al. (2009). Nevertheless, the 83 µg C/L total decrease in carbon concentration is only slightly higher than the lower limit of detection (LOD) of the LC-OCD analysis. The LOD concentration of biopolymers, calculated as the average concentration of demineralised water (68 µg C/L) plus 3 times the standard deviation (2 µg C/L), is 74 µg C/L (www.doc-labor.de). Moreover, the LOD is likely higher in saline waters due to the interference of salts. Therefore, it is possible that the decrease in concentration of organic materials in the batch solution after 4-day incubation is below the LOD of LC-OCD analysis.

In summary, the bacterial growth potential tests demonstrated that AOM (3.5kDa - 0.2µm) from marine algae can enhance growth of bacteria in seawater. AOM, especially the biopolymer fraction, is not easily biodegradable by bacteria based on the change in organic carbon observed within the 4-day incubation period. AOM like TEPs in natural waters can be degraded by bacteria in a matter of hours to several months (Passow, 2002). Although organic material consumed by bacteria may have originated from filtered seawater, it is apparent that the AOM substantially enhanced biological activity in the water as the net growth of bacteria have increased with higher concentration of added AOM. Increased bacterial activity has been reported in algal bloom situations especially at the senescent period of the bloom where most AOM are released (Grossart, 1999; Cole, 1982). It might be possible that high molecular weight AOMs such as biopolymers may eventually be degraded by bacteria when the readily biodegradable organic matter in the water is already depleted. However, further investigations are necessary to verify this theory.

9.3.3 Effect of AOM in the feedwater

The effect of AOM in the feedwater on biological fouling development in cross-flow membrane system was investigated in capillary UF set-up. Four parallel filtration experiments were performed to compare cake resistance and pressure drop increase in the membrane system when fed with solutions comprising different compositions of nutrients and AOM, namely:

1. 90% (artificial seawater + 0.5 mg C/L AOM) + 10% (natural seawater)
2. 90% (artificial seawater + CPN nutrients) + 10% (natural seawater)
3. 90% (artificial seawater + 0.5 mg C/L AOM + PN nutrients) + 10% (natural seawater)
4. 90% (artificial seawater + 0.5 mg C/L AOM + CPN nutrients) + 10% (natural seawater)

The feedwater solutions with nutrients were initially spiked with either CPN or PN nutrients (C = 0.2 mg C/L, P = 0.01 mg P/L, N = 0.04 mg N/L) to accelerate bacterial growth. Since the feedwater was continuously recycled and nutrients were eventually depleted by bacteria, additional nutrients were added every 7 days. All experiments were operated at room temperature (20±2 °C), flux of 15 L/m².h and cross-flow velocity of 0.15 m/s to simulate the conditions in a reverse osmosis system during spring period in the Netherlands. The cake and feed channel pressure drop in the 4 experiments, within 20 days of monitoring, are compared in Figure 9-8.

The capillary membranes used in the experiments have an average clean water resistance of 7.9 x 10¹² (± 15%) and feed channel pressure drop of 78 mbar (± 5%) at cross-flow velocity of 0.15 m/s. The capillary UF membrane is a low molecular weight cut-off UF (7 kDa) and its membrane resistance is about 5-7 times higher than in most commercial UF membranes (100-300 kDa). A low MWCO UF was selected for the biofouling experiments so that most of AOM biopolymers from the feed water can be rejected by the membrane (Villacorte et al., 2013).

As shown in Figure 9-8, the increase in cake resistance and feed channel pressure drop were not substantially different for the 4 experiments during the first 8 days of operation. After day 8, an exponential increase of both cake resistance (up to 350%) and pressure drop (up to 500%) was recorded for the membrane module fed with both AOM and CPN nutrients. An exponential increase (up to 100%) in cake resistance was also observed for the membrane fed with CPN nutrients only, but no significant increase in pressure drop. No remarkable increase in cake resistance and pressure drop was observed for membranes fed only with AOM and membranes fed with AOM and PN nutrients.

In general, the parallel biofouling experiments demonstrated that AOM introduced in nutrient-rich seawater can substantially (4x in this study) enhance biological fouling. However when introduced under nutrient limiting conditions, rapid biological fouling may not occur even if AOM is abundantly present in the feed water. It was also illustrated that AOM was not effectively utilised by bacteria as an alternative source of biodegradable carbon during the 18-day period. This is in agreement with the bacterial growth potential test experiments.

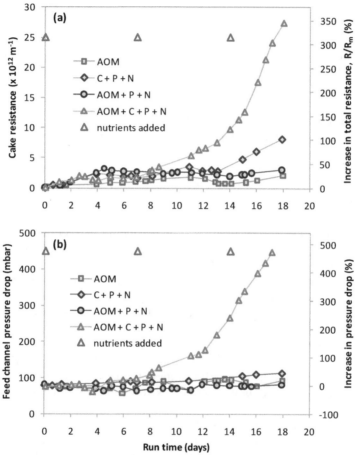

Figure 9-8: Change in hydraulic resistance (a) and feed channel pressure drop (b) through 1-m long cross-flow capillary membranes fed with and without AOM (0.5 mgC/L) or nutrients.

9.3.4 Effect of AOM deposits on the membrane surface

When sticky AOM material (e.g., TEPs) deposit on membrane surfaces, they may serve as conditioning layer which allow more effective attachment of bacteria and accelerate biofilm development (Berman and Holenberg, 2005). This hypothesis was investigated by performing parallel experiments in cross-flow capillary membranes pre-fouled with different concentrations of AOM. The experiments were performed in two phases:

1. In the first phase (AOM fouling), 3 membrane modules were feed with artificial seawater (ASW) spiked with different concentrations of AOM (0.2, 0.4 and 0.8 mgC/L). The membrane system was operated (J=15 L/m^2.h; CFV=0.15 m/s) until an apparent increase (>15%) in total resistance was observed in all of the 3 modules.

2. In the second phase (biofouling), the feedwater of pre-fouled modules were replaced with ASW solution containing CPN nutrients (0.2 mg C/L, 0.01 mg P/L, 0.04 mg N/L) and bacteria. For comparison, a clean membrane (blank) was also operated with the same feed water. The operational settings (J and CFV) were similar to the first phase to minimize detachment of deposited AOM. The filtrate and concentrate water were continuously re-circulated back to the feed tank; hence, additional nutrients were added to the feedwater every 1-2 days to maintain growth of bacteria.

Two sets of experiments were performed based on the above-mentioned procedure. In the first set of experiments, the feed solution in the biofouling phase was prepared containing 10% untreated seawater with its natural bacterial community. In the second set of experiments, the feed solution of the biofouling phase was inoculated with ~1 million cells/ml of cultured marine bacteria (*V. harveyi*) instead of untreated seawater. The results of the two sets of experiments are presented in Figure 9-9.

The AOM fouling phase in the first set of experiments was performed for 2.5 days (Figure 9-9a). Within this period the total resistance increased by about 60% for the membrane fed with 0.8 mg C/L AOM while a similar increase (~30%) was observed for membranes coated with 0.2 and 0.4 mg C/L. For the second set of experiments, the AOM fouling phase was performed longer (~5 days) due to slower rate of increase in cake resistance (Figure 9-9b). The increase in total resistance after 5 days of operation was 18%, 24% and 32% for membranes fouled with 0.2, 0.4 and 0.8 mg C/L of AOM, respectively. The substantial difference in observed fouling rates between the two sets of experiments was attributed to possible variations of the resistance and/or stickiness of the different batches of AOM used in the experiments. The reason for such variations is still subject to further investigations.

Biofouling started to kick-off 5 days after untreated natural seawater was added to the feed solution (Figure 9a) while biofouling immediately kick-off when *Vibrio harveyi* was added to the feedwater (Figure 9-9b). The possible reason for the difference in fouling rate is that *V. harveyi* is already adapted to the ASW matrix as it was previously grown in similar medium, while the local seawater bacterial population needed some time to adjust to the artificial seawater medium and water temperature. It may also indicate that *V. harveyi* are much more effective in utilising available nutrients than bacteria present in the seawater sample used in the experiment. Moreover, *V. harveyi* are relatively larger (up to 10 µm in length; see Annex Figure 9-15) than most marine bacteria and it has been reported to produce substantial amounts of TEP-like EPS (Bramhachari and Dubey, 2006), which probably make them more effective in forming biofilm on the surface of the membrane. Preliminary tests of sample taken from a batch culture of *V. harveyi* verified the substantial presence of bacterial TEPs (results not shown).

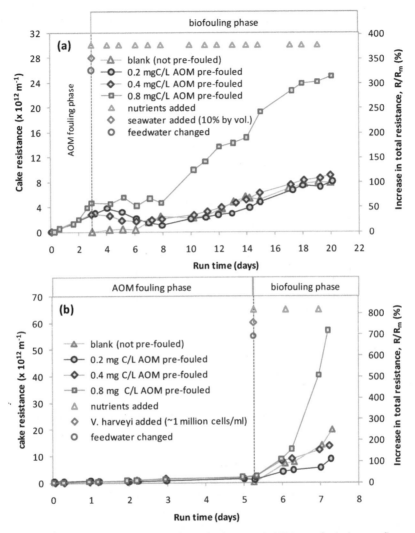

Figure 9-9: Change in hydraulic resistance through clean and AOM pre-fouled cross-flow capillary membranes. (a) Biofouling tests with bacteria from seawater. (b) Biofouling tests with model bacteria *V. harveyi*.

In biofouling experiments with mixed seawater bacteria (Figure 9-9a), the final cake resistance in the blank membrane (not pre-fouled) is similar to that of membranes pre-fouled with 0.2 and 0.4 mg C/L of AOM. However, experiments with membranes pre-fouled with 0.8 mgC/L showed ~3 times higher final cake resistance.

In biofouling experiments with *V. harveyi* (Figure 9-9b), the final cake resistance of membranes pre-fouled with 0.2 and 0.4 mgC/L of AOM were 40% and 55% lower than the blank, respectively. The membrane pre-fouled with 0.8 mg C/L of AOM was remarkably higher (180%) than the blank. The substantial difference between low (<0.4 mg C/L) and high (>0.4 mg C/L) concentrations of TEP may be due to variations in coverage area of deposited AOM. The relatively higher blank cake resistance in the biofouling test with *V. harveyi* can be attributed to natural variability in the attachment and/or growth of bacteria. Overall, the results of two batches of experiments demonstrate that a substantial accumulation of AOM can accelerate biofouling in cross-flow membranes.

An autopsy of the fouled capillary membranes was performed after biofouling experiments with *V. harveyi*. Concentrations of organic carbon, $TEP_{0.4\mu m}$ and *V. harveyi* from extracted biofilm samples are presented in Figure 9-10.

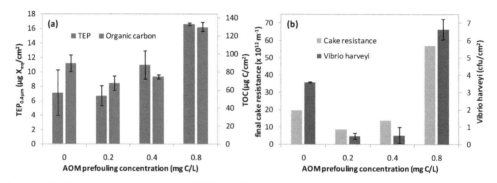

Figure 9-10: Average concentrations of TEP, total organic carbon and *V. harveyi* in biofilm samples extracted from capillary membranes used in the biofouling tests with *V. harveyi* bacteria in comparison with the final cake resistance.

Among the three parameters measured, *V. harveyi* concentration best coincides with the final cake resistance of the 4 experiments. The highest recorded concentration of *V. harveyi* bacteria (~6.5 x 10^6 cells/cm²) is within the range of concentration reported in severely biofouled RO membrane modules (Vrouwenvelder *et al.*, 2008). The highest concentration of *V. harveyi* also corresponds to the highest concentrations of TEP and TOC in the biofilm, an indication that the bacteria are releasing substantial concentrations of TEP-like EPS. The accumulation of these materials may have likely caused the rapid increase in biofilm resistance.

9.3.5 Effect of AOM deposits in spiral wound membrane

The feed channel diameter (0.8 mm) of the capillary membranes used in the biofouling experiments (discussed in Sections 9.3.3 and 9.3.4) are comparable to a spiral wound membrane. However unlike capillary membranes, spiral wound membranes have a feed spacer which can substantially affect the accumulation of biofilm in the feed-concentrate channel (Vrouwenvelder *et al.*, 2009a; 2010). The spacers can also provide additional surface area for AOM and bacterial attachment, whereby biofilm formation will not only form on the membrane but also on the spacers.

Further investigations were performed on the effect of AOM on biofouling specifically in spiral wound membranes based on the changes of feed channel pressure drop. Parallel experiments were performed in four MFS cells with spiral wound membranes and spacers, operated under similar hydraulic condition (CFV=0.2 m/s) in SWRO. The reproducibility of hydraulic conditions in 4 MFS cells was investigated for different cross-flow velocities (0.05-0.4 m/s). It was found that the reproducibility of results is acceptable only for cross-flow velocities ≥0.2 m/s (see Annex Figure 9-16) for the results). Furthermore, biofouling reproducibility tests were also performed for the 4 MFS cells where the cells were fouled by feeding saline solution containing *V. harveyi* and nutrients from the same feed water tank (see Annex Figure 9-17). It was observed that the pressure drop increases in MFS cells 1 and 4 were similar but substantially higher (~70%) than both MFS cells 2 and 3. Hence, the subsequent biofouling experiments were performed in duplicates where MFS 1 and 2 were set as the control/blank cells (clean membrane) and MFS 3 and 4 as the positive cells (pre-fouled with AOM).

The biofouling experiments with AOM were performed in 3 phases, namely:

1. Phase 1. Saline AOM solutions containing 1 mg C/L of biopolymers were fed through MFS 3 and 4 until >15% increase in pressure drops was observed for both cells.

2. Phase 2. The feed solution of MFS 3 and 4 were replaced with artificial seawater solution spiked with ~1 million cells/ml of *V. harveyi*. At the same time, the operation of MFS 1 and 2 were started and feed with similar feed solution as MFS 3 and 4. The feedwater was re-circulated in the system for ~24 hours to allow attachment of bacteria on the membrane/spacers.

3. Phase 3. The contents of the feed tanks were replaced with clean saline solution spiked with CPN nutrients (0.1 mg C/L, 0.01 mg P/L, 0.02 mg N/L). To minimise depletion of nutrients, additional nutrients were added to the feedwater regularly. The feedwater tanks were replenished every 3-4 days with clean artificial seawater to minimise bacterial growth in the tank.

The results of the 21-day monitoring of pressure drop in all three phases of the experiments are presented in Figure 9-11.

The pressure drop increase after AOM fouling (Phase 1) for ~3 days were 23% and 17% for MFS 3 and 4, respectively. Within 24 hours after introducing *V. harveyi* bacteria to the system (Phase 2), the pressure drop increased in MFS 3 and 4 by 15% and 9%, respectively but no apparent increase was observed for the blank cells (MFS 1 and 2). During the biofouling phase (Phase 3), rapid fouling was observed for MFS 3 and 4 while a slower fouling rate was observed for MFS 1 and 2. Pressure drop in MFS 3 have increased by ~900% 7 days after *V. harveyi* was introduced to the system while a similar increase in MFS 4 was only observed 18 days after the bacteria was introduced. In comparison, the control cells (MFS 1 and 2) only showed ~250% increase in pressure drop after 18 days the bacteria was introduced.

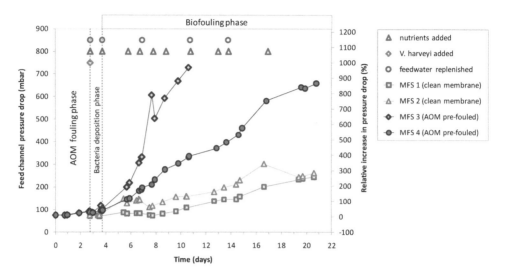

Figure 9-11: Pressure drop across the feed channel of MFS cells fouled and pre-fouled with AOM (1 mg C/L). MFS 3 and 4 were operated starting at the beginning of the experiment (day 0) and were fouled with 1 mg C/L AOM in feedwater for 3 days while MFS 1 and 2 were operated starting at day 3 with clean membranes/spacers.

In general, the MFS biofouling experiments demonstrated that pre-fouling of spiral wound membrane/spacer with AOM can substantially accelerate biofouling in the system. The remarkable difference in fouling rate between MFS 3 and 4 despite being operated similarly can be attributed to the reproducibility issues of the MFS cells (see Annex Figure 9-17). However, considering that a higher pressure drop increase was observed in MFS 3 (23%) than MFS 4 (17%) during AOM fouling phase may indicate that more AOM were deposited in MFS 3 than in MFS 4. Since the accumulated AOM may serve as a good attachment site for bacteria, more accumulation of AOM may have led to better attachment of *V. harveyi* resulting in faster development of biofilm.

As shown in Figure 9-11, the experiment was terminated earlier for MFS 3 (day 10) to avoid risk of leaking in the plastic tubing due to high pressure. Just before MFS 3 was taken out for autopsy (day 10), photographs of the 4 MFS cells were taken from the transparent top of the cells for comparison. As shown on Figure 12, biofilm accumulation is apparent on the membrane and spacers of AOM pre-fouled membranes (MFS 3 and 4) while the control membranes/spacers appear to have no visible accumulation of biofilm. Also, no visible accumulation was observed during the AOM fouling phase despite an apparent increase in pressure drop, likely because of the transparent nature of AOMs. The cell with the most biofilm accumulation was MFS 3, where streamers and blobs of biofilm can be observed obstructing various portions of the feed channel. MFS 4 showed biofilm in similar location but relatively fewer accumulations than MFS 3. In both MFS 3 and 4, biofilm accumulation was concentrated behind intersections of spacer mesh and appeared to be consistent between the feed and outlet side of the cell. Such observation is similar to what was reported in previous experimental and modelling studies (Vrouwenvelder *et al.,*

2009a,b; 2010). In MFS 1 and 2, only small accumulations of biofilms were observed near the feed side while the rest of membranes/spacers appeared to have no visible biofilm. The visual observations (Figure 9-12) were relatively consistent with the pressure drop readings presented in Figure 9-11.

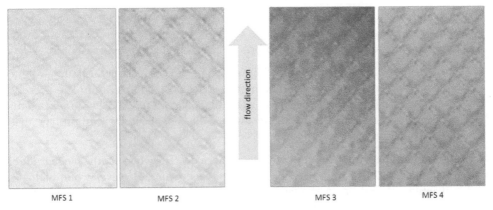

Figure 9-12: Photographic images of membranes and spacers as observed from the transparent top of 4 MFS cells during the 10th day of the parallel biofouling experiments.

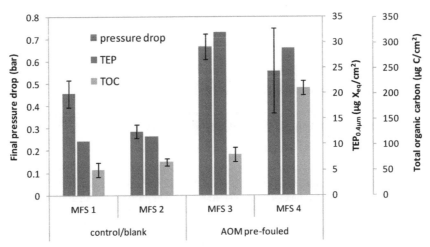

Figure 9-13: TEP and TOC concentrations in biofilm extracted from membranes and spacers in the MFS biofouling experiments in comparison with the final cake resistance. For MFS 1, 2 and 4, biofilm samples were extracted on day 21 while samples were extracted on day 10 for MFS 3.

Biofilm samples extracted from 4 MFS after the end of each MFS experiments were analysed for TOC and $TEP_{0.4\mu m}$ concentrations (Figure 9-13). Both TOC and $TEP_{0.4\mu m}$ concentrations were consistently higher for MFS membranes/spacers pre-fouled with AOM than the control. Among the two parameters measured, TEP concentration was more consistent with the final pressure drop recorded by the MFSs than TOC. This may further indicate that the biofilm formed by *V. harveyi* comprised substantial concentrations of TEP-like EPS and that these may have been mainly responsible for the increase in pressure drops.

9.4 Conclusions

- In clean membranes, better adhesion of AOM was observed on polyamide RO membrane than on polyethersulfone UF membrane. However, AOM can attach more strongly on AOM-fouled membranes than on clean membranes.

- The accumulation of AOM during algal bloom can substantially enhance biological growth in seawater. The increase in bacterial activity in the presence of AOM can be due to the ability of AOM biopolymers (e.g., TEPs) to absorb and concentrate nutrients from the feedwater so that bacteria can utilise these nutrients more effectively.

- The biofouling rate was substantially accelerated when both biodegradable nutrients and AOM were present in the RO feed water. However when biological nutrients are limiting in the feedwater, AOM were not observed to accelerate the occurrence of biological fouling.

- The pre-conditioning of RO membrane as a result of initial fouling by AOM can also accelerate biological fouling.

References

Allison, D. G., & Sutherland, I. W. (1987). The role of exopolysaccharides in adhesion of freshwater bacteria. Journal of General Microbiology, 133(5), 1319-1327.

Austin B. and Zhang X-H. (2006) Vibrio harveyi: a significant pathogen of marine vertebrates and invertebrates. Letters in Applied Microbiology 43, 119–124.

Baker, J. S. and Dudley, L. Y. (1998) Biofouling in membrane systems—A review. Desalination 118(1-3), 81-89.

Bar-Zeev E, Berman-Frank I, Girshevitz O, Berman T. (2012) Revised paradigm of aquatic biofilm formation facilitated by microgel transparent exopolymer particles. PNAS 109(23), 9119-24.

Bar-Zeev, E., Berman-Frank, I., Liberman, B., Rahav, E., Passow, U. and Berman, T. (2009) Transparent exopolymer particles: Potential agents for organic fouling and biofilm formation in desalination and water treatment plants. Desalination & Water Treatment 3, 136–142.

Berman, T., and Holenberg (2005) M., Don't fall foul of biofilm through high TEP levels. Filtration & Separation 42(4), 30-32.

Berman, T., R. Mizrahi & C. G. Dosoretz (2011) Transparent exopolymer particles (TEP): A critical factor in aquatic biofilm initiation and fouling on filtration membranes. Desalination, 276, 184-190.

Bramhachari P.V. and Dubey S.K. (2006) Isolation and characterization of exopolysaccharide produced by Vibrio harveyi strain VB23. Letters in Applied Microbiology 43, 571–577.

Caron , D.A., Garneau, M.E., Seubert, E., Howard M.D.A.,, Darjany L., Schnetzer A., Cetinic, I., Filteau, G., Lauri, P., Jones, B. and Trussell, S. (2010) Harmful algae and their potential impacts on desalination operations off southern California. Water Research 44, 385–416.

Characklis, W. G. (1981) Bioengineering report: Fouling biofilm development: A process analysis. Biotechnology and Bioengineering 23: 1923–1960.

Childress, A. E., & Elimelech, M. (1996). Effect of solution chemistry on the surface charge of polymeric reverse osmosis and nanofiltration membranes. Journal of Membrane Science, 119(2), 253-268.

Cole, J. J. (1982). Interactions between bacteria and algae in aquatic ecosystems. Annual Review of Ecology and Systematics 13, 291-314.

de la Torre, T., Lesjean, B., Drews, A., and Kraume, M. (2008) Monitoring of transparent exopolymer particles (TEP) in a membrane bioreactor (MBR) and correlation with other fouling indicators. Water Science and Technology 58 (10), 1903–1909.

De Vittor, C., Larato, C., & Umani, S. F. (2009). The application of a plug-flow reactor to measure the biodegradable dissolved organic carbon (BDOC) in seawater. Bioresource Technology, 100(23), 5721-5728.

Discart, V., Bilad, M. R., Vandamme, D., Foubert, I., Muylaert, K., & Vankelecom, I. F. J. (2013). Role of transparent exopolymeric particles in membrane fouling: Chlorella vulgaris broth filtration. Bioresource Technology, 129, 18-25.

Dreszer, C., Vrouwenvelder, J. S., Paulitsch-Fuchs, A. H., Zwijnenburg, A., Kruithof, J. C., & Flemming, H.C.. (2013) Hydraulic resistance of biofilms. Journal of Membrane Science 429, 436-447.

Flemming, H. C. & G. Schaule (1988) Biofouling on membranes - A microbiological approach. Desalination, 70, 95-119.

Flemming, H. C. (2002) Biofouling in water systems–cases, causes and countermeasures. Applied Microbiology and Biotechnology 59, 629-640.

Flemming, H.C., Schaule, G., Griebe, T., Schmitt, J., & Tamachkiarowa, A. (1997) Biofouling - the achilles heel of membrane processes. Desalination, 113(2-3), 215-225.

Frank, B. P., & Belfort, G. (2003). Polysaccharides and sticky membrane surfaces: Critical ionic effects. Journal of Membrane Science, 212(1-2), 205-212.

Grossart, H.P. (1999). Interactions between marine bacteria and axenic diatoms (cylindrotheca fusiformis, nitzschia laevis, and thalassiosira weissflogii) incubated under various conditions in the lab. Aquatic Microbial Ecology 19(1), 1-11.

Huber S. A., Balz A., Abert M. and Pronk W. (2011) Characterisation of aquatic humic and non-humic matter with size-exclusion chromatography – organic carbon detection – organic nitrogen detection (LC-OCD-OND). Water Research 45 (2), 879-885.

Hurwitz, G., Guillen, G. R., & Hoek, E. M. V. (2010). Probing polyamide membrane surface charge, zeta potential, wettability, and hydrophilicity with contact angle measurements. Journal of Membrane Science, 349(1-2), 349-357.

Kang S.T., Subramani A., Hoek E.M.V., Deshusses M.A., Matsumoto M.R. (2004) Direct observation of biofouling in cross-flow microfiltration: mechanisms of deposition and release. Journal of Membrane Science 244 (1–2), 151–165.

Kennedy, M. D., Muñoz-Tobar, F. P., Amy, G. L. and Schippers, J. C. (2009) Transparent exopolymer particle (TEP) fouling of ultrafiltration membrane systems. Desalination & Water Treatment 6(1-3) 169-176.

Li Q. and Elimelech M. (2004) Organic Fouling and Chemical Cleaning of Nanofiltration Membranes: Measurements and Mechanisms. Environ. Sci. Technol. 38 (17), 4683–4693.

Myklestad S., Haug A., Larsen B. (1972) Production of carbohydrates by the marine diatom Chaetoceros affinis var. willei (Gran) Hustedt. II. Preliminary investigation of the extracellular polysaccharide. Journal of Experimental Marine Biology and Ecology, 9 (2) , 137-144.

Nguyen, T.; Roddick, F.A.; Fan, L. (2012) Biofouling of Water Treatment Membranes: A Review of the Underlying Causes, Monitoring Techniques and Control Measures. Membranes 2, 804-840.

Pankratz, T., 2008. Red tides close desal plants. Water Desalination Report 44, 1.

Passow, U. (2002) Transparent exopolymer particles (TEP) in aquatic environments, Progress In Oceanography 55(3-4), 287-333.

Passow, U. and Alldredge, A. L. (1995) A Dye-Binding Assay for the Spectrophotometric Measurement of Transparent Exopolymer Particles (TEP). Limnology and Oceanography 40(7), 1326-1335.

Pieracci, J., Crivello, J. V., & Belfort, G. (1999). Photochemical modification of 10kDa polyethersulfone ultrafiltration membranes for reduction of biofouling. Journal of Membrane Science, 156(2), 223-240.

Prest, E.I., Hammes, F., Kötzsch, S., van Loosdrecht, M.C.M, Vrouwenvelder, J.S. (2013) Monitoring microbiological changes in drinking water systems using a fast and reproducible flow cytometric method, Water Research In Press, doi: 10.1016/j.watres.2013.07.051.

Ralston J., Larson I., Rutland M., Feiler A., Kleijn M. (2005) Atomic force microscopy and direct surface force measurements (IUPAC Technical Report), Pure and Applied Chemistry, 77, 2149-2170.

Ramus, J. (1977) Alcian Blue: A Quantitative Aqueous Assay for Algal Acid and Sulfated Polysaccharides. Journal of Phycology 13, 345-348.

Ricq, L., Pierre, A., Bayle, S., & Reggiani, J.C. (1997). Electrokinetic characterization of polyethersulfone UF membranes. Desalination 109(3), 253-261.

Schock, G. & A. Miquel (1987) Mass transfer and pressure loss in spiral wound modules. Desalination 64, 339-352.

Schurer R., Tabatabai A., Villacorte L., Schippers J.C., Kennedy M.D. (2013) Three years operational experience with ultrafiltration as SWRO pre-treatment during algal bloom. Desalination & Water Treatmen 51 (4-6), 1034-1042.

Schurer, R., Janssen, A., Villacorte, L., Kennedy, M.D. (2012) Performance of ultrafiltration & coagulation in an UF-RO seawater desalination demonstration plant. Desalination & Water Treatment 42(1-3), 57-64.

SLMB (2012) Determining the total cell count and ratios of high and low nucleic acid content cells in freshwater using flow cytometry. Analysis method 333.1, The Swiss Food Book (Schweizerische Lebensmittelbuch). Federal Office of Public Health, Switzerland.

Spruijt E., van den Berg S.A., Cohen-Stuart M.A., and van der Gucht J. (2012) Direct Measurement of the Strength of Single Ionic Bonds between Hydrated Charges. ACS Nano 6 (6), 5297–5303.

Van Nevel, S., Hennebel, T., De Beuf, K., Du Laing, G., Verstraete, W., & Boon, N. (2012). Transparent exopolymer particle removal in different drinking water production centers. Water Research, 46(11), 3603-3611.

Villacorte, L. O., Kennedy, M. D., Amy, G. L., and Schippers, J. C. (2009a) The fate of Transparent Exopolymer Particles (TEP) in integrated membrane systems: Removal through pretreatment processes and deposition on reverse osmosis membranes. Water Research 43 (20), 5039-5052.

Villacorte, L. O., Kennedy, M. D., Amy, G. L., and Schippers, J. C. (2009b) Measuring Transparent Exopolymer Particles (TEP) as indicator of the (bio)fouling potential of RO feed water. Desalination & Water Treatment 5, 207-212.

Villacorte, L. O., Schurer, R., Kennedy, M., Amy, G. and Schippers, J.C. (2010a) Removal and deposition of Transparent Exopolymer Particles (TEP) in seawater UF-RO system. IDA Journal 2 (1), 45-55.

Villacorte, L. O., Schurer, R., Kennedy, M., Amy, G., Schippers, J.C. (2010b) The fate of transparent exopolymer particles in integrated membrane systems: a pilot plant study in Zeeland, The Netherlands. Desalination & Water Treatment 13, 109-119.

Villacorte, L.O., Ekowati, Y., Winters, H., Amy, G.L., Schippers, J.C. and Kennedy, M.D. (2013) Characterisation of transparent exopolymer particles (TEP) produced during algal bloom: a membrane treatment perspective. Desalination & Water Treatment 51 (4-6), 1021-1033.

Vrouwenvelder, J.S., Graf von der Schulenburg, D.A., Kruithof, J.C., Johns, M.L., Van Loosdrecht, M.C.M. (2009a) Biofouling of spiral wound nanofiltration and reverse osmosis membranes: a feed spacer problem. Water Research 43: 583-594.

Vrouwenvelder, J.S., Hinrichs, C., Van der Meer, W.G.J., Van Loosdrecht, M.C.M., Kruithof, J.C. (2009b). Pressure drop increase by biofilm accumulation in spiral wound RO and NF membrane systems: Role of substrate concentration, flow velocity, substrate load and flow direction. Biofouling 25:543-555.

Vrouwenvelder, J.S., Manolarakis, S.A., van der Hoek, J.P., van Paassen, J.A.M., Van der Meer, W.G.J., Van Agtmaal, J.M.C., Prummel, H.D.M., Kruithof, J.C., Van Loosdrecht, M.C.M. (2008) Quantitative Biofouling Diagnosis in Full Scale Nanofiltration and Reverse Osmosis Installations. Water Research 42: 4856-4868.

Vrouwenvelder, J.S., Picioreanu, C., Kruithof, J.C., Van Loosdrecht, M.C.M. (2010) Biofouling in spiral wound membrane systems: Three-dimensional numerical model based evaluation of experimental data. Journal of Membrane Science 346, 71–85.

Vrouwenvelder, J.S., Van Paassen, J.A.M., Wessels, L.P., Van Dam, A.F., Bakker, S.M. 2006. The membrane fouling simulator: A practical tool for fouling prediction and control. Journal of Membrane Science 281: 316-324.

Weinrich L.A., Schneider O.D. and LeChevallier M.W. (2011) Bioluminescence-Based Method for Measuring Assimilable Organic Carbon in Pretreatment Water for Reverse Osmosis Membrane Desalination Appl. Environ. Microbiol 77 (3), 1148-1150.

Winters H. and Isquith I. R. (1979). In-plant microfouling in desalination. Desalination, 30(1), 387-399.

Winters, H., Isquith, I., Arthur, W. A., & Mindler, A. (1983). Control of biological fouling in seawater reverse osmosis desalination. Desalination, 47(1-3), 233-238.

Yamamura, H., Kimura, K., Okajima, T., Tokumoto, H., & Watanabe, Y. (2008). Affinity of functional groups for membrane surfaces: Implications for physically irreversible fouling. Environmental Science and Technology, 42(14), 5310-5315.

Annexes: Supporting information

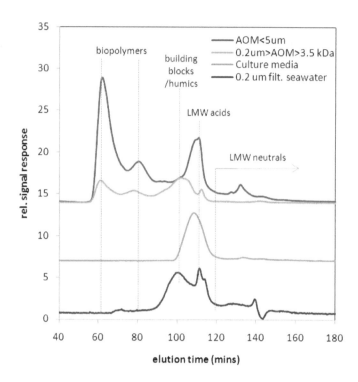

Figure 9-14: LC-OCD chromatograms of AOM samples, algae culture media and filtered seawater used in the bacterial growth potential tests.

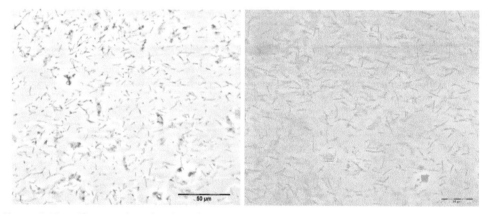

Figure 9-15: Photographs of *Vibrio harveyi* taken through Olympus BX51 microscope, 200x magnification using a black and white camera (left) and coloured camera (right).

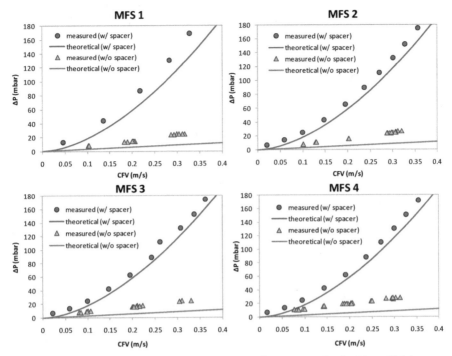

Figure 9-16: Measured and theoretical pressure drops of the 4 MFS feed with artificial seawater at various cross-flow velocities with and without spacers. Theoretical calculations were based on the method by Schock & Miquel (1987).

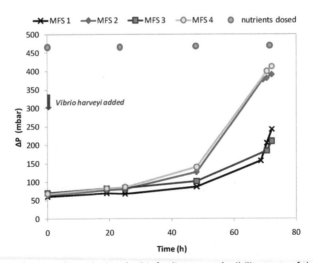

Figure 9-17: Pressure drop readings during the biofouling reproducibility tests of the 4 MFS feed from the same feed tank with artificial seawater and nutrients.

10

General conclusions

Contents

10.1 Conclusions 280

    10.1.1 Algal bloom and algogenic organic matter (AOM) 281
    10.1.2 TEP method developments 281
    10.1.3 TEP and biopolymer removal by pre-treatment 282
    10.1.4 UF fouling by algal-derived biopolymers 283
    10.1.5 Role of algal cells in UF fouling 283
    10.1.6 Biological fouling enhanced by AOM 284

10.2 General outlook 284

References 285

10.1 Conclusions

Seawater desalination using reverse osmosis (SWRO) technology has been rapidly growing in terms of installed capacity, plant size and global application. A currently emerging threat to SWRO is the seasonal proliferations of microscopic algae in seawater known as algal blooms. These natural phenomena can potentially affect every coastal area in the world where SWRO installations are located. Recent algal bloom incidents in the Middle East have resulted in temporary closure of various SWRO plants due to clogging and poor effluent quality of the pre-treatment system (mainly granular media filtraters) and/or as a drastic measure to avoid irreversible fouling problems in the downstream RO membranes. Furthermore, more extra large plants (>500,000 m³/day) are expected to be installed over the coming years but frequent chemical cleaning (>1/year) of the SWRO would be unfeasible for such plants. This has redirected the focus of the desalination industry to improving the reliability of the pre-treatment system in minimising the fouling in SWRO. To maintain stable operation in SWRO plants during algal blooms, pre-treatment using ultrafiltration (UF) membranes has been proposed. However, some concerns have also been expressed regarding the rate of fouling in UF membranes during direct pre-treatment of seawater suffering from algal bloom.

The potential problems which may occur in membrane-based desalination plants (UF pre-treatment followed by SWRO) during treatment of algal bloom impaired seawater are: (1) particulate fouling in UF due to accumulation of algal cells and their detritus, (2) organic fouling in UF or RO due to accumulation of AOM, and (3) biological fouling in RO initiated and/or enhanced by AOM. These potential issues were addressed in this study by performing theoretical and experimental analyses. Furthermore, various tools/methods were successfully developed and/or further improved to study the characteristics and membrane fouling potential of algal bloom impacted water, namely:

- development of a new method to measure transparent exopolymer particles (TEP) in freshwater and seawater to as small as 10kDa and further verified to minimise the effect of salinity and lowering the level of detection in seawater;

- development of a simple model to assess the deposition of algal cells through an inside-out capillary UF membrane and their impact to the hydraulic ;

- interaction force measurement using AFM to assess the stickiness of AOM and their affinity to UF and RO membranes;

- application of LC-OCD, F-EEM, FTIR, and staining with sugar specific lectins coupled with CLSM to indentify the composition of AOM produced by bloom-forming algae;

- application of flow cytometry to investigate the bacterial growth potential of algal biopolymers; and

- application of membrane fouling simulators (MFS) and cross-flow capillary UF (7 kDa MWCO) to investigate and demonstrate the possible role of AOM in biological fouling in RO membranes.

The above-mentioned theoretical and analytical tools/methods were assessed and/or implemented in the various segments of this study, the major findings and conclusions of which are discussed in the following sections.

10.1.1   Algal bloom and algogenic organic matter (AOM)

The nature and behaviour of algal-derived organic material (AOM) produced by three common species of bloom-forming algae (AT: Alexandrium tamarense, CA: Chaetoceros affinis and MSp: Microcystis sp.) were characterized by employing various characterisation techniques. The observed growth behaviour, peak cell concentration and AOM concentration vary substantially between algal cultures. CA produced 2-3 times more TEP than AT and 6-9 times more than MSp. The AOM produced by the 3 species were mainly biopolymers (polysaccharides and proteins) while some refractory organic matter (e.g., humic-like substances) and/or low molecular weight biogenic substances were also observed. Polysaccharides with associated fucose and sulphated functional groups were ubiquitous in the biopolymer fraction. The stickiness of AOM in terms of adhesive and cohesive forces varies between source species (CA>AT≥MSp). In general, cohesion forces between AOM materials are generally stronger than the adhesion forces between AOM and clean UF or RO membranes.

The AOM retention by microfiltration (MF) and ultrafiltration (UF) membranes were also measured and compared for membranes with different pore sizes. The biopolymer fraction of AOM was substantially removed by MF/UF membranes but not the low molecular weight fractions (e.g., humic-like substances). MF membranes with pore sizes of 0.4 μm and 0.1μm retained 14-47% and 42-56% of biopolymers, respectively. UF membranes with molecular weight cut-offs of 100kDa and 10kDa retained 65-83% and 83-97% of biopolymers, respectively.

To illustrate the adverse effect of algal bloom on membrane filtration, the membrane fouling potential based on the modified fouling index (MFI-UF) was measured in batch cultures and extracted AOM. The membrane fouling potential (MFI-UF) of the 3 batch cultures correlated well with TEP concentration ($R^2>0.85$) but no correlation was observed with algal cell concentration. The membrane fouling potential of AOM depends strongly on the source algal species (CA>MSp>AT). A higher specific membrane fouling potential (MFI per mg C/L biopolymers) were measured for AOM fractions between 10-100 kDa than the larger fractions (>100kDa).

10.1.2   TEP method developments

Over the years, a number of methods have been developed to measure transparent exopolymer particles (TEPs) in seawater and freshwater based on Alcian blue staining and spectrophotometric techniques. However, critical analysis of these methods identified several questions regarding their reproducibility, calibration, limit of detection and range of TEP sizes taken into account. To address this issue investigations were performed focusing on improving the methods developed by Passow and Alldredge (1995) and Thornton *et al.* (2007), because these are, in principle, applicable in both saline and freshwater sources.

The two existing methods showed tendency to overestimate TEP concentrations when measuring brackish and saline water samples due to coagulation of Alcian blue dye in the presence of dissolved salts. To minimise this problem, a membrane rinsing procedure was introduced to remove saline moisture from the membrane after retention of TEP. The rinsing step was effective in lowering the analytical blank and the detection limit of both methods.

Based on the technique introduced by Passow and Alldredge (1995), a modified/improved method was proposed for measuring TEP larger than 0.4 μm ($TEP_{0.4μm}$) or larger than 0.1 μm ($TEP_{0.1μm}$). The method was verified with respect to minimising interference of salinity, improving the reproducibility and lowering the level of detection.

A new method was developed, partially based on the technique of Thornton *et al.* (2007) on which TEP is measured by retention on 10 kDa MWCO membranes. The advantage of the new method is it can potentially measure smaller colloidal TEPs down to 10 kDa. The protocol of the new method was also tested and developed in a way that the effect of salinity is minimised and the limit of detection is known. This method also enables the measurement of different fractions of TEP by making use of membranes having different pore sizes.

To allow comparison between the two improved methods, an integrated calibration protocol was developed using Xanthan gum as the standard. $TEP_{0.4μm}$ and TEP_{10kDa} measurements were successfully applied to monitor TEP production in marine algal cultures and TEP removal over the treatment processes of several water treatment plants.

10.1.3   TEP and biopolymer removal by pre-treatment

The fate of TEPs over the treatment processes of several RO plants was assessed using the newly developed and improved methods (Chapter 5) to measuring particulate and colloidal TEPs ($TEP_{0.4μm}$ and TEP_{10kDa}).

TEP monitoring (1-3 years) in seawater and freshwater sources in 2 RO plants revealed that high TEP concentrations occurred mainly during the spring algal bloom season (March-May). UF (with and without coagulation) and conventional coagulation followed by flotation-filtration or sedimentation-filtration were effective in removing both $TEP_{0.4μm}$ and TEP_{10kDa}. Comparable reductions were also observed with the biopolymer concentration (measured by LC-OCD) and the membrane fouling potential (in terms of MFI-UF).

Data collected from 5 plants showed that the TEP_{10kDa} concentration correlated better with MFI-UF than $TEP_{0.4\mu m}$, biopolymers or humic acid concentrations, an indication that colloidal TEPs can likely cause more (serious) fouling in membrane systems than other forms of organic matter.

Membrane autopsies of RO elements taken from 2 plants revealed substantial accumulation of TEPs and bacteria in RO membranes and spacers. Some of these TEPs may have been produced locally by biofilm bacteria.

10.1.4   UF fouling by algal-derived biopolymers

UF fouling caused by AOM from a common bloom-forming alga (*Chaetoceros affinis*) in seawater was investigated at various levels of solution pH, ionic strength and ionic composition.

The UF fouling propensity of AOM measured based on MFI-UF and the observed non-backwashable fouling rate increased with increasing biopolymer concentration. Lowering the pH of seawater containing AOM increased the specific fouling potential (MFI per mg C/L of AOM). Substantial variations in MFI were also observed when varying the ionic strength in the feedwater. Among the major mono- and di-valent cations in seawater, Ca^{2+} ions demonstrated a significant influence on the fouling propensity of AOM, possibly due to bridging effects.

A model was developed based on MFI-UF and backwashability to project the time before chemical cleaning is needed in the UF system. Results showed an exponential increase in cleaning frequency with a linear increase in TEP concentration.

For comparison, fouling by model biopolymers (SA: sodium alginate and XG: xanthan gum) were also investigated under similar conditions with AOM. The fouling propensity (MFI-UF and non-backwashable fouling) of AOM on UF was substantially higher than with model biopolymers at biopolymer concentrations typical in seawater algal blooms. Furthermore, the changes in solution chemistry have a more pronounced effect on the fouling propensity of SA and XG than AOM, an indication that these model biopolymers do not reliably simulate the fouling propensity of micro-algal biopolymers in seawater.

10.1.5   Role of algal cells in UF fouling

Theoretical and experimental investigations were performed to determine the fouling and plugging potential of micro-algal cells on the operation of dead-end capillary UF membrane during severe algal bloom situations.

The membrane fouling potential of 16 species of bloom-forming micro-algae, calculated based on the cake filtration model (expressed as MFI-UF), were 100 to 10,000 times lower than what was observed in filtration experiments with two laboratory grown algal species (*Chaetoceros affinis* and *Microcystis sp.*). This observation was mainly attributed to the presence of AOM produced by algae (e.g., transparent exopolymer particles, TEP). Experimental data showed some correlation ($R^2 < 0.5$) between algae concentration and membrane fouling potential while a better correlation ($R^2 > 0.8$) was observed for TEP concentration and fouling potential. Experiments with a UF test unit demonstrated severe non-backwashable fouling which is attributed to AOM itself and/or AOM attached to algal cells.

Based on theoretical analyses, the transport and deposition of algal cells through capillary membrane are mainly influenced by inertial lift and shear-induced diffusion. The larger the algal cell, the higher is the possibility that it will accumulate at the end or in the middle of the capillary (depending on the design). Capillary plugging due to accumulation of algal cells in feed channels can cause increase in membrane and cake resistance due to localised increase in flux of the remaining active membrane filtration area. Such potential problems can be addressed by shortening the run length of filtration cycles, lowering the flux and introducing a small cross-flow (bleed) at the dead-end of the capillary.

10.1.6 Biological fouling enhanced by AOM

The possible role of AOM in biofouling of reverse osmosis (RO) membranes treating algal bloom impaired seawater was investigated based on lab-scale experiments with artificial seawater containing AOM from bloom-forming marine alga (*Chaetoceros affinis*).

Accelerated (8-20 days) biofouling experiments, simulating similar conditions in seawater RO systems, were performed using cross-flow capillary membranes and membrane fouling simulators (MFS) to demonstrate the effect of AOM on biofouling development. It was illustrated in these experiments that when biodegradable nutrients (CNP) are readily available in the feed water, the substantial presence of AOM - either in the feedwater or on the membrane/spacers - can accelerate the occurrence of biofouling in SWRO. However, when nutrients are not readily available in the feed water, direct organic fouling by AOM may occur but with a much lower fouling rate than when nutrients (N,C,P) are readily available. These findings were supported by the results of growth potential tests (using flow cytometry) whereby the net growth of bacteria in natural seawater increased with increasing AOM concentration. Furthermore, adhesion force measurement (using atomic force microscopy) demonstrated that AOM can strongly adhere to RO membranes and that adhesion become stronger when the membrane is already fouled with AOM. Consequently, an effective removal of AOM from the RO feed water is necessary to minimise biofouling in SWRO plants affected by algal blooms.

10.2 General outlook

This study demonstrated that developing better analytical tools are essential in elucidating algal blooms in seawater sources and their adverse impact on the operation of SWRO plants. It also highlighted the importance of developing effective pre-treatment processes to remove AOM from the raw water (e.g., UF) and reduce the fouling potential of the feed water for downstream SWRO membranes. Since algal blooms in seawater generally occur within a short period of time (e.g., few days to weeks) and are also difficult to predict, it is essential that TEP, MFI-UF and algae concentration are regularly monitored in the raw water of membrane-based desalination plants so that timely corrective measures can be implemented in the pre-treatment system during the onset of an algal bloom.

Furthermore, the analytical parameters developed/applied in this study together with the knowledge obtained from the outcome of this study may help future engineers/operators in developing better membranes (e.g., low molecular weight cut-off UF), more robust pre-treatment systems and effective operation strategies (e.g., lower membrane flux, coagulant dosing, etc.) to maintain stable operation in SWRO plants during severe algal blooms.

References

Passow and Alldredge, 1995 Passow, U. and Alldredge, A. L. (1995) A Dye-Binding Assay for the Spectrophotometric Measurement of Transparent Exopolymer Particles (TEP). *Limnology and Oceanography* **40**(7), 1326-1335.

Thornton, D. C. O., Fejes, E. M., DiMarco, S. F. and Clancy, K. M. (2007) Measurement of acid polysaccharides (APS) in marine and freshwater samples using alcian blue. *Limnology and Oceanography: Methods* **5**, 73–87.

Samenvatting

Omgekeerde osmose (RO) is momenteel de toonaangevende en geprefereerde zeewaterontzilting technologie, vooral in gebieden waar zoetwater bronnen beperkt zijn. Echter, zeewater RO (SWRO) systemen zijn gevoelig voor operationele problemen vooral te wijten aan membraanvervuiling. Om membraanvervuiling te minimaliseren, is voorbehandeling van zeewater stroomopwaarts van het SWRO systeem nodig. De meeste SWRO installatie, vooral in de regio van het Midden-Oosten, installeren een voorbehandeling met coagulatie gevolgd door granulair media filtratie (GMF). De laatste jaren wordt lage druk membraan zoals ultrafiltratie (UF) steeds vaker gebruikt als een voorkeursalternatief voor GMF.

Er is steeds meer bewijs dat de seizoensgebonden verspreiding van microscopische algen in het zeewater genaamd algenbloei een belangrijke oorzaak is van membraanvervuiling in SWRO installaties. Deze natuurlijke verschijnselen kunnen mogelijk optreden in de meeste kustgebieden van de wereld waar SWRO installaties staan. Recente ernstige algenbloei uitbraken in het Midden-Oosten hebben verstopping van GMF voorbehandeling systemen veroorzaakt en ook geleid tot onaanvaardbare kwaliteit veroorzaakt (hoge densiteit voor slib index, SDI>5) van het effluent. Dit laatste leidde uiteindelijk tot een tijdelijke sluiting van een aantal SWRO zuiveringen voornamelijk te wijten aan zorgen van onomkeerbare vervuiling in de downstream SWRO systeem. De slechte prestaties van GMF tijdens algenbloei heeft de focus van de ontzilting industrie verschoven naar de toepassing van UF als voorbehandeling voor SWRO. Niettemin is een uitgebreid onderzoek naar de negatieve effecten van algenbloei nog steeds nodig om mogelijke vervuilingproblemen naar behoren te beperken in zowel UF en RO membranen.

Het doel van deze studie is om de negatieve gevolgen van zeewater algenbloei in de werking van UF en RO-membraan systemen te begrijpen. Bovendien is dit onderzoek gericht op de ontwikkeling/verbetering van methoden om de eigenschappen te onderzoeken en de membraanvervuiling potentieel van algen en organische stof van algen (AOM) te meten. Het uiteindelijke doel is om ingenieurs/exploitanten beter begrip evenals betrouwbare evaluatie-instrumenten te verschaffen om robuuste processen en effectieve operationele strategieën te ontwikkelen om stabiele werking in membraan-gebaseerde ontziltingsinstallaties tijdens algenbloei te houden.

De potentiële problemen die kunnen optreden bij membraan-gebaseerde ontziltingsinstallaties (UF voorbehandeling gevolgd door SWRO) tijdens de behandeling van water met algenbloei zijn: (1) deeltjes vervuiling in UF gevolg van ophoping van algen en hun bezinksel, (2) biologische vervuiling in UF en RO als gevolg van ophoping van AOM, en (3) biologische vervuiling in RO geïnitieerd en/of verbeterd door AOM. Deze mogelijke problemen werden in dit onderzoek geadresseerd door een combinatie van theoretische en experimentele analyses.

De eerste stap om beter inzicht te krijgen in de effecten van algenbloei op membraanfiltratie was om karakterisering studies uit te voeren met batch culturen van algemene bloeivormende soorten zeewater (*Chaetoceros affinis* , *Alexandrium tamarense*) en zoetwater (*Microcystis sp.*) algen. De 3 soorten bloeivormende algen toonde verschillende groeipatroon en AOM productie in verschillende fasen van hun levenscyclus. De AOM geproduceerd door de 3 algen soorten werden geëxtraheerd en gekarakteriseerd met verschillende microscopische, choromatografische en spectrofotometrische technieken. De belangrijkste bevinding is dat AOM voornamelijk biopolymeren (bijvoorbeeld polysacchariden, proteïnen) bevatten terwijl de resterende fracties persistente organische stoffen (bijv. humusachtige stoffen) bevatten en/of laagmoleculaire biogene stoffen. Polysacchariden met bijbehorende sulfaat en fucose functionele groepen waren alomtegenwoordig in biopolymeren geproduceerd door de 3 algensoorten. Bovendien waren deze biopolymeren in staat te kleven aan schone UF en RO membranen en veel sterker op al door biopolymeren vervuilde membranen. Ook werd vastgesteld dat deze klevende AOM fractie hoofdzakelijk bestaat uit transparante exopolymeer deeltjes (TED).

Verdere onderzoeken werden uitgevoerd om de aanwezigheid van TED in membraan-gebaseerde ontzilting systemen te volgen. Hiertoe werden 2 bestaande methoden om TED te meten eerst gemodificeerd of verder verbeterd voor hun toepasbaarheid voor het bestuderen van verschillende voorbehandelingen en membraanvervuiling. Diverse verbeteringen werden ingevoerd voornamelijk om de inmenging van opgeloste zouten in het monster te minimaliseren, om de detectie ondergrens te verlagen, om zowel deeltjes en colloïdale TED's (tot 10 kDa) te meten en een reproduceerbare kalibratie methode te ontwikkelen met een standaard TED (Xanthaan gom). Succesvolle toepassing van beide methoden (met modificaties en verbeteringen) werd aangetoond voor de monitoring van TED accumulatie in algenculturen en TED verwijdering in de voorbehandelingen van 4 RO installaties.

Lange termijn (~4 jaar) TED monitoring in het ruwe water van een SWRO installatie in Nederland vertoonde pieken van TED concentraties meestal in de lente algenbloei (maart-mei) periode, die voornamelijk werden gedomineerd door diatomeeën en/of Phaeocystis. Verdere monitoring van het voorkomen van TED in de zuiveringsprocessen van 4 RO installaties liet aanzienlijke vermindering van TED's, biopolymeren en membraanvervuiling potentiële (in termen van de gemodificeerde vervuilingindex-ultrafiltratie, MFI-UF) zien na UF (met en zonder coagulatie), coagulatie-sedimentatie-snelle zandfiltratie en coagulatie-opgeloste luchtflotatie behandelingen. Anderzijds, laboratoriumschaal membraan afstoting experimenten toonden aan dat volledige verwijdering van algen afgeleide biopolymeren mogelijk is door UF membranen met lager moleculegewicht cut-off (≤10 kDa). De resultaten van zowel laboratoriumschaal en full-scale installatie experimenten toonden over het algemeen significante correlatie tussen TED concentraties en MFI-UF aan, wat een indicatie is dat TED's waarschijnlijk een belangrijke rol spelen in de vervuiling van UF voorbehandeling systemen en mogelijk in SWRO systemen als deze niet effectief verwijderd worden in het voorbehandelingproces.

De rol van algen op de vervuiling van capillaire UF membranen werd onderzocht op basis van theoretische en experimentele analyse. Resultaten tonen aan dat membraanvervuiling van UF tijdens filtratie van door algenbloei getroffen wateren vooral het gevolg is van ophoping van TED op membraanporiën en -oppervlakken, in plaats van de accumulatie van algen cellen. Dit werd bevestigd door de significante correlatie ($R^2>0,8$) tussen MFI-UF en TED concentratie en niet met algencel concentratie ($R^2<0,5$).

Aanvullende onderzoeken zijn uitgevoerd om te controleren of algencellen ernstige verstopping kan veroorzaken in capillaire membranen. Theoretisch kan algencel transport door een van-binnen-naar-buiten aangedreven capillaire UF hoofdzakelijk beïnvloed door inertieel optillen, die verstopping kunnen veroorzaken omdat daarmee een grotere afzetting van algen in het doodlopende deel van het capillair. Hydraulische berekeningen suggereren dat ernstige verstopping als gevolg van significante accumulatie van grote algen cellen (>25 µm) aanzienlijke toename van membraan en cake resistentie kan veroorzaken als gevolg van lokale toename van de flux tijdens constante flux filtratie. Deze potentiële problemen kunnen worden aangepakt door het verkorten van de periode tussen filtratiecycli, verlagen van het membraan flux en/of introduceren van een dwarsstroming (bleed) aan het doodlopende einde van het capillair.

Door biopolymeren van bloei-vormende zeewieren vervuilde capillaire UF membranen werden verder onderzocht in relatie tot de chemische samenstelling van het voedingswater. Het verlagen van de pH van het voedingswater kan leiden tot een verhoging van het UF membraan vervuilingspotentieel van biopolymeren. Het variëren van de ionsterkte van het voedingswater kan ook leiden tot aanzienlijke variaties van vervuilingspotentieel. De rol van mono- en di-waardige kationen op de vervuilingspotentieel van biopolymeren is ook onderzocht door het variëren van het ion samenstelling van het voedingswater matrix. Onder de belangrijkste kationen rijk aanwezig in zeewater, Ca^{2+} ionen heeft een significante invloed op de tendens tot vervuiling van AOM qua MFI-UF en niet-terugspoelbare vervuiling. Bovendien is het effect van de chemische samenstelling op vervuilingsgedrag van algen biopolymeren wezenlijk anders dan wat werd waargenomen bij model polysacchariden (bv. natriumalginaat en xanthaan gom) die traditioneel gebruikt worden in UF vervuiling experimenten. Dit kan erop wijzen dat deze model polysacchariden niet betrouwbaar zijn in het simuleren van de vervuiling neiging van algen biopolymeren in zeewater.

De laatste fase van het onderzoek was gericht op het aantonen van de belangrijke rol van AOM op biofilm ontwikkeling in SWRO systemen door het uitvoeren van versnelde biologische aangroei experimenten (8-20 dagen) in dwars-stroom capillaire UF-membranen (10 kDa MWCO) en spiraalvormig gewonden RO membranen (met behulp van MFS). In deze experimenten is geïllustreerd dat wanneer voedingsstoffen niet beperkt zijn in het voedingswater, de aanzienlijke concentratie AOM - in het voedingswater of op het membraan/spacers - kan het optreden van biofouling in SWRO versnellen. Echter, als voedingsstoffen beperkt zijn in het voedingswater, kan directe organische vervuiling door AOM optreden, maar in een veel geringere mate van vervuiling, dan wanneer vervuiling nutriënten (N, C, P) direct beschikbaar zijn. De bevindingen worden ondersteund door de resultaten van het groeipotentieel proeven (met flowcytometrie) waarbij de netto groei van zeewater bacteriën in natuurlijk zeewater een lineaire toename met AOM concentratie aantoonde. Bijgevolg, effectieve verwijdering van AOM van RO voedingswater is nodig om biofouling te minimaliseren in SWRO installaties aangetast door algenbloei.

Over het algemeen, toonde deze studie aan dat betere analytische parameters en tools essentieel zijn in het ophelderen van de nadelige effecten van algenbloei in zeewater bronnen op de werking van membraan-gebaseerde ontziltingsinstallaties (UF-RO). Het benadrukte ook het belang van het ontwikkelen van effectieve voorbehandeling processen om AOM te verwijderen uit het ruwe water en het membraanvervuiling potentieel van het voedingswater voor downstream SWRO systeem te verminderen. Aangezien zeewater algenbloei algemeen plaatsvindt binnen een korte tijdsperiode (bijvoorbeeld enkele dagen tot weken) en deze ook moeilijk te voorspellen zijn, is het essentieel dat MFI-UF, TED en algen concentraties regelmatig worden gemonitord in het ruwe en voorbehandelde water van SWRO installaties, zodat tijdig corrigerende maatregelen in het voorbehandeling systeem kan worden geïmplementeerd tijdens het begin van een algenbloei.

Abbreviations

AB	alcian blue
AEx	anion exchange
AFM	atomic force microscopy
AOC	assimilable organic carbon
AOM	algal organic matter
APS	acid polysaccharides
ASP	amnesic shellfish poisoning
ASW	artificial seawater
AT	*Alexandrium tamarense*
ATP	adenosinetriphosphate
AZP	azaspiracid shellfish poisoning
BDOC	biodegradable dissolved organic carbon
BOD	biological oxygen demand
BW	backwashable fouling
CA	Chaetoceros affinis
CEB	chemically enhanced backwashing
CEx	cation exchange
CFP	ciguatera fish poisoning
CIP	cleaning in place
CiR	chemically irreversible fouling
CLSM	confocal lasser scanning microscopy
CR	chemically reversible fouling
DAF	dissolved air flotation
DLS	dynamic light scattering
DMF	dual media filtration
DSP	diarrhetic shellfish poisoning
ED	electrodialysis
EOM	extracellular (algal) organic matter
EPS	extracellular polymeric substances
FEEM	fluorescence excitation-emission matrices
FTF	filter-transfer-freeze
FTIR	fourier transform infrared
GAC	granular activated carbon
GMF	granular media filtration
HAB	harmful algal bloom
IMS	integrated membrane systems
IOM	Intracellular (algal) organic matter

IR	infrared
LC-OCD	liquid chromatography-organic carbon detection
LMW	low molecular weight
LOD	limit of detection
MF	microfiltration
MFI-UF	modified fouling index - ultrafiltration
MFS	membrane fouling simulator
MS	microstrainer
MSp	*Microcystis sp.*
MWCO	molecular weight cut-off
N	nitrogen
nBW	non-backwashable fouling
NF	nanofiltration
NOM	natural organic matter
NSP	neurotoxic shellfish poisoning
OC	organic carbon
OCD	organic carbon detector
ON	organic nitrogen
OND	organic nitrogen detector
P	phosphorous
PA	polyamide
PC	polycarbonate
PES	polyethersulfone
pH	hydrogen potential
PS	polystyrene
PSP	paralytic shellfish poisoning
PV	pressure vessel
RC	regenerated cellulose
RO	reverse osmosis
RSF	rapid sand filtration
SA	sodium alginate
SDI	silt density index
SEM	scanning electron microscopy
Si	silicate
SWRO	seawater reverse osmosis
TDS	total dissolved salts
TEP	transparent exopolymer particles
TMP	trans-membrane pressure
TOC	total organic carbon
UF	ultrafiltration
UPW	ultrapure water
UVD	UV detector
XG	xanthan gum

Publications and Awards

Journal publications

1. Schurer R., Tabatabai A., Villacorte L., Schippers J.C., Kennedy M.D. (2013) Three years operational experience with ultrafiltration as SWRO pre-treatment during algal bloom. *Desalination & Water Treatmen* 51 (4-6), 1034-1042.

2. Villacorte, L.O., Ekowati, Y., Winters, H., Amy, G.L., Schippers, J.C. and Kennedy, M.D. (2013) Characterisation of transparent exopolymer particles (TEP) produced during algal bloom: a membrane treatment perspective. *Desalination & Water Treatment* 51 (4-6), 1021-1033.

3. Schurer, R., Janssen, A., Villacorte, L., Kennedy, M.D. (2012) Performance of ultrafiltration & coagulation in an UF-RO seawater desalination demonstration plant. *Desalination & Water Treatment* 42(1-3), 57-64.

4. de Ridder, D.J., Villacorte, L., Verliefde, A.R.D., Verberk, J.Q.J.C., Heijman, S.G.J., Amy, G.L. and van Dijk, J.C. (2010) Modeling equilibrium adsorption of organic micropollutants onto activated carbon. *Water Research* 44 (10), 3077-3086.

5. Villacorte, L. O., Schurer, R., Kennedy, M., Amy, G. and Schippers, J.C. (2010) Removal and deposition of Transparent Exopolymer Particles (TEP) in seawater UF-RO system. *IDA Journal* 2 (1), 45-55.

6. Villacorte, L. O., Schurer, R., Kennedy, M., Amy, G., Schippers, J.C. (2010) The fate of transparent exopolymer particles in integrated membrane systems: a pilot plant study in Zeeland, The Netherlands. *Desalination & Water Treatment* 13:109-119.

7. Villacorte, L. O., Kennedy, M. D., Amy, G. L., and Schippers, J. C. (2009) The fate of Transparent Exopolymer Particles (TEP) in integrated membrane systems: Removal through pretreatment processes and deposition on reverse osmosis membranes. *Water Research* 43 (20), 5039-5052.

8. Villacorte, L. O., Kennedy, M. D., Amy, G. L., and Schippers, J. C. (2009) Measuring Transparent Exopolymer Particles (TEP) as indicator of the (bio)fouling potential of RO feed water. *Desalination & Water Treatment* 5, 207-212.

Journal articles submitted/in preparation

1. Villacorte L.O., Tabatabai S.A.A., Amy G., Schippers J.C. and Kennedy M.D., (2013) Impact of algal blooms on seawater reverse osmosis: membrane fouling and pre-treatment options. *In preparation for Desalination.*

2. Villacorte L.O., Ekowati Y., Neu T., Kleijn M., Winters H., Amy G., Schippers J.C. and Kennedy M.D. (2013) Characterisation of organic matter released by bloom-forming dinoflagellate, diatom and cyanobacteria. *In preparation for Water Research.*

3. Li S., Winters H., Villacorte L.O., Ewas A.-H., Kennedy M.D., Amy G.L. (2013) Characterization and comparison of transparent exopolymer particles (TEP) produced by bacteria and algae. *Submitted to Water Research.*

4. Villacorte, L. O., Ekowati, Y., Calix-Ponce H.N., Schippers, J.C., Amy, G. and Kennedy, M. (2013) Measuring transparent exopolymer particles in freshwater and seawater: limitations, method improvements and applications. *In preparation for Water Research.*

5. Villacorte, L. O., Ekowati, Y., Calix-Ponce, H.N., Schurer, R., Kennedy, M., Amy, G. and Schippers, J.C. (2013) Monitoring Transparent Exopolymer Particles (TEP) in integrated membrane systems. *In Preparation for Water Research.*

6. Villacorte L.O., Karna B., Ekowati Y., Kleijn M., Amy, G., Schippers, J.C. and Kennedy, M.D. (2013) Fouling of ultrafiltration membranes by algal biopolymers in seawater. *In preparation for Journal of Membrane Science.*

7. Villacorte L. O., Gharaibeh M., Schippers J.C. and Kennedy M.D. (2013) Fouling potential of algae in inside-out driven capillary UF membrane. *In Preparation for Journal of Membrane Science.*

8. Villacorte L.O., Nyambi Z., Keseilius V., Calix-Ponce H.N., Ekowati Y., Prest E., Vrouwenvelder J.S., Winters H., Amy G., Schippers J.C. and Kennedy M.D. (2013) Biofouling in cross-flow membranes facilitated by algal organic matter. *In Preparation for Water Research.*

Publications in conference proceedings

1. Villacorte, L., Micor, K., Calix, H.N., Amy, G., Schippers J.C., and Kennedy, M.D. (2012) Fouling of UF membranes during algal bloom: the role of transparent exopolymer particles (TEP). In Proc. of IWA World Water Congress and Exhibition, Busan, South Korea, 16-21 September.

2. Villacorte, L., Schippers, J.C., Amy, G., and Kennedy, M.D. (2012) Membrane filtration of algal bloom impacted waters: Investigating the fouling potential of algal cells and transparent exopolymer particles (TEP). In Proc. of EDS Conference on Membranes in Drinking and Industrial Water Production, Leeuwarden, Netherlands, 10-12 September 2012.

3. Ekowati, Y., Villacorte, L., Winters, H., Schippers, J.C., Amy, G., and Kennedy, M.D. (2012) Characterisation, rejection and fouling potential of algal organic matter from 'red tide' algae. In Proc. of EDS Conference on Membranes in Drinking and Industrial Water Production, Leeuwarden, Netherlands, 10-12 September 2012.

4. Villacorte, L., Ekowati, Y., Salinas-Rodriguez, S.G., Winters, H., Amy, G., Schippers, J.C., Kennedy, M.D. (2012) MF/UF rejection and membrane fouling potential of transparent exopolymer particles (TEP) produced by 'Red Tide' algae.

In Proc. of EDS Conference and Exhibition on Desalination for the Environment: Clean water and energy, Barcelona, Spain, 22-26 April 2012.

5. Schurer, R., Tabatabai, A., Villacorte, L., Schippers J.C. and Kennedy M.D. (2012) Three years operational experience with ultrafiltration as SWRO pre-treatment during algal bloom. In Proc. of EDS Conference and Exhibition on Desalination for the Environment: Clean water and energy, Barcelona, Spain, 22-26 April 2012.

6. Amy G., Salinas S., Villacorte L., Ha C.W., Hamad J., Yangalli V., Kennedy M.D., Croue J.P. (2011) Application of Innovative Natural Organic Matter (NOM) Characterization Protocols to Membrane Fouling Assessment and Control: Seawater, Wastewater, and Freshwater Sources. In Proc. of 6th IWA Specialist Conference on Membrane Technology for Water & Wastewater Treatment, Aachen, Germany, 4-7 October.

7. Villacorte L. O., Karna B., Berenstein D. E., Calix H. N., Schurer R., , Amy G., Schippers J. C., Kennedy M. D. (2011) Non-backwashable fouling of UF membranes caused by transparent exopolymer particles (TEP) in seawater. In Proc. of 6th IWA Specialist Conference on Membrane Technology for Water & Wastewater Treatment, Aachen, Germany, 4-7 October.

8. Villacorte, L. O., Berenstein, D.E., Calix, H.N., Vrouwenvelder H., Amy, G., Schippers, J.C., Kennedy, M.D. (2011) Membrane fouling due to marine algal bloom: the role of transparent exopolymer particles (TEP). In Proc. of IDA World Congress & Exposition, Perth, Western Australlia, 4-9 September.

9. Calix, H.N., Villacorte, L. O., Amy, G., Schippers, J.C. and Kennedy, M.D. (2011) Identification of transparent exopolymer particles (TEP) on fouled membranes In Proc. of IDA World Congress & Exposition, Perth, Western Australlia, 4-9 September.

10. Villacorte, L. O., Berenstein, D.E., Calix, H.N., Amy, G. Schippers, J.C., Kennedy, M.D. (2011) Fouling of UF/MF of Membranes during an Algal Bloom: the role of transparent exopolymer particles (TEP) in seawater. In Proc. of the AMTA Annual Conference & Exposition, Miami Beach, FL, USA, 19-23 July.

11. Calix, H.N., Villacorte, L. O., Amy, G., Schippers, J.C. and Kennedy, M.D. (2011) Visualization of transparent exopolymer particles (TEP) on fouled membranes. In Proc. of the AMTA Annual Conference & Exposition, Miami Beach, FL, USA, 19-23 July.

12. Villacorte, L. O., Kennedy, M., Amy, G. and Schippers, J.C. (2010) Filtration resistance of foulant layers containing transparent exopolymer particles (TEP) during membrane filtration. EDS EuroMed 2010: Desalination for Clean Water and Energy, Tel-Aviv, Israel, 3-7 October.

13. Villacorte, L. O., Schurer, R., Kennedy, M., Amy, G. and Schippers, J.C. (2010) Organic Fouling of Ultrafiltration Membranes by Transparent Exopolymer Particles (TEP) in Seawater. IWA World Water Congress and Exhibition, Montreal, Canada, 19-24 September.

14. Villacorte, L. O., Schurer, R., Kennedy, M., Amy, G. and Schippers, J.C. (2010) Fouling of Ultrafiltration Membranes by Transparent Exopolymer Particles (TEP) in Seawater. In Proc. of the AMTA Annual Conference & Exposition, San Diego, CA, USA, 12-15 July. *(Best Student Paper Presentation Award)*

15. R. Schurer, A. Janssen, L. O. Villacorte, M. Kennedy (2010) Evides Pilot Seawater Desalination: efficacy of coagulant in ultrafiltration pre-treatment in integral UF-RO membrane desalination. In Proc. of the Conference on Membranes for Drinking and Industrial Water, Trondheim, Norway, 27-30 June, IWA-EDS.

16. Villacorte, L. O., R. Duraisamy, Kennedy, M., Amy, G. and Schippers, J.C. (2010) Transparent Exopolymer Particles: a promising tool for organic/biological fouling studies in membrane systems? In Proc. of the Conference on Membranes for Drinking and Industrial Water, Trondheim, Norway, 27-30 June, IWA-EDS.

17. Villacorte, L. O., Schurer, R., Kennedy, M., Amy, G. and Schippers, J.C. (2009) Removal and deposition of Transparent Exopolymer Particles (TEP) in seawater UF-RO system. In Proc. of the IDA World Congress for Desalination and Water Reuse, Dubai, UAE, 7-12 November.

18. Villacorte, L. O., Kennedy, M. D., Amy, G. L., and Schippers, J. C. (2009) Removal and deposition of Transparent Exopolymer Particles (TEP) in integrated membrane systems treating surface water and secondary effluent. In Proc. of the IWA Membrane Technology Conference, Beijing, China, 1-3 September.

19. Villacorte, L. O., Schurer, R., Kennedy, M., Amy, G. and Schippers, J.C. (2009) The fate of transparent exopolymer particles in integrated membrane systems: a pilot plant study in Zeeland, The Netherlands. In Proc. of EDS Congress on Desalination for the Environment, Clean Water and Energy, Baden-Baden, Germany, 17-20 May.

20. Villacorte, L. O., Kennedy, M. D., Amy, G. L., and Schippers, J. C. (2008) Measuring dissolved Transparent Exopolymer Particles (TEP) as indicator of the (bio)fouling potential of RO feed water. In Proc. of EDS EuroMed 2008: Desalination for Clean Water and Energy Cooperation among Mediterranean Countries of Europe and the MENA Region, Dead Sea, Jordan, 9–13 November. *(Best Student Paper Presentation Award)*

21. Amy, G., Maeng, S.K., Jekel, M., Ernst, M., Villacorte, L.O., Yangali, V., Kim, T.U., Reemtsma, T. (2008) Advanced water/wastewater treatment process selection for organic micropollutant removal: a quantitative structure-activity relationship (QSAR) approach. In: International Water Week - Water Convention, Singapore, 23-27 June.

Awards and scholarship

1. **Best oral presentation award** (the Netherlands, September 2013) UNESCO-IHE Annual PhD Symposium 2013, UNESCO-IHE Delft.

2. **Outstanding student paper oral presentation award** (USA, July 2010) AMTA Annual Conference & Exposition, San Diego, CA, USA, American Membrane Technology Association (AMTA).

3. **Best student paper oral presentation award** (Jordan, November 2008) EuroMed Conference, Desalination Cooperation among Mediterranean Countries of Europe and the MENA region, Dead Sea, Jordan, European Desalination Society (EDS).

4. **MSc fellowship** (Netherlands, 2006-2008). NUFFIC-NFP and UNESCO-IHE.

Curriculum Vitae

LOREEN OPLE VILLACORTE

Oct. 2013 - present	: Lecturer in Water Supply Engineering, Environmental Engineering and Water Technology Department, UNESCO-IHE Institute for Water Education, the Netherlands.
May 2009 – Jan. 2014	: PhD research fellow at UNESCO-IHE, Wetsus Centre of Excellence for Sustainable Water Technology and TUDelft, the Netherlands.
Oct. 2008 – Apr. 2009	: Researcher at UNESCO-IHE, the Netherlands.
Oct. 2006 – Sept. 2008	: Master of Science in Water Supply Engineering at UNESCO-IHE, the Netherlands.
Jan. 2005 – Sept. 2006	: Water Supply Engineer at the Water Resources Centre, University of San Carlos, the Philippines.
Apr. 2004 – Dec. 2004	: Hydrologist at the Water Resources Centre (WRC), University of San Carlos, the Philippines.
Jan. 2004 – Mar. 2004	: Field Engineer of Central Cebu WaterREMIND project funded by the Royal Netherlands Embassy to the Philippines, USC-WRC, the Philippines.
Jun. 1998 – Mar. 2003	: Bachelor of Science in Civil Engineering, University of San Carlos – Technological Centre, Cebu, the Philippines.
20 June 1981	: Born in Cebu City, the Philippines.

This work was performed in the cooperation framework of Wetsus, centre of excellence for sustainable water technology (www.wetsus.nl). Wetsus is co-funded by the Dutch Ministry of Economic Affairs and Ministry of Infrastructure and Environment, the European Union Regional Development Fund, the Province of Fryslân and the Northern Netherlands Provinces. The authors like to thank the participants of the research theme "Biofouling" for the fruitful discussions and their financial support.

T - #0390 - 101024 - C304 - 240/170/16 - PB - 9781138026261 - Gloss Lamination